A DICTIONARY OF STATISTICAL TERMS

Prepared for the International Statistical Institute by

MAURICE G. KENDALL, M.A., SC.D.
Chairman, Scientific Control Systems Ltd

WILLIAM R. BUCKLAND, B.SC., PH.D.
Divisional Director, Economist Intelligence Unit Ltd.

A DICTIONARY OF STATISTICAL TERMS

Published for the International Statistical Institute
Third Edition
Revised and Enlarged

HAFNER PUBLISHING COMPANY, INC.
New York, N.Y.

Oliver & Boyd
Tweeddale Court
14 High Street
Edinburgh EH1 1YL
A Division of Longman Group Ltd

First published 1957
Second Edition 1960
Reprinted 1966 1967
Third Edition 1971
Reprinted 1972

ISBN 0 05 002280 6

© **1971 The International Statistical Institute**

Printed in Great Britain by T. and A. Constable Ltd
Hopetoun Street, Edinburgh

74.12 - 1510

In the U.S.A.
Hafner Publishing Co., Inc.
866 Third Avenue
New York, N.Y. 10022

FOREWORD

The first edition of this *Dictionary of Statistical Terms* was prepared by Dr Kendall and Dr Buckland for the International Statistical Institute, with financial assistance from UNESCO, and was published in 1957. The text was in English, with Glossaries in French, German, Italian and Spanish.

In a field like statistics that is of relatively recent growth, is expanding rapidly, and has fruitful applications in science, in government, in business and industry, and in everyday affairs, it is no surprise that the first edition soon established itself as a highly important addition to the resources of the statistical profession. My own uses of the Dictionary, for instance, are those of a professor of statistics who might be presumed, or at least expected, to know his field well. I have found the Dictionary invaluable in looking up unexplained terms that I come across in reading—the author evidently was taught the term somewhere and assumes that everybody knows it. Another frequent use is in statistical consultation. The client and I both use the term but with different meanings. There is confusion until I reach for the 'Kendall & Buckland' and we can agree on a common understanding and get on with the problem. Considering how large the field has become by 1970, I expect many similar uses of this revised and enlarged third edition.

The Institute and the statistics profession owe a massive debt to Drs Kendall and Buckland for their work in preparing these editions. Their combination of authoritative knowledge of theory and wide experience in applications could not be bettered for this purpose. It remains for me to add the thanks of the Institute to those which the compilers give in their Preface to the colleagues who helped them in the work. The publishers, Oliver & Boyd, have done their usual competent job of presentation of the product.

W. G. Cochran
President, I.S.I.

PREFACE TO THE THIRD EDITION

This new edition is both revised and extended; revised in the light of helpful comments received from colleagues all over the world, and extended to cover the many new concepts which have emerged in the last fifteen years. So far as has been practicable this present edition takes account of literature in both book and journal form up to the end of 1968. The pace of advance may be judged from the fact that the number of items has increased from some 1700 in the first edition to over 2500 in this.

In general the principles to which we have worked are similar to those stated in the Preface to the first edition, but we alone are responsible for the definitions. On this occasion we have endeavoured to give an author and date for the introduction of the more recent concepts and the actual references will be readily identified through the three-volume *Bibliography of Statistical Literature* by Kendall & Doig (published by Oliver & Boyd) or the issues of *Statistical Theory & Methods Abstracts* (previously published by Oliver & Boyd; from 1971 published by Longman Journals Division).

With regard to the glossaries of equivalent terms (in French, German, Italian and Spanish), which were an integral part of the first edition and formed a special separate publication at the second edition, we find that the position has changed during the intervening years. Accordingly, it was agreed that for this revised and extended edition the ISI would arrange only for an English-Russian/Russian-English glossary of equivalent terms. This has been prepared by Professor S. Kotz for separate and nearly simultaneous publication by Oliver & Boyd.

A task of this kind is essentially a collaborative venture and we have received much help from our professional colleagues especially in the drafting of lists of new terms. We also wish to acknowledge the considerable assistance given by our research associate, Ronald A. Fox, B.Sc., who made the initial survey of book literature and drafted the relevant definitions. We would also gratefully acknowledge support on the inescapable administrative and editorial tasks given throughout the project by Mrs Mary Sydenham.

As before, we shall be pleased for any reader to call our attention to errors, omissions or obscurities.

London
June, 1970

M. G. K.
W. R. B.

EXTRACTS FROM THE PREFACE TO THE FIRST EDITION

We undertook this work early in 1951 at the invitation of the International Statistical Institute. A description of the problems which we encountered and of the methods by which we tried to solve them is given in a paper* read before the Rome meeting of the Institute in September 1953. We repeat here only as much as is necessary to explain some of the features of the following work. . . .

The function of a dictionary, in our view, is to provide an explanation of terms in current use, whether they are intrinsically desirable or not. We have omitted a few elementary terms which are self-explanatory. Several colleagues have urged us to go rather further and to omit expressions which are open to criticism. We sympathise with their viewpoint, but we thought it our duty to keep a tight rein on our inclination to omit expressions which are confusing or redundant, and have contented ourselves with a statement of opinion about those which we felt should be allowed to lapse into disuse.

We have tried to keep our notation and usage consistent, but we have not attempted any standardisation of symbolism or nomenclature. It is very possible that a work of this kind, however much we disclaim intention, may become normative. We recognise this and have taken our responsibilities very seriously. At the same time we have tried not to standardise more than consistency required. In our experience the intensity of desire for a uniform symbolism which would cover the whole domain of theoretical statistics varies inversely with a knowledge of the extent and complexity of that domain. . . .

There are some terms current in foreign languages, especially in Italian, with no equivalent in English or with different meanings from the literal English equivalent. A number of Italian terms were kindly provided by Professor Gini and have been included in English translation in the main body of the Dictionary. They are distinguished from ordinary English terms by an asterisk. We are responsible for the English translation and equivalents.

* M. G. Kendall (1954), 'The Projected Dictionary of Statistical Terms' *Bulletin de l'Institut International de Statistique*, **34**, Part 2, 629.

CONTENTS

α-Error In the theory of testing hypotheses an error incurred by rejecting a hypotheses when it is actually true: more usually known as **Error of the First Kind** or Type I Error. In Quality Control it is equivalent to **Producer's Risk.**

α-Index See **Pareto Curve.**

Abbe-Helmert Criterion A test of randomness in a time-series based upon the fact that in a random series auto-correlation coefficients of order k, for all k greater than zero, should vanish. The criterion tests the significance of the first serial correlation ($k = 1$) by the use of the large-sample formula var $r_1 = 1/(n-1)$.

*Abnormal Curve (Curva Anormale) A frequency curve with **Abnormality**, that is to say, differing from a normal curve with the same median and dispersion about the median. If the frequency curve crosses this corresponding normal curve more than once in at least one of the two branches on either side of the median, it is said to be complex abnormal (*anormale complessa*). If it is symmetrical and in each branch crosses the normal curve only once it is simple abnormal (*anormale semplice*).

*Abnormality (Anormalità) In Italian usage, a measure of dissimilarity from the normal curve, somewhat similar to a measure of **Kurtosis.** For the normal curve the **Mean Difference** with repetition, $\varDelta_R$, is related to the mean deviation about the **Median**, δ, by the relation $\varDelta_R = \delta\sqrt{2}$. Frequency distributions for which this is realised are said to have neutral abnormality (*anormalità neutra*); if $\varDelta_R < \delta\sqrt{2}$ the distribution is hyponormal (*iponormale*) and if $\varDelta_R > \delta\sqrt{2}$ it is hypernormal (*ipernormale*).

It is possible to consider separately the two halves of the distribution lying on either side of the median. Abnormality in both halves is called *anormalità bilaterale*; abnormality in only one half is called *anormalità uni-laterale*. If both halves have the same type of abnormality there is said to be *anormalità uniforme*, e.g. if both have neutral abnormality the distribution is said to possess *anormalità neutra uniforme*. If the two halves have different types there is said to be *anormalità disforme*; and if one half has neutral abnormality there is said to be *anormalità neutra disforme*.

*Abnormality, Index of A measure of **Abnormality.** Let $x_1, ..., x_m$ be a set of values and $\mu_l, ..., \mu_m$ the **Cograduated** values of a normal distribution with the same median and mean deviation about the median. One such index is

$$N_1 = \frac{1}{m} \sum_{i=1}^{m} | x_i - \mu_i |.$$

A quadratic index of similar type is

$$N_2 = \frac{1}{m} \sum_{i=1}^{m} (x_i - \mu_i)^2.$$

Relative indices may be obtained by division by the maximum values which these indices may have, namely twice the mean deviation about the median in the case of N_1 and $\sqrt{2}$ times the root-mean-square about the median in the case of N_2.

Abrupt Distribution A continuous frequency distribution is said to be abrupt at a finite terminal point of its range if the frequency and the first derivative of the frequency are not both zero at that point.

Absolute Deviation The difference, taken without regard to sign, of a variate value from some given value which may be a constant, a parent value or a sample value. This concept enters into the **Mean Deviation.**

Absolute Difference The difference, taken without regard to sign, between the values of two variables; and in particular of two random variables. This concept enters into the **Mean Difference.**

Absolute Error The absolute error of an observation x is the absolute deviation of x from its 'true' value.

Absolute Frequency The actual frequency of a variate, as distinct from the relative frequency, namely the ratio of the frequency to the total frequency of all variate values.

Absolute Measure This term is occasionally used to describe a measure of variate values which is independent of origin and scale; for example, the moment ratio μ_4/μ_2^2 or the corresponding sample statistic m_4/m_2^2 The usage is not entirely satisfactory owing to possible confusion with measures, such as the mean deviation, which are based on absolute quantities, i.e. quantities taken regardless of sign.

Absolute Moments The moments of a frequency distribution in which the deviations about a fixed point are taken without regard to sign; that is to say, the rth absolute moment about a value a is

$$\nu_r' = \int_{-\infty}^{\infty} | x - a |^r dF(x).$$

As for ordinary moments, the absolute moments about an arbitrary value are usually denoted by a prime, those about the mean without a prime, e.g.

$$\nu_r = \int_{-\infty}^{\infty} | x - \mu_1' |^r dF(x).$$

For even values of r the ordinary and absolute moments are identical.

Absolutely Unbiassed Estimator See **Unbiassed Estimator.**

Absorbing Barrier Certain additive or **Random Walk Processes** represent the motion of a particle in one or more dimensions; and in certain cases limitations may be imposed on the motion in the form of barriers which, once reached, 'absorb' the particle and end the motion (as distinct from reflecting it). The boundary lines terminating a **Sequential Sampling Process** are of this type.

Absorbing Markov Chain An important class of non-ergodic Markov chains; namely those that possess **Absorbing States**.

Absorbing Region A generalisation from one dimension of the idea of an **Absorbing Barrier**.

Absorbing State A state in a stochastic process which, once reached, cannot be quitted. It is also known as an **Absorbing Barrier**, but this term is generally used where the process terminates when the barrier is reached.

Accelerated Stochastic Approximation A method proposed by Kesten (1958) for increasing the speed of convergence in a **Robbin's-Munro, Kiefer-Wolfowitz** or other type of stochastic approximation. The method depends upon the number of changes of sign, in relation to the size of the step from one approximation to the succeeding one.

Acceleration by Powering Certain arithmetic processes for extracting the characteristic roots of a matrix by iterative methods can be shortened (i.e. the number of iterations required can be reduced) by operating on a power of the matrix. The calculations and the convergence of the iterative process are then said to be accelerated by powering.

Acceptable Quality Level (AQL) The proportion of effective units in a batch which is regarded as desirable by the consumer of the batch; the complement of the proportion of defectives which he is willing to tolerate.

Acceptable Reliability Level A concept, analogous to **Acceptable Quality Level**, in which the characteristic is expressed in terms of failure rate per unit time, e.g. 1000 hours.

Acceptance Boundary An alternative name for **Acceptance Line**.

Acceptance Control Chart A simple form of quality control chart in which lots are accepted or rejected according to whether plotted values of the mean value of the lot or batch fall inside or outside the specified acceptance control limits.

Acceptance Error An alternative name for β-**error**.

Acceptance Inspection The inspection of items to determine whether they are acceptable, that is to say, conform to standards required by the intending user.

Acceptance Line In sequential analysis the graph of the **Acceptance Number** as ordinate against the sample number as abscissa. It is also known as the **Acceptance Boundary**. There is a corresponding rejection line.

Acceptance Number In acceptance inspection schemes, and in sequential analysis generally, the number of defective items (dependent on the sample number) which, if attained, requires the acceptance of the batch under examination. It is usually accompanied by a rejection number, which is the number of defectives requiring the rejection of the batch. If the number of defectives at any one stage is above the acceptance and below the rejection number, sampling is continued.

Acceptance Region In the theory of testing hypotheses, a region in the **Sample Space** such that, if a sample point falls within it, the hypothesis under test is accepted.

Accumulated Deviation The graduation of a grouped frequency distribution provides, for each grouping interval, a 'theoretical' or 'expected' frequency. The difference of this quantity and the observed quantity taken with regard to sign and cumulated for increasing values of the variates starting from the least, is called the accumulated deviation. It is equivalent to the difference between the observed and theoretical distribution functions of the grouped distributions.

Accumulated Process A stochastic process derived from the cumulated sum of random variables. For example, the total damage during time t of a sequence of thunderstorms forms an accumulated process.

Accuracy Accuracy in the general statistical sense denotes the closeness of computations or estimates to the exact or true values. In a more specialised sense the word also occurs as meaning (*a*) in relation to an estimator, *unbiassedness* (see **Unbiassed Estimator**); (*b*) in relation to the reciprocal of the standard error, the **Precision**. Neither usage can be recommended.

For intrinsic accuracy, see **Intrinsic**.

Adaptive Optimisation A system of control for physical processes proposed by Box & Jenkins (1962), based upon statistical adaptation as a series of observations proceeds through time. It has a particular application in the field of statistical quality control.

Addition, of Variates Let $x_1, x_2,..., x_n$ be n variates with a joint distribution. The univariate quantity z with distribution $\Pr(z \leqslant z_0 = x_1 + x_2 + ... + x_n)$ is said to be the sum of the n variates, which are said to be added to yield z. The usage sometimes gives rise to confusion; for example $x_1 + x_2 = x_1 + x_3$ does not involve, as it would for ordinary algebraic quantities, that if $x_1 + x_2$ and $x_1 + x_3$ follow the same law of distribution, x_2 and x_3 have the same distribution. [See also **Convolution**.]

2

Additive Model A model in which the factors influencing the dependent variable have an additive effect. In older usage the factors were considered as statistically independent, but modern usage does not impose this limitation, e.g. a linear regression equation is additive even if the regressors are dependent. [See also **Additivity of Means**.]

Additive Properties of χ^2 If two independent variates u_1 and u_2 are distributed as χ^2 with v_1 and v_2 degrees of freedom respectively, their sum u_1+u_2 is distributed as χ^2 with v_1+v_2 degrees of freedom. This is sometimes called the additive property of χ^2.

Additive (Random Walk) Process A stochastic process with independent increments, that is to say, a process $\{x_t\}$ is additive if, for $t_1<t_2<...<t_n$, the differences $x_{t_2}-x_{t_1}$, $x_{t_3}-x_{t_2}$, etc. are independent. The expressions 'differential process', 'process with independent increments' are equivalent, but are usually confined to the case when the parameter t is continuous. When the increments are discrete and finite, the process may also be said to be additive; a synonym in this case is 'Random Walk' process.

Additivity of Means In a multiple-factor experiment, a hypothesis often considered is that the effects of the factors are independent and additive. The hypothesis is then described (perhaps rather loosely) as additive. The same idea is expressed by speaking of the means of factor-effects over a number of observations as additive.

Admissible Decision Function In the general theory of statistical decision functions an admissible decision function is one for which it can be shown that there is no decision function which is uniformly better. The criterion defining 'better' is usually stated in terms of **Risk Functions** but other definitions are possible.

Admissible Hypothesis Generally, a hypothesis is said to be admissible when it is possible within the conditions of the problem. More specifically, if there is a distribution of known mathematical form depending on k unknown parameters θ then a statistical hypothesis can be stated by specifying any k set of θ's which is *a priori* possible. Such hypotheses form the set of admissible hypotheses.

Admissible Numbers In the terminology of A. H. Copeland a (real) number is said to be admissible if its digits obey the conditions required by von Mises' theory for a random series. For example, if it is written in the scale of 10 the frequency of occurrence of the digits 0 to 9 under any systematic method of selection (independent of the actual values) tends to the limit 1/10. [See **Irregular Kollektiv**.]

Admissible Strategy A strategy such that no other strategy exists with at least as good an outcome for each possible state of nature and with a better one for some particular state.

Admissible Test A test of a hypothesis is said to be admissible if, in respect of a particular class of alternatives, there is no other test with **uniformly greater power**; e.g. if a uniformly most powerful test exists, no other test is admissible.

Affinity A concept introduced by Bhattacharya (1943) as a measure of divergence or distance between two statistical populations. It provides, for example, a generalisation of **Mahalanobis Distance.**

Age Dependent Birth and Death Process A **Birth and Death Process** where the birth (fission) and death (extinction) rates are not constant over time but change in a manner which is dependent upon the age of the individual or homogeneous group concerned. This concept is closely related to the **Hazard Rate** or **Force of Mortality.**

Age Dependent Branching Process See **Branching Process.**

Age Specific Death Rate A term generally used in life analysis of human or biological populations to denote the probability of dying in the next unit time period of an entity that had survived for a stated period of time. Substitution of word 'failure' for 'death' renders the term useful in connection with physical systems or components.

Aggregation A word used to denote the compounding of primary data into an aggregate, usually for the purpose of expressing them in a summary form. For example, national income and price index-numbers are aggregative, as contrasted with the income of an individual or the price of a single commodity.

Aggregative Index An index-number which is constructed by aggregating a number of items (as distinct, for example, from picking out a representative item). In price indices, if p_o typifies the prices in the base period and p_n those for the current period and q_o and q_n typify the quantities in the base and current periods respectively, then the two principal forms of aggregative index are:

$$\frac{\Sigma(p_nq_o)}{\Sigma(p_oq_o)} \text{ and } \frac{\Sigma(p_nq_n)}{\Sigma(p_oq_n)}$$

where the summation takes place over the commodities to be aggregated. [See **Laspeyres' Index, Paasche's Index**.]

Aggregative Model The statistical study of an economic system usually involves setting up a model expressing known relations between, or hypotheses concerning, the variables under study. When these 'variables' are themselves constructed from groups of individual variables, as when a price index-number is substituted for a set of prices, the model is said to be aggregative.

Agreement, Coefficient of This coefficient relates to the situation where m observers provide paired comparisons

3

for n objects. A coefficient of agreement between the verdicts of the m observers is given by

$$u = \frac{8\Sigma}{m(m-1)(n-1)n} - 1,$$

where Σ is the sum of the number of agreements between pairs of judges. The coefficient of agreement is a generalisation of Kendall's coefficient of rank correlation (τ), to which it reduces when $m = 2$. [See **Kendall's Tau**.]

Aitken Estimator An exact least squares estimator, due to Aitken (1948), which minimises the generalised variance for linear estimation of a vector of parameters θ.

Aleatory Variable A **Random Variable or Variate**. There being no word similar to 'random' in Romance languages, the word is usually translated as aléatoire (French) or aleatorio (Italian) and works written in those languages have influenced some English writers to introduce an English equivalent 'aleatory'. There is no need for this.

Algorithm A word of changing meaning in recent years. Formerly (and still in some other languages) it was almost equivalent to 'formula'. More recently, and especially in computation, it has come to mean an explicit relation which permits of the calculation of an assigned quantity by iterative processes converging on the true value; and, slightly more generally, any explicit relation which leads to the desired quantity in however protracted a manner. For example, if x_n is an 'estimated' value of $\sqrt{2}$, the algorithm

$$x_{n+1} = \tfrac{1}{2}\left(x_n + \frac{2}{x_n}\right)$$

will lead to an improved estimate.

Alias This term is used in two branches of statistical analysis. In connection with design of experiments, when a factorial design is only fractionally replicated certain comparisons do not distinguish between some of the treatment combinations; these are said to be aliases.

In connection with the spectrum analysis of time series, an analogous concept arises because a single width of the interval of observation does not permit a distinction between certain angular frequencies.

Alienation, Coefficient of If r is the product-moment correlation between two variates the coefficient of alienation k is $\sqrt{(1-r^2)}$. It is equal to the square root of the **Coefficient of Non-determination**. The term occurs mainly in psychology and is apparently due to T. L. Kelley (1919).

Allocation, of a Sample The way in which sample numbers are assigned to various parts of a population by the sampling plan; e.g. for a stratified population it may be decided to allocate the total sample number to the strata in proportion to the numbers of individuals in those strata.

Allokurtic See **Kurtosis**.

Allowable Defects In quality control, the number of allowable defects in a sample is the critical number (designated by the particular scheme of sampling inspection) such that, if this number is exceeded, the whole of the remainder of the inspection-lot must be examined or the lot rejected out of hand. [See also **Acceptance Number**.]

Almost Certain A probability dependent on a parameter n may tend to unity as n tends to infinity. The event to which it relates is then said to be 'almost certain' in the limit. More generally, if the measure of an event is unity and its converse has measure zero but does not relate to an empty set, the event is 'almost' certain.

Almost Stationary A term proposed by Granger & Hatanaka (1964) to cover the situation where the series resulting from wide interval sampling of a non-stationary time series is itself stationary.

Alphabet A term used in information theory instead of 'sample space' where the outcome of a trial can be one of a finite number of results denoted by the letters A, B, C, ... For example, if the outcome is success or failure, denoted by A and B respectively (or 0 and 1), the experience is drawn from a space or 'alphabet' consisting of two letters.

Alternating Renewal Process A special case of the two-state **Semi-Markov Process** in which the two types of intervals appear alternately.

Alter Periodogram A form of analysis suggested by Dinsmore Alter in 1937 for investigating periodicities in time-series. It depends on the behaviour of the sum of *absolute* differences of terms k time-intervals apart, for varying values of k. If the squares of differences were taken instead of absolute differences the resulting form would be simply related to the **Correlogram**.

Alternative Hypothesis In the theory of testing hypotheses, any admissible hypothesis alternative to the one under test.

Amount of Information See **Information**.

Amount of Inspection In quality control, the number of items (size of sample) taken from each lot and inspected according to the particular sampling plan being employed.

Amplitude In relation to a time series, the amplitude of a fluctuation is the value of the ordinate at its peak or trough taken from some mean value or trend line. Sometimes the difference between values at peak and trough is referred to as 'amplitude'.

Amplitude Ratio Some time series exhibit seasonal movements which are regular in phase but vary in amplitude from year to year. The actual amplitude in any year, expressed as a proportion of the average amplitude taken over a long period is called the amplitude ratio. It affords a measure of departure from normal seasonal variation.

Analogue Computer A device which simulates some mathematical process or relationship, and hence one in which the results of the process can be observed as physical quantities, such as voltage or current. [See **Digital Computer**.]

Analysis of Covariance See **Covariance Analysis**.

Analysis of Dispersion A generalisation of the **Univariate Analysis of Variance** providing a theory for multivariate tests of significance. It is sometimes known as multivariate analysis of variance.

Analysis of Variance See **Variance Analysis, Variance Component**.

Analytic Regression A regression relation where the independent variables (regressors) are formed by polynomials, trigonometric sums or other analytic expressions.

Analytic Survey A (sample) survey where the primary purpose of the design is the comparison between sectors or subgroups of the population sampled.

Analytic Trend See **Trend**.

Ancillary Information This phrase is mostly used in the customary sense of information which is additional or supplementary to the main body of information available. It also occurs in the specialised sense of 'information' conveyed by **Ancillary Statistics**. [See **Supplementary Information**.]

Ancillary Statistic According to the theory of inference associated with the name of R. A. Fisher the likelihood function contains all the information provided by the sample about the unknown parameters of the population which it may be desired to estimate. If there are no **Sufficient Estimators** some loss of information in estimation is inevitable; but it can be reduced by taking additional functions of the variates which can be combined with the **Maximum Likelihood Estimator**. Such functions are called ancillary statistics.

Anderson-Darling Statistic A modified version (W_n^2) of the **Cramer-von Mises** ω^2 **Test Statistic**

Angular Transformation A variate transformation expressing a variate y in terms of a variate x by a trigonometrical formula such as the **Arc Sine Transformation**.

Anomic See **Clisy**.

Antimode The variate value, if any, for which a frequency distribution has a minimum. The expression is usually confined to the case where the minimum is not zero and is a true minimum, e.g. the zero tails of a normal distribution at infinity are not antimodes in this sense.

*Antiseries (Antiserie) In Italian usage, given a series $x_1, x_2, ..., x_n$, with weights $p_1, p_2, ..., p_n$, the antiseries is defined as the reciprocals $1/x_1, 1/x_2, ..., 1/x_n$ taken with weights $x_1 p_1, x_2 p_2, ..., x_n p_n$.

Antithetic Transforms Transformations of variates which will bring them into the scope of mutually compensated variations as arise in **Antithetic Variates**.

Antithetic Variates A concept introduced by Hammersley & Morton (1956) in connection with control variates in Monte Carlo analysis. Antithetic variates are such that the set of estimators involved mutually compensate the variations.

Aperiodic State See **Period (Stochastic Process)**.

Approximation Error In general, an error due to approximation in numerical calculations as distinct, for example, from an error of observation. More particularly, a rounding error. [See **Rounding**.]

Arbitrary Origin In the calculation of the moments of a frequency distribution, it is often desirable to calculate the moments about some convenient, though arbitrary, origin before transforming them to moments about the arithmetic mean as the origin. Moments about an arbitrary origin are usually written with a prime: μ_r', as distinct from those about the mean which are written without a prime: μ_r.

Arc Sine Distribution A distribution, occurring in the theory of recurrent events, of the form

$$F = \frac{2}{\pi} \arc \sin \sqrt{x}, \quad 0 \leqslant x \leqslant 1.$$

Arc Sine Transformation A transformation of a variate x into a variate y by some relation of the type $y = \arc \sin (x+k)$ where k is chosen as convenient. The object is usually to make the variance of y more 'stable' than that of x, i.e. to render it more nearly constant for different populations from which x might arise.

Area Comparability Factor In the analysis of vital statistics it sometimes occurs that, whereas the population and deaths at each age are known for the whole country and its localities for the census year, in subsequent years only the total deaths and populations of the localities are known. The problem of adjusting the local crude **Death**

Rates in these later years for comparisons between localities is met by using an Area Comparability Factor.

A common form of this factor is obtained by dividing the average Age Specific Death Rate for the whole country in the census year by a similar average for the locality. The corrected death rate for any given locality is obtained by multiplying the crude death rate by the Area Comparability Factor.

Area Sampling A method of sampling used when no complete frame of reference is available. The total area under investigation is divided into small sub-areas which are sampled at random or by some restricted random process. Each of the chosen sub-areas is then fully inspected and enumerated, and may form a frame for further sampling if desired. The term may also be used (but is not to be recommended) as meaning the sampling of a domain to determine area, e.g. under a crop.

Arfwedson Distribution A discrete probability distribution proposed by Arfwedson (1951) for an urn sampling problem where drawings are random with replacement. An urn containing N numbered balls is sampled n times; the probability of achieving v different balls ($v = 1, \ldots, n$) is

$$F(N, n, v) = \binom{N}{v} \frac{f(n, v)}{N^n}$$

where $f(n, v)$ is the number of series containing v selected numbers.

Arithmetic Distribution A discontinuous distribution for which the variate values are equidistant and can be represented as $a \pm \lambda b$, $\lambda = 0, 1, 2, \ldots$.

In the multivariate analogue the expression 'lattice distribution' is used.

Arithmetic Mean The arithmetic mean of a set of values $x_1, x_2, \ldots, x_n$ is their sum divided by their number, namely $\frac{1}{n} \sum_{j=1}^{n} x_j$. It is often denoted by a bar, e.g. $\bar{x}$.

For a continuous distribution with distribution function $F(x)$ the arithmetic mean is the integral $\int_{-\infty}^{\infty} x \, dF$ and is usually denoted by μ_1 or μ_1' [see **Moments**]. This integral, interpreted in the Riemann-Stieltjes sense, may also be regarded as defining the arithmetic mean in the case of a discontinuous distribution.

In current English usage the word 'arithmetic' is frequently omitted so that where a 'mean' is mentioned the arithmetic mean is to be understood.

Armitage's Restricted Procedure See **Restricted Sequential Procedure**.

Array In the most general sense, an explicit display of a set of observations. More usually, the term denotes some special arrangement of the observations, e.g. in order of magnitude. A *frequency array* is an array of frequencies according to variate values, that is to say, a frequency distribution. The term 'array' is often used for the individual frequency distributions which form the separate rows and columns of a bivariate frequency table.

Arrival Distribution The probability distribution of the number of items of a defined type which arrive at a given point per unit time. The concept is particularly appropriate to service situations where congestion can lead to queuing: the arrivals can be at a service point or the end of the queue.

Ascertainment Error See **Non-Sampling Error**.

Assignable Variation That part of the variation in, for example, an industrial process which can be attributed to specific causes; such as poorly trained operators, faulty machine settings, sub-standard raw material, etc.

Associable Design A concept proposed by Shah (1960) that, if there are s partially balanced incomplete block designs each with v treatments and b blocks, these will be deemed associable designs if the orthogonal matrix $\mathbf{L}$ is also **Canonical**, i.e. that $\mathbf{L}'\mathbf{N}_i\mathbf{N}_j'\mathbf{L}$ is diagonal for all $i, j = 1, 2, \ldots, s$.

Associate Class The concept of associate class in certain experiment designs, e.g. partially balanced incomplete blocks, was introduced by Bose & Nair (1939). The complete definition is somewhat lengthy, furthermore there are five types of such relationships between treatments which are known as **Association Schemes**. The most important of these is the **Group Divisible Designs**.

Association In the most general sense, the degree of dependence, or independence, which exists between two or more variates whether they be measured quantitatively or qualitatively. More narrowly, the term is mostly used to denote the relationship between variates which are simply dichotomised, namely in a 2×2 table as distinct from **Contingency**, which measures relationship in an $m \times n$ table of attributes, and **Correlation**, which measures relationship in a classification according to specified ranges of variate values.

If, in a two-fold table, the frequencies of the attributes (A, B) (not-A, B), $(A,$ not-$B)$ and (not-A, not-$B)$ are respectively a, b, c, d, the association between A and B is said to be positive if

$$a > \frac{(a+b)(a+c)}{a+b+c+d}$$

within sampling limits, and negative in the contrary case; if the inequality becomes an equality the attributes are independent.

Association, Coefficient of A measure of the degree of association between two attributes. In the notation of the previous item, one such coefficient (due to Yule, 1900) is

$$Q = \frac{ad-bc}{ad+bc}.$$

Another coefficient is

$$V = \frac{ad-bc}{|\{(a+b)(a+c)(b+d)(c+d)\}^{\frac{1}{2}}|}.$$

A further coefficient of a similar kind

$$Y = \frac{1-\sqrt{(bc/ad)}}{1+\sqrt{(bc/ad)}}$$

is called the coefficient of colligation.

The concept of the coefficient of association has been developed by Goodman & Kruskal (1954, 1959 and 1960) especially for the more general case of the $m \times n$ table. [See also **Contingency**.]

Association Scheme In the design of partially balanced incomplete block experiments, the arrangement which shows which treatments do, or do not, appear together in a block is called the association scheme.

Assumed Mean An **Arbitrary Origin** or working mean for the calculation of moments. The expression is not to be recommended.

Asymmetrical Distribution A distribution which is not symmetrical, that is to say, for which there is no central value a such that $f(x-a) = f(a-x)$, $f(x)$ being the frequency function. [See **Skewness**.]

Asymmetrical Factorial Design See **Symmetrical Factorial Design**.

Asymmetrical Test See **One-sided Test**.

***Asymmetry (Asimmetria)** See ***Symmetry**.

Asymptotic Bayes Procedure A term proposed by Whittle (1965) for the concept, where c is the cost of experimentation, of

$$\lim_{c \to 0} \left\{ \frac{F(S,d)}{F(S)} \right\} = 1$$

and $F(S, d)$ is the expected future loss from using sequential procedure d from a base of information S.

Asymptotic Distribution The limiting form of a frequency or probability distribution dependent on a parameter, such as sample-number or time, as that parameter tends to limit, usually infinity.

Asymptotic Efficiency The efficiency of an estimator in the limit as the sample size increases. [See **Efficient Estimator**.]

Asymptotic Normality A distribution dependent on a parameter n, usually a sample number, is said to be asymptotically Normal if, as n tends to infinity, the distribution tends to the Normal form.

Asymptotic Relative Efficiency The limiting value, as the sample sizes tend to infinity, of the ratios encountered in the various forms surrounding the concept of **Relative Efficiency**.

Asymptotic Standard Error The standard error of a statistic nearly always depends on the sample number n. If, as n tends to infinity, the standard error S is asymptotically equivalent to a quantity Q, i.e. is such that S/Q tends to unity, then Q may be termed the asymptotic standard error. As so defined, Q is not unique but in the vast majority of cases S^2 is expressible as a term in n^{-1} plus terms of lower order in n, and the first is taken as the square of Q.

In many cases standard error depends on (unknown) parent parameters. It is customary in large sample theory to estimate these parameters from the sample itself and to substitute the estimates in formulae for standard error. The results are, in general, exact only asymptotically.

Asymptotically Efficient Estimator See **Efficient Estimator**.

Asymptotically Locally Optimal Design A term relating to the design of bio-assay experiments (Andrews & Chernoff, 1955) where the doses cannot be known exactly and it is desirable to minimise the asymptotic variance of the response parameter.

Asymptotically Most Powerful Test A test of significance which remains **Uniformly Most Powerful** as the sample size tends to infinity.

Asymptotically Stationary A concept of limiting probability associated with certain stochastic processes, in particular Markov chains with stationary transition probabilities. Given a process $\{x_n\}$, a stationary process $\{x_n^*\}$ exists such that

$$\lim_n \Pr\{(x_n, x_{n+1} \ldots) \epsilon B\} = \overset{*}{\Pr}\{(x_n^*, x_{n+1}^* \ldots) \epsilon B\}$$

for every $B\epsilon\mathscr{B}_\infty$.

Asymptotically Subminimax A minimax decision procedure which is desirable for some small sample problems but undesirable for large samples was termed asymptotically minimax by Robbins (1951).

Asymptotically Unbiassed Estimator See **Unbiassed Estimator**.

Attack Rate In medical statistics, the ratio between the number of new cases of sickness and the population at risk in a unit time period.

Attenuation Where observations on bivariate material are subject to errors of measurement the true correlation between the variates will be obscured, usually being underestimated. The correlation is then said to be attenuated. In 1904 Spearman advanced the following formula which would correct this attenuation

$$r'_{xy} = \frac{r_{xy}}{\{r_{xx} r_{yy}\}^{\frac{1}{2}}}$$

where r_{xy} is the geometric mean of correlations between independent determinations of x and y; r_{xx} and r_{yy} are the means of correlation between independent determinations of x and of y and r'_{xy} is the corrected correlation. [See **Reliability Coefficient**.]

*Attraction, Index of (Indice di Attrazione) An Italian index of concordance, between qualitative variables. If the marginal frequencies in an $s \times s$ table are f_i. and $f_{\cdot j}$ $(i, j, = 1, 2, ..., s)$ and the frequency in the ith row and jth column is f_{ij}, the index is

$$\left(n \sum_{i=1}^{s} f_{ii} - \sum_{i=1}^{s} f_i. f_{\cdot j}\right) \Big/ \left(n \sum_{i=1}^{s} f'_{ii} - \sum_{i=1}^{s} f_i. f_{\cdot j}\right)$$

where n is the total frequency and f' is the smaller of $f_i., f_{\cdot j}$. This definition applies if the numerator is positive. In the contrary case the denominator is replaced by $\sum f_i. f_{\cdot j} - \sum f''_{ii}$, where f'' is zero if $f_i. + f_{\cdot j} < n$ and $f_i. + f_{\cdot i} - n$ in other cases.

The index of resemblance (*rassomiglianza*) is defined as

$$\frac{n \sum f_{ii} - \sum f_i. f_{\cdot j}}{\{(n^2 - \sum f_i.^2)(n^2 - \sum f_{\cdot j}^2)\}^{\frac{1}{2}}}$$

Attribute A qualitative characteristic of an individual, usually employed in distinction to a variable or quantitative characteristic. Thus, for human beings sex is an attribute but age is a variable. Very often attributes are dichotomous, each member of a population being allotted to one of two groups according to whether he does or does not possess some specified attribute; but manifold classification can also be carried out on the basis of attributes, as when individuals are classified as belonging to various blood-groups.

*Atypical Characteristic (Carattere Atipico) See ***Characteristic**.

Auto-catalytic Curve See **Growth Curve**.

Autocorrelation The internal correlation between members of series of observations ordered in time or space.

Autocorrelation Coefficient If ξ_t is a stationary stochastic process with mean m and variance σ^2 the autocorrelation coefficient of order k is defined by

$$\rho_k = \rho_{-k} = \frac{1}{\sigma^2} \mathscr{E}(\xi_t - m)(\xi_{t+k} - m)$$

where the expectation relates to the joint distribution of ξ_t and ξ_{t+k}.

In a slightly more limited sense, if x_t is the **Realisation** of a stationary process with mean m and variance σ^2 the autocorrelation coefficients are given by a similar formula where the expectation is to be interpreted as

$$\lim_{n_2 - n_1 \to \infty} \frac{1}{(n_2 - n_1)} \sum_{j=n_1}^{n_2} (x_{t+j} - m)(x_{t+j+k} - m).$$

In a more limited sense still, the expression is applied to the correlations of a finite length of the realisation of a series. Terminology on the subject is not standardised and some writers refer to the latter concept as **Serial Correlation**, preferring to denote the sample value by the Latin derivative 'serial' and retaining the Greek derivative 'auto' for the whole realisation of infinite extent.

Analogous expressions, omitting division by σ^2, provide *autocovariances*. [See also **Correlogram**.]

Autocorrelation Function The autocorrelation function of a stationary stochastic process is the **Autocovariance** divided by the variance, e.g. for a series with zero mean and range $a \leq t \leq b$, defined at each time point, is given by

$$\rho(\tau) = \frac{1}{b-t-a} \int_a^{b-t} u(t)\, u(t+\tau) dt \Big/ \frac{1}{a-b} \int_a^b u^2(t) dt.$$

The limits a and b may be infinite subject to the existence of the integrals or sums involved.

The numerator of this expression is called the auto-covariance function.

Autocovariance See **Autocorrelation**.

Autocovariance Function For any **Stationary Process** the function $\gamma(k) = C(x_{t+k}, x_t)$ where C represents the co-variance of the terms in the bracket is known as the autocovariance function. It has an obvious generalisation to continuous processes when, subject to existence, k may be continuous.

Autocovariance Generating Function A function of a variable z which, when expanded as a power series in z, yields the autocovariance of a **Stationary Process**.

Autonomous Equations In econometrics, an equation which describes the behaviour of one particular group or sector of the economy, e.g. a demand equation describing only buyer's behaviour, is said to be autonomous if it is affected only by changes in the behaviour of that particular group or sector. Equations of this kind are sometimes termed 'structural equations' but this is not to be recommended, since such equations may include variates from the whole system.

Autoregression The generation of a series of observations whereby the value of each observation is partly dependent upon the values of those which have immediately preceded it, i.e. each observation stands in a regression relationship with one or more of the immediately preceding terms. A scheme of autoregression may be regarded as a **Stochastic Process** of a conditional kind.

Autoregressive Model An econometric model based upon the **Autoregressive Process** but also containing lagged versions of some or all of the **Endogenous Variables** appearing in the model specification.

Autoregressive Process A stochastic process suggested by

Yule (1921) for the representation of a system oscillating under its own internal forces which, being damped, are regenerated by a stream of random external shocks. The realisation of such a scheme in the form of a series defined at equidistant points of time may be expressed as

$$u_{t+j} = f(u_t, u_{t+1}, u_{t+2}, ..., u_{t+j-1}) + \epsilon_{t+j} \qquad (1)$$

where ϵ is a random variable and f represents a functional relationship. In most practical applications this is taken as linear, e.g.

$$u_{t+2} = \alpha u_{t+1} + \beta u_t + \epsilon_{t+2} \qquad (2)$$

and the name derives from the fact that this may be regarded as a regression of u_{t+2} on u_{t+1} and u_t.

The expression is now used to denote any process of type (2) even if it is not stationary.

Autoregressive Series A series generated by an **Autoregressive Process**; the realisation of an autoregressive process.

Autoregressive Transformation If there is autocorrelation in the error term of an autoregressive process it is sometimes possible to transform the original variates to new variates such that the autoregressive scheme in the transformed variates has an uncorrelated error term. This procedure is known as an autoregressive transformation.

Auto Spectrum A spectrum resulting from the analysis of a single time series. [See also **Cross Spectrum**.]

Average A familiar but elusive concept. Generally, an 'average' value purports to represent or to summarise the relevant features of a set of values; and in this sense the term would include the median and the mode. In a more limited sense an average compounds all the values of the set, e.g. in the case of the arithmetic or geometric means. In ordinary usage 'the average' is often understood to refer to the arithmetic mean.

For Average Deviation, etc., see **Mean Deviation**, etc.

Average Amount of Inspection In quality control, the average number of items inspected per lot. The expression is used either where the sample size is not fixed, as in sequential analysis, or where all lots not accepted with fixed sample size are separately inspected *in toto* and rectified.

Average Article Run Length The **Average Run Length** is the average number of *samples* taken before a signal showing the need for action is given. The average article run length is the average number of *items* sampled before action is taken.

Average Corrections (for grouping) Corrections to moments for grouping can be regarded in two ways, according as the end-points of the grouping mesh are treated as located at random on the variate scale or not. If they are located at random, correction terms can be derived as the deviations of 'grouped' moments from the true values averaged over all possible positions of the grouping mesh. These are known as average corrections. There is apt to be confusion with the case where the mesh is not treated as randomly located, owing to the fact that the average corrections have the same form as **Sheppard's Corrections** which relate to a fixed grouping but require for their validity conditions of high contact on the terminals of the distribution.

Average Critical Value Method A method for assessing the relative efficiency of statistical tests in regression analysis of time series due to Geary (1966). Broadly speaking the average critical value is equivalent to a power function value of one half for large sample sizes.

Average Deviation A synonym of **Mean Deviation**, not to be recommended.

Average Extra Defectives Limit A method of insuring the effectiveness of continuous sampling plans by adjusting a process that has gone out of control. This effectiveness increases as the number of defectives that slip through uninspected decreases before the plan adjusts to the deterioration in quality.

Average Inaccuracy A measure proposed by Theil (1965) in a regression context, defined as the expected sum of squares of the squared estimation errors as a fraction of the expected sum of squares of the disturbances estimated.

Average of Relatives An average, usually in the form of an index number, of a set of *relatives*, that is to say values obtained as the ratio of a magnitude in the given period to the corresponding magnitude in the base period. In price index numbers, the price relatives are usually weighted by the values either of the base period or of the given period. Where the weights used are the values in the base year the formula reduces to that of **Laspeyres**. If values in the given period are used with a harmonic average the formula reduces to that of **Paasche**.

Average Outgoing Quality Level See **Average Quality Protection**.

Average Outgoing Quality Limit (AOQL) The greatest percentage of defective items that can be found as an average of outgoing quality. A sampling inspection plan may also be designed with the AOQL as a parameter in the sense of it being the maximum percentage defective that can be accepted.

Average Quality Protection In quality control, a procedure which aims at keeping the proportion of defective items in deliveries of a manufactured product (after inspection and rectification if necessary) at or below some specified limit. This limit, usually expressed as a percentage, is called the **Average Outgoing Quality Limit**; it may be given in terms of defective or effective units, e.g. as either 5 per cent defective or 95 per cent effective. The actual proportion is called the average outgoing quality *level*. In cases where the intention is to control the proportion of defectives in each delivered lot (as distinct from the average of a number of lots) the procedure is called lot quality protection, and the limit is the lot tolerance limit or lot tolerance per cent defective. [See also **Consumer's Risk, Rectifying Inspection.**]

Average Run Length (ARL) The average run length of a sampling inspection scheme at a given level of quality is the average number of samples of *n* items taken in the period between the time when the process commences to run at the stated level and that at which the scheme indicates a change from acceptable to rejectable quality level is likely to have occurred. This concept must not be confused with **Average Article Run Length**.

Average Sample Number Curve The graph of the **Average Sample Number Function**, with the function as ordinate against the parameter as abscissa.

Average Sample Number (ASN) Function In sequential analysis, the expected or average sample number required to reach a decision, considered as a function of the parameter concerning which the decision is to be made; e.g. in quality control for defective items, the average number inspected per batch, for acceptance of a batch, as a function of the proportion of defects produced by the manufacturing process.

Average Sample Run Length An alternative name sometimes used for **Average Run Length** in order to distinguish it from the concept of **Average Article Run Length**.

Axonometric Chart A chart devised for the purpose of representing a solid on a plane surface; a **Stereogram**.

β-**Error** In the theory of testing hypotheses, an error incurred by accepting a hypothesis when it is incorrect. This kind of error is also referred to as **Consumer's Risk**, an **Error of Second Kind** or a Type II error.

Bachelier Process See **Brownian Motion Process.**

Backward Equations See **Kolmogorov Equations.**

Bagai's Y_1 Statistic A variant of the **Wilks-Lawley U-statistic** proposed by Bagai (1962) in the form

$$Y = |\mathbf{A}| / |\mathbf{C}|$$

where A and C are independent sums of product matrices based upon sample observations with degrees of freedom n_1 and n_2 respectively.

Bahadur Efficiency An approximate measure of asymptotic relative performance for sequences of test statistics proposed by Bahadur (1960).

Balanced Confounding In the design of factorial experiments it is sometimes possible to arrange for different components of interaction to be **Partially Confounded** to the same extent. The confounding of these components of interaction is then said to be balanced.

Balanced Differences In a systematic sample of an ordered series the selected units are not located at random and therefore no fully valid estimate of sampling error is possible in general. One method of overcoming this difficulty is to construct artificial strata by dividing the series into 'blocks' of equal length and to regard the members falling within a block as having been chosen at random within that block.

If there are only two members in the block an estimate of error is based on their difference. If there are more it may be based on more complicated linear functions designed to eliminate systematic error, e.g. for seven members there would be used 'balanced differences' of the form:

$$d = \tfrac{1}{2}y_1 - y_2 + y_3 - y_4 + y_5 - y_6 + \tfrac{1}{2}y_7.$$

The numbers of terms is arbitrary but seven or nine will eliminate most of any systematic component of variation.

Similar ideas are applicable to systematic sampling in more than one dimension.

Balanced Factorial Experimental Design A generalisation by Shah (1958, 1960) of work by Bose (1947). The following conditions must be satisfied, (a) each treatment is replicated the same number of times, (b) each block has same number of plots, (c) estimates of contrasts for different interactions are uncorrelated, and (d) complete balance is achieved over each interaction. This latter is so only if all normalised contrasts in an interaction are estimated with the same variance.

Balanced Incomplete Block See **Incomplete Block.**

Balanced Lattice Square See **Square Lattice.**

Balanced Sample If the mean value of some characteristic is known for a population and the value of the characteristic can be ascertained for each member of a sample, it is possible to choose the sample so that the mean value of the characteristic in it approximates to the parent mean. Such a sample is said to be balanced.

The object of balancing is to obtain a sample which is representative of the parent in respect of some other characteristic for which the parental value is not known. Authorities differ about the value and validity of the method.

Ballot Theory Various generalisations and extensions of a theorem by Bertrand (1887): if a candidate (A) scores a votes and candidate (B) scores b votes, $a > \mu b$ and μ is non-negative integer, then the probability throughout the counting that the votes for (A) are always greater than μ times the votes for (B) is $\mathrm{Pr} = (a - \mu b)/(a + b)$.

Band Chart When a complex quantity, that is to say, a magnitude which is the sum of certain component parts, is recorded for successive intervals of time it is often convenient to show the movements of the total on a chart which also shows, for each point of observation, partition into components. The movement of the intervals representing the components describes bands across the chart which may be coloured or cross-hatched to assist the visual interpretation.

Bar Chart The graphical representation of frequencies or magnitudes by rectangles drawn with lengths proportional to the frequencies or magnitudes concerned. There are various complications which can be incorporated into this simple concept. For example, component parts of a total can be shown by sub-dividing the length of the bar. Two or three kinds of information can be compared by groups of bars each one of which is shaded or coloured to aid identification. Figures involving increases and decreases can be shown by using bars drawn in opposite directions, above and below a zero line.

Bartholomew's Problem Parameter estimation in life testing procedures was originally associated with observations from a single exponential distribution. Bartholomew considered, under the assumption that all items that are put on test come from the same exponential life distribution, the problem of obtaining the maximum likelihood estimate of the mean life when data are available giving the dates of installation on life test of n items of equipment, the life test being terminated at some time T without reference to the experiment.

Bartlett and Diananda Test An extension (1950) of **Quenouille's Test** for the fitting of autoregression schemes to time series.

Bartlett Relation A differential equation relationship proposed by Bartlett (1949) for a class of stochastic processes (including the Markov Process) between the characteristic function of the vector variate $\mathbf{x}$ and the derivate cumulant function.

Bartlett's Collinearity Test A test for direction and collinearity of latent roots and vectors in principal component analysis. It was proposed by Bartlett (1951) and developed by Kshirsagar & Gupta (1965). The question at issue is whether some of the roots and associated latent vectors are distinguishable. [See also **Multicollinearity**.]

Bartlett's Decomposition The **Wishart Distribution** was shown by Bartlett (1933) to be comprised of k chi-squared variables and $\frac{1}{2}k(k-1)$ standard Normal variables which are independent within as well as between these groups. The decomposition depends upon expressing the matrix form of the Wishart distribution as the product of a (lower) triangular matrix and its transpose.

Bartlett's Test An approximate test for the homogeneity of a set of variances from a number of independent Normal samples, given by Bartlett in 1937.

Bartlett's Test of Second-Order Interaction A test of significance proposed by Bartlett (1935) for the presence of a second-order interaction in a $2 \times 2 \times 2$ contingency table. It is based upon the ratio of the cross-product ratios used to determine the first order interaction.

Base A number or magnitude used as a standard of reference. It may occur as a denominator in a ratio or percentage calculation. It may also be the magnitude of a particular time series from which a start is to be made in the calculation of a new relative series—an **Index Number** —which will show the observations as they accrue in the future in relation to that of the **Base Period**.

Base Line The horizontal line on a graph corresponding to some convenient basic measurement of the variable represented on the ordinate scale. The base is often taken to be zero.

Base Period The period of time for which data used as the base of an index number, or other ratio, have been collected. This period is frequently one of a year but it may be as short as one day or as long as the average of a group of years. The length of the base period is governed by the nature of the material under review, the purpose for which the index number (or ratio) is being compiled and the desire to use a period as free as possible from abnormal influences in order to avoid bias.

Base Reversal Test This is the same as the **Time Reversal Test**.

Base Weight The weights of a weighting system for an index number computed according to the information relating to the base period instead, for example, of the current period.

It is usual in writing formulae to denote the information from the base period with a suffix o and that for the given period with a suffix n. A price index weighted according to prices and quantities of the base period, i.e. base weights, might be

$$I_{on} = \frac{\Sigma \, p_o q_o \left(\dfrac{p_n}{p_o} \right)}{\Sigma \, p_o q_o},$$

where p_o, q_o are the base weights and summation takes place over the commodities composing the index.

Basic Cell A term proposed by Mahalanobis to denote the smallest area for which a variate may be considered to have a sufficiently precise meaning. [See also **Quad**.]

Batch Variation In quality control, the variations in a product which is made or examined in batches, as distinct from one which is produced or examined continuously. The batch variation may be made up of variation within each batch, due to the ordinary process of manufacture, and variation between batches, which may also be due to the quality of raw materials used. [See also **Interclass Variance, Intraclass Variance**.]

Bates-Neyman Model A model proposed for the study of slight and severe accidents in which the statistical form is known as the **Multivariate Negative Binomial Distribution**.

Battery of Tests A term used in applied psychology for a group of tests to which subjects are submitted. The usual objects of subsequent analysis are either to provide predictions of each subject's aptitude for one or more occupations on the basis of a weighted combination of series in the tests, or else to provide variables whose correlations may subsequently be analysed into group or general factors common to the tests.

Bayes' Estimation The estimation of population parameters by the use of methods of inverse probability and in particular of **Bayes' Theorem**. If $\Pr(\theta|H)$ denotes the prior probability of θ then the posterior probability of θ is given by:

$$\Pr(\theta|x_1, x_2, ..., x_n, H) = \Pr(\theta|H)\Pr(x_1, x_2, ..., x_n|\theta H).$$

θ is estimated by choosing that value which maximises the posterior probability. If **Bayes' Postulate** in invoked $\Pr(\theta|H)$ is constant and the method is equivalent to the maximisation of the likelihood $\Pr(x_1, x_2, ..., x_n|\theta H)$.

Bayes' Postulate A postulate concerning the prior probabilities of a set of hypotheses (cf. **Bayes' Theorem**) to the general effect that in the absence of information to the contrary all prior probabilities are to be assumed equal. The postulate (or something equivalent) is necessary if **Bayes' Theorem** is to be applied in situations where there is no information concerning the prior probabilities. The postulate is the critical point in a theory of inference based on inverse probability and is usually acceptable only to those for whom probability is not a limiting frequency.

Bayes' Risk The minimised value of the expected risk in a **Bayes' Solution**.

Bayes' Solution In the terminology of statistical decision-functions, a Bayes' solution is a decision function which minimises the average risk relative to some probability distribution.

Bayes' Strategy A Bayes' strategy is one for which the available decision rules can be linearly ordered in terms of the **Bayes' Risk**. The decision rule with the smallest risk is the Bayes' decision rule.

Bayes' Theorem This theorem of T. Bayes (1763) states that if $q_1, q_2, ..., q_n$ are a set of mutually exclusive events, the probability of q_r, conditional on prior information H and on some further event p, varies as the probability of q_r on H alone times the probability of p given q_r and H, namely

$$\Pr(q_r|pH) \propto \Pr(q_r|H)\Pr(p|q_rH).$$

If $q_1, q_2, ..., q_n$ are exhaustive the constant of proportionality is

$$1/\{\sum_{r=1}^{n}\Pr(q_r|H)\Pr(p|q_rH)\}.$$

In the main application of the theorem, p is an observed event and the q's are hypotheses explaining the event. The three terms in the above expression are then called, in order, the **posterior probability**, the prior probability and the **Likelihood**. The theorem enables the probabilities of the explaining hypotheses to be determined; this use is called the method of **inverse probability**.

The principal difficulty lies in determining, or even defining, the prior probabilities and the resolution of the difficulty by **Bayes' Postulate** has occasioned much controversy. This does not, however, affect the theorem, which is a simple consequence of the product law of probability.

Bayesian Inference A form of inference which regards parameters as being random variables possessed of prior distributions reflecting the accumulated state of knowledge. In its methods it makes use of **Bayes' Theorem** and its extensions.

Bayesian Probability Point A critical value, analogous to a **Confidence Point or Limit**, derived according to the philosophy of **Bayes' Theorem** and the use of the prior distribution(s) of the parameter(s) under estimation.

Beall-Rescias Generalisation of Neyman's Distribution The probability generating function of Neyman's types A, B and C distributions with parameters λ_1 and λ_2 suggest the general form

$$G(z) = \exp\left\{\lambda_1\Gamma(\beta+1)\sum_{i=0}^{\infty}\frac{\lambda_2^{i+1}(z-1)^{i+1}}{\Gamma(\beta+i+2)}\right\}.$$

This is actually a probability generating function and defines Beall-Rescias' generalisation of Neyman's distribution with parameters $\beta, \lambda_1, \lambda_2; 0 < \beta < \infty, 0 < \lambda_1 < \infty, 0 < \lambda_2 < \infty$ over the range $\{0, 1, 2, ...\}$.

Behrens-Fisher Test A test of significance for the difference between the means of random samples from two Normal populations with unequal variances. It is based on the concept of **Fiducial Inference** and has been the subject of considerable controversy.

Behrens' Method A method for estimating the **Median Effective Dose** of a stimulus based upon **Quantal Responses**. It is closely allied to the **Reed-Münch Method**. Although put forward by Behrens in 1929 it was independently proposed by Dragstedt and Lang in 1928; for this reason it is sometimes referred to as the Dragstedt-Behrens method. It is of restricted validity.

Bell-shaped Curve A symmetrical frequency curve, usually of a continuous frequency distribution, which shows a marked similarity to a vertical section through a bell.

Berge's Inequality An inequality of the Tchebychev type for two correlated variates proposed by Berge (1937) as follows:

$$\Pr\{\,|\,x_1\,|\,\geqslant k\sigma_1 \text{ or } |\,x_2\,|\,\geqslant k\sigma_2\} \leqslant \{1-(1-\rho^2)^{\frac{1}{2}}\}/k^2)$$

for all $k > 0$ and where ρ is the correlation parameter.

Berksonian Line A term, derived from a method proposed by Berkson, to decide the regression line between two variables where the values of the independent variables are set at pre-assigned levels instead of being measured.

Bernoulli Distribution Another name for the **Binomial Distribution**.

Bernoulli Numbers The Bernoulli number of order r, B_r, is defined as the numerical coefficient of $t^r/r!$ in the expansion of $t/(e^t-1)$ as a power series in x.

Explicitly, $B_0 = 1$, $B_1 = -\frac{1}{2}$, $B_2 = \frac{1}{6}$, $B_4 = -\frac{1}{30}$, $B_6 = \frac{1}{42}$, $B_3 = B_5 = B_7 = \ldots = 0$.

There are variations in the notation and some writers give $B_1 = -\frac{1}{2}$, $B_2 = \frac{1}{6}$, $B_3 = -\frac{1}{30}$ ....

Bernoulli Polynomial The Bernoulli polynomial $B_r^{(n)}(x)$, of order n and degree r is defined as the coefficient of $t^r/r!$ in the expansion of $\left(\dfrac{t}{e^t-1}\right)^n e^{xt}$.

Bernoulli's Theorem A theorem propounded by James Bernoulli in the fourth part of his *Ars Conjectandi* which was published in 1713 after his death in 1705. Effectively it is a proposition in pure mathematics to the effect that the observed proportional frequency in random drawings of individuals from a population of attributes with constant probability p converges to p in probability; or, to put it another way, the proportional frequency in the binomial distribution $(q+p)^n$ lying within a range $\pm\epsilon\sigma$ of the mean value p tends to unity with increasing n however small ϵ may be, σ being the standard deviation $\sqrt{(pq/n)}$.

Bernoulli himself seems to have regarded the theorem as something beyond a mathematical proposition, perhaps a justification of a frequency theory of probability.

Bernoulli Trials Sequences of events, such as those given by coin tossing or dice throwing, in which successive trials are independent and at each trial the probability of appearance of a 'successful' event remains constant; the distribution of successes is then given by the **Binomial** or **Bernoulli Distribution**.

Bernoulli Variation A sampling situation in which members are chosen from a population of attributes such that the probability of occurrence is constant; hence the sampling distribution of occurrences in samples of fixed size is binomial. The term is used in contradistinction to **Lexis** and **Poisson Variation**.

Bernstein's Inequality An inequality of the **Bienaymé-Tchebychev** type. If a distribution has mean a and variance σ^2 and if the absolute moment of order r, v_r, exists and obeys the inequality

$$v_r \leqslant \tfrac{1}{2}\sigma^2 r\,!\,h^{r-2},$$

where h is some constant, then

$$\Pr\{\,|\,x-a\,|\,>t\sigma\} \leqslant 2\exp\left(\frac{-t^2\sigma^2}{2\sigma^2+2ht\sigma}\right).$$

Bernstein's Theorem A form of the **Central Limit Theorem** for dependent variates, given by S. Bernstein in 1927.

Berry's Inequality This inequality proves that

$$\Pr(S \geqslant t\sigma) < 1-\Phi(t)+1\cdot88\frac{M}{\sigma}$$

where $\Phi(t)$ is the distribution function for a unit Normal random variable and S is the sum of random variables σ_i^2. The x_i are bounded so that

$$|\,x_i-\mathscr{E}(x_i)\,| \leqslant M_i \text{ and Max } (M_i) = M.$$

Bessel Function Distribution A frequency distribution which involves Bessel functions. For example, the distribution of the covariance of two Normal correlated variates involves a Bessel function of the second kind with imaginary argument.

Best Asymptotically Normal Estimator The original name given by Neyman (1949) to a class of estimator which later became known as **Regular Best Asymptotically Normal Estimator**.

Best Critical Region See **Critical Region**.

Best Estimator The estimation of population parameters from information provided by the sample raises the question whether there is a 'best' estimator. The answer depends mainly on the criteria which are laid down as to the 'goodness' of an estimator. If there is a criterion which distinguishes one of two estimators as better than the other and if there exists an estimator which is better than any other, it is said to be the best.

Various criteria have been suggested, e.g. that of

Sufficiency, Minimum Variance or **Closeness**. It is not always true that a 'best' estimator exists.

Best Fit   See **Goodness of Fit**.

Best Linear Unbiassed Estimator   A linear function of the order statistics available in the sample of information of the variable under analysis for which the variance is minimum and which is also **Unbiassed**.

Beta Coefficients   This expression occurs in two distinct senses: (a), in elementary statistics, to denote the coefficients in regression equations, which are often represented by the letter β, e.g. the linear equation
$$y = \beta_0 + \beta_1 X_1 + \beta_2 X_2 + \ldots + \beta_p X_p + \epsilon;$$
(b) to denote the **Moment-Ratios**, especially those used by K. Pearson (1895) to describe **Skewness** and **Kurtosis**.

Beta Distribution   The term beta distribution is usually applied to the form
$$dF = \frac{x^{\alpha-1}(1-x)^{\beta-1}}{B(\alpha, \beta)}\, dx, \qquad 0 \leqslant x \leqslant 1; \alpha, \beta > 0.$$
A second form, sometimes known as a beta distribution of the second kind, is
$$dF = \frac{y^{\alpha-1}}{B(\alpha,\beta)(1+y)^{\alpha+\beta}}\, dy, \quad 0 \leqslant y \leqslant \infty; \alpha, \beta > 0.$$
This is easily transformed into the first type by putting $x = 1/(1+y)$. It has also been referred to as an 'inverted' beta distribution.

Distributions of the first kind are a special case of the **Pearson Type I Distribution** and those of the second kind are a special case of the **Pearson Type VI Distribution**.

Beta Probability Plot   The graph of cumulated distribution function as ordinate against the variate as abscissa for beta (Type I or Type VI) distributions. For Type I the variate has finite range. For Type VI it is infinite and is used to test a variance ratio. It does not involve specially ruled paper. The plots can be of two kinds: (a) ordered quantities in the unit range against quantities of the appropriate beta distribution of the first kind (Pearson Type I) or (b) ordered ratios of mean squares against quantities of the appropriate beta distribution of the second kind (Pearson Type VI).

Between-groups Variance   See **Interclass Variance**.

Bhattacharyya Bounds   A system of lower bounding values to variance of an estimator where no minimum value exists, due to Bhattacharyya (1946). The bounds depend on a series of derivatives of the likelihood function.

Bhattacharyya's Distance   A measure of the 'distance' between two populations. If the frequency functions are $f(x)$ and $g(x)$ the distance is given by
$$\text{arc cos} \int_{-\infty}^{\infty} \{f(x)g(x)\}^{\frac{1}{2}}\, dx.$$

Analogous expressions can be given for discontinuous and multivariate distributions.

Bias   Generally, an effect which deprives a statistical result of representativeness by *systematically* distorting it, as distinct from a random error which may distort on any one occasion but balances out on the average.

For bias in estimation see **Unbiassed Estimator**.

Biassed Estimator.   See **Unbiassed Estimator**.

Biassed Sample   A sample obtained by a biassed sampling process, that is to say, a process which incorporates a systematic component of error, as distinct from random error which balances out on the average. Non-random sampling is often, though not inevitably, subject to bias, particularly when entrusted to subjective judgement on the part of human beings.

Biassed Test   A test is said to be biassed if it gives a lower probability of rejecting the hypothesis under test (H_0) when the alternative hypothesis (H_1) is true than when H_0 is true. Expressed in another way if the hypothesis under test is $\theta = \theta_0$ and the **Power Function** of the test has a minimum value at a point $\theta \neq \theta_0$ then the test is biassed.

Bienaymé-Tchebychev Inequality   An inequality derived by Bienaymé (1853) and rediscovered by Tchebychev (1867); it is a special case of the more general **Tchebychev Inequality**. The inequality is generally stated in the form
$$\Pr\{\,|\,x-a\,| > t\sigma\} \leqslant 1/t^2$$
where $\mathscr{E}(x) = a$ and $\mathscr{E}(x-a)^2 = \sigma^2$ exist and $t > 1$. That is to say, the probability that a variate will differ from its mean by more than t times its standard deviation is at most $1/t^2$ for any probability distribution. The limits so placed are in general rather crude but the inequality is a valuable one in the theory of stochastic convergence.

Bifactor Model   A model of factor structure, due to Holzinger, which is an extension of the **Simple Two-Factor Model**. It is supposed that a battery of tests can be analysed into a general factor and a number of mutually exclusive group factors, e.g.

Test		a	b	c	d
1	.	. x	x		
2	.	. x	x		
3	.	. x		x	
4	.	. x		x	
5	.	. x		x	
6	.	. x			x
7	.	. x			x

The factor 'a' is the general factor with factors 'b', 'c', and 'd' associated with mutually exclusive groups of tests; '*b*', for example, occurring in tests 1 and 2 but not elsewhere.

Bilateral Exponential This distribution is the convolution of an exponential distribution and its Mirrored Distribution; the distribution of $x_1 - x_2$ when x_1 and x_2 are independent and have a common exponential distribution.

Bimodal Distribution A frequency distribution with two Modes.

Binary Experiment An experiment E for which the possible outcomes x fall into one of two distributions is termed a binary experiment (Birnbaum, 1961).

Binary Sequence In general, any sequence each number of which can take one of two possible values. In a probabilistic context a series of numbers each of which represents the outcome of an (independent or dependent) binomial trial and hence where the two possible outcomes can be represented as 0 or 1.

Binomial Distribution If an event has probability p of appearing at any one trial, the probability of r appearing in n independent trials is $\binom{n}{r} q^{n-r} p^r$, where $q = 1 - p$. This is the term involving p^r in the binomial expansion of $(q+p)^n$, which, since it arrays the various probabilities for $r = 0, 1, ..., n$, is known as the binomial distribution. It is also known as the Bernoulli distribution after James Bernoulli who gave it in his (posthumous) *Ars Conjectandi* in 1713.

Binomial Index of Dispersion A statistic for testing whether a set of samples is homogeneous with respect to some common attribute.

If there are k samples of sizes $n_1, ..., n_k$ with proportions $p_1, ..., p_k$ and p is the mean proportion for all members together, i.e.

$$p = (\sum_{i=1}^{k} n_i p_i)/ \sum_{i=1}^{k} n_i,$$

the index of dispersion is

$$\sum_{i=1}^{k} n_i(p_i-p)^2/\{p(1-p)\}.$$

The significance of the index, as denoting departure from homogeneity, may be tested in the χ^2 distribution with $k-1$ degrees of freedom.

The index is a particular case of the **Lexis Ratio**.

Binomial Probability Paper A graph-paper with a grid which is specially designed to facilitate the analysis of enumeration data, i.e. data in the form of proportions or percentages from a binomial population. Both the rectangular coordinate axes are graduated in terms of the square-root of the variable.

Binomial Variation Another name for **Bernoulli Variation**.

Binomial Waiting Time Distribution An alternative name for the **Negative Binomial Distribution**. It refers to the realisation of a success for the kth time after $k+x-1$ independent binomial trials.

Bipolar Factor In factor analysis, a factor which is positively correlated with some variates (tests) but negatively correlated with others. When such a factor is identified with some recognisable quality, before or after rotation, it is regarded as expressing a property which may have a negative as well as a positive intensity; e.g. cowardice as opposed to bravery, cowardice being regarded as a quality in itself and not as mere absence of bravery.

Bipolykays The bivariate k-statistics of a sample; an extension of the concept of **Polykays** given by Hooke (1956).

Birnbaum-Raymond-Zuckerman Inequality An inequality in multivariate analysis giving the probability that deviations from means lie within a hyperellipse

$$\Pr\left\{ \sum \frac{(x_i - \mu_i^2)}{\sigma_i^2 k_i^2} \geqslant t^2 \right\} \leqslant 1/t^2 \sum 1/k_i^2.$$

If the σk's are all equal the surface becomes a hypersphere.

An improvement due to Berge (1938) made use of the correlations between variates.

Birth and Death Process A stochastic process which attempts to describe the growth and decay of a population the members of which may die or give birth to new individuals. The types mainly studied have relatively simple laws of reproduction and mortality.

Birth, Death and Immigration Process A simple extension of **Birth and Death Process** which takes into account immigration into the system from outside. Emigration may be treated synonymously with 'death'.

Birth Process A stochastic process describing the population of a system in which individual members may give birth to new members. The expression is often confined to the case where the variate (population) increases only by jumps of amount $+1$, the probability of a jump from n to $n+1$ in time dt being asymptotically $\lambda_n dt$. Here λ_n may also depend on t. [See also **Poisson Process, Branching Process**.]

Birth Rate The crude birth rate of an area is the number of births actually occurring in that area in a given time period, divided by the population of the area as estimated at the middle of the particular time period. The rate is usually expressed in terms of 'per 1000 of population'.

This crude birth rate is capable of considerable refinement according to the various specific viewpoints adopted in the analysis of vital statistics. For example, it may be adjusted to allow for changes in the proportion of the female population in the child-bearing age groups. [See **Fertility Rate.**]

Biserial Correlation Originally, a coefficient designed to measure the correlation of two qualities, one of which is represented by a measurable variate, the other a simple dichotomy according to the presence or absence of an attribute. The coefficient usually employed is Pearson's biserial η.

Later (1909) Pearson extended the connotation of 'biserial' to the case where both characteristics were dichotomies and proposed a coefficient known as biserial **r**. This has not come into use, being replaced by **Tetrachoric Correlation.**

Bispectrum The Fourier transform of the third order moment function of a stationary random process. It is the method of analysing quadratic effects as the spectrum is used for linear problems.

Bit An abbreviation for binary digit, which is a digit of a number written in the scale of two; for example the number 2 is expressed as 10, 4 as 100 and 8 as 1000.

In modern communication theory the term also refers to a single piece (bit) of information conveyed by an electrical impulse.

Bivariate Binomial Distribution An extension of the binomial distribution to the case where a member can exhibit success or failure in each of two attributes. If the probabilities of success for the attributes are denoted by p_{11}, p_{10}, p_{01} and p_{00} the probability of x successes of the first and y of the second in a sample of s is given by

$$\binom{s}{x} \sum_i \left\{ \binom{s-x}{y-i} \binom{x}{i} p_{11}^i \, p_{10}^{x-i} \, p_{01}^{y-i} \, p_{00}^{s-x-y+i} \right\}$$

where the summation takes place up to $i =$ the smaller of x, y.

Bivariate Distribution The distribution of a pair of variates x_1 and x_2, written for the continuous case as
$$dF = f(x_1, x_2) \, dx_1 \, dx_2.$$

For discontinuous or grouped data the distribution may be set out in a rectangular array known as a bivariate or correlation table.

Bivariate Logarithmic Distribution A generalisation of the logarithmic distribution arising as the limit of the (0, 0) truncated **Bivariate Negative Binomial.** Alternatively termed bivariate logarithmic series distribution.

Bivariate Logarithmic Series Distribution See **Bivariate Logarithmic Distribution.**

Bivariate Multinomial Distribution See **Bivector Multinomial Distribution.**

Bivariate Negative Binomial Distribution A straightforward generalisation of the **Negative Binomial Distribution** to bivariate material.

Bivariate Normal Distribution Two variates x_1 and x_2 with means m_1 and m_2 and variances σ_1^2 and σ_2^2 are distributed in the bivariate normal form if their distribution function is given by

$$dF = \frac{1}{2\pi\sigma_1\sigma_2(1-\rho^2)^{\frac{1}{2}}} \exp \left[-\frac{1}{2(1-\rho^2)} \right.$$
$$\left. \left\{ \left(\frac{x_1-m_1}{\sigma_1}\right)^2 - \frac{2\rho(x_1-m_1)(x_2-m_2)}{\sigma_1\sigma_2} + \left(\frac{x_2-m_2}{\sigma_2}\right)^2 \right\} \right] dx_1 \, dx_2,$$

where ρ is a parameter not greater than unity in absolute value. For any fixed x_1 (or x_2) the variate x_2 (or x_1) is normally distributed. The parameter ρ is (or ought to be) called the correlation parameter.

Bivariate Pareto Distribution A distribution, in two different forms, where both marginal distributions are univariate Pareto distributions (Mardia, 1962).

Bivariate Pascal Distribution Consider an experiment of observing individuals one after another for two characters A_1 and A_2 until exactly k individuals possessing both the characters are observed. Let x_i denote the number of individuals possessing character A_i observed in the experiment, $i = 1, 2$. Then (x_1, x_2) has the bivariate Pascal distribution. This distribution is a special case of the bivariate negative binomial distribution.

Bivariate Sign Test A bivariate analogue of the **Sign Test** proposed by Hodges (1955) which can be regarded as a two-dimensional walk (Klotz, 1959).

Bivector Multinomial Distribution Suppose an individual can exhibit one set of s_1 attributes and also one of another set of s_2 attributes. Let p_{ij} be the probability that an individual exhibits the ith attribute of the first set and the jth attribute of the second set where $\sum_{i, j} p_{ij} = 1$. In a sample of n the probabilities of the various possible combinations are given by the expansion of
$$(\sum p_{ij})^n.$$
This is the bivector or bivariate multinomial distribution. It may be regarded as a univariate multinomial in the $s_1 s_2$ combinations.

Blackwell's Theorem A theorem concerning minimum variance estimation stated by Blackwell (1947) and also by Rao (1949). If a minimum variance estimator exists, it is always a function, e.g. conditional expected value, of the sufficient estimator.

Blakeman's Criterion In the regression of y on x the correlation ratio (η^2) measures the variation of means of arrays: the total sum of squares of deviations of means of arrays from the hypothetical regression line is given by $(\eta^2 - R^2)/(1 - R^2)$ where R is the correlation coefficient between the variable y and its value Y yielded by the regression line. A comparison of this quantity with its standard error is called Blakeman's Criterion. A test of non-linearity of regression is nowadays usually carried out by variance-analysis.

Block The name given, mainly in experimental design, to a group of items under treatment or observation. For example, a block may comprise a group of contiguous plots of land, all the animals in a litter, the results obtained by a single operator, or meteorological observations on a series of days at a given place. The general purpose of dividing all the material in an experiment into blocks (or of regarding it as so divided by the circumstances of the case) is to isolate sources of heterogeneity; the items in a block being so far as possible homogeneous and uncontrolled variation in the experimental material being measured by comparisons between blocks.

The variation in the experimental observations is usually divided (by variance-analysis) into effects due to differences between blocks and effects due to variation within blocks; these being known as interblock and intra-block effects (variances, etc.) respectively. The expressions 'interblock' and 'intrablock information' occur both in the general sense and in the specialised sense of **Information**. Thus for estimates of certain treatment comparisons the 'information' may be the reciprocal of the most efficient estimator if it exists. [See also **Incomplete Block, Randomised Block**.]

Block Diagram A block diagram is made up of vertically placed rectangles situated adjacent to each other on a common base line. Where the characteristic to be depicted is quantitative the height of the rectangles is usually taken to be proportional to this quantitative variable. When this kind of diagram is used to portray a frequency distribution it takes the name of **Histogram**.

Blum Approximation A multi-dimensional extension proposed by Blum (1954) of the **Robbins-Munro** and **Kiefer-Wolfowitz Stochastic Approximation Procedures**.

Bock's Three Component Model A model proposed by Bock (1958) for allowing for differences between judges in the method of **Paired Comparisons**. The observed merits of object A_i in its comparison with A_j by judge k could be represented by

$$Y_{ik(j)} = V_i + \omega_{ik} + Z_{ik(j)};$$
$$(i, j, = 1, 2, ..., t; i \neq j : k = 1, 2, ..., n)$$

where the brackets around j indicate that this subscript serves merely as a label. V_i represents the merits of object A_i and ω_{ik} and $Z_{ik(j)}$ are random variables, the former being components peculiar to specific objects and judges.

Boole's Inequality An inequality developed by Boole in 1854 to give the limits to the frequencies in certain logically defined classes in terms of the frequencies in other classes. It has application in probability theory. One simple form states that if $A_1, A_2, ..., A_k$ are compatible events, the probability that at least one occurs is not greater than the sum of the probabilities that each occurs independently of the occurrence of the others:

$$1 - \Pr(\text{not-}A_1, \text{not-}A_2, ..., \text{not-}A_k) \leqslant$$
$$\Pr(A_1) + \Pr(A_1) + \Pr(A_2) + ... + \Pr(A_k).$$

Borel-Cantelli Lemmas Two lemmas, given in a simple case by Borel and more generally by Cantelli (1917) occurring in the theory of probability. The first lemma, in brief, says that if we have a series of events $A_1, A_2, ...,$ with probabilities $p_1, p_2, ...$ (not necessarily independent) and $\sum_{i=1}^{\infty} p_i$ converges, then from some point onwards it is virtually certain that only a finite number of any A_k will occur. The second says that if the events are independent and Σp_i diverges, then it is virtually certain that an infinite number will occur. The lemmas are useful in providing proofs for the **Laws of Large Numbers**.

Borel-Tanner Distribution In queueing theory the distribution, under certain conditions, of the number of customers served before the queue vanishes for the first time. The two effective conditions are r customers in the queue at zero time and constant service time and probability of new customers. The original distribution ($r = 1$) is due to Borel (1948) and generalised by Tanner (1953) for $r > 1$.

Bose Distribution The distribution of the variance ratio when the variates are correlated was given by R. C. Bose (1935).

Bose-Einstein Statistics In statistical mechanics one possible basic assumption concerning states and energy levels is equivalent to supposing that r distinguishable particles are distributed among n cells ($r < n$) in such a way that each of the n^r arrangements is equally probable. This gives rise to the Maxwell-Boltzmann statistics. If the particles are indistinguishable there are $\binom{n+r-1}{r}$ distinguishable arrangements and if these are taken as equally probable there result Bose-Einstein statistics. As a particular case, if not more than one particle may appear in any cell there are $\binom{n}{r}$ equally probable arrangements; and these form the basis of Fermi-Dirac statistics.

Bounded Completeness If, in the definition of **Completeness** $\mathscr{E}\{h(x)\} = 0$ implies that $h(x) = 0$ only for all bounded $h(x)$, $f(x \mid \theta)$ is called boundedly complete.

Bowley Index See **Marshall-Edgeworth-Bowley Index**.

Box-Jenkins Model A model for predicting non-stationary time series with or without seasonal variations. It depends on autoregression of the mth difference of the series, with a moving average residual.

Branching Markov Process See **Branching Process**.

Branching Poisson Process An alternative designation proposed by Lewis (1964) for the **Poisson Clustering Process**. However, as the principal features of interest in the ordinary concept of a **Branching Process** are unimportant or meaningless in this connection the usage does not seem to be appropriate.

Branching Process A stochastic process describing the growth of a population in which the individual members may have offspring, the lines of descent 'branching-out' as new members are born. Sometimes referred to as a chain-reaction process. If the life-time distribution of the individual members, i.e. the waiting-time in a particular state, follows the negative exponential form the (continuous time) process is called a branching Markov process. For a general form of life-time distribution the term age-dependent branching process is used.

Branching Renewal Process A **Renewal Process** in which each of a series of primary events generates a subsidiary series of events. The complete process is the superposition of the events in the primary and subsidiary processes.

Brandt-Snedecor Method A name sometimes given to one of the formulae for calculating χ^2 from a $2 \times n$ table. If the frequencies in the ith column of the table are a_i and b_i and $p_i = a_i/(a_i+b_i)$, $q_i = 1-p_i$, $\bar{p} = \Sigma a_i/\Sigma(a_i+b_i)$, $\bar{q} = 1-\bar{p}$, the summation taking place over the n columns,

$$\chi^2 = \frac{1}{\bar{p}\bar{q}}\left\{\Sigma(a_i p_i) - \frac{(\Sigma a_i)^2}{\Sigma(a_i+b_i)}\right\}.$$

Bravais Correlation Coefficient An obsolete synonym for the correlation parameter in bivariate Normal variation; and, by extension, to the **Product-Moment Correlation Coefficient** which estimates it.

Brownian Motion Process An additive **Stochastic Process** in a real variate x_t defined at time t such that $x_t - x_s$ is Normally distributed with zero mean and variances $\sigma^2 \mid t-s \mid$, where σ^2 is a constant.

Brown's Method A method of predicting time series, using the basic idea of **Exponential Weights**, proposed by Brown (1959) and further modified (1963). The method is designed to allow for first and second order polynomial trends. See also **Discounted Least Squares**.

Bruceton Method An alternative designation for the 'Up-and-down' Method or the **Staircase Method**.

Bulk Sampling The sampling of materials which are available in bulk form. That is to say, it is the population which is in bulk; the term does not mean the drawing of a sample in bulk. Examples of such sampling would be the sampling of a shipment of coal for ash-content, or of tobacco for moisture content.

Bunch-Map Analysis The technique of the bunch-map is central to the method of **Confluence Analysis** proposed by Frisch (1934). In order to guard against the appearance of unexpected relationships between the variables of a multivariate system he proposed to examine all possible subsets of regression coefficients in the complete set. If these subsets are drawn on standard diagrams, the representation of any set of regression coefficients (obtained by minimising the sum of squares in the direction of each variate in turn) produces a pencil of beams diverging from the origin—a 'bunch' of lines.

The main object of this presentation is to judge the effect on a set of variates of introducing a new variate. A new variate is judged useful if it tightens the bunch or changes the general slope of the bunch, and the coefficients obtained by minimising the sum of squares in the direction of the new variate yield lines lying between those already derived. On the other hand, if the introduction of a new variate scatters the beams the variate is said to be detrimental.

Burkholder Approximation An extension by Burkholder (1956) to cover certain asymptotic properties of stochastic approximation procedures due to **Robbins & Munro** and **Kiefer & Wolfowitz**.

Burr's Distribution A frequency distribution introduced by Burr (1942) where the cumulative distribution has a simple algebraic form
$$F(x) = 1-1/(1+x^c)^k \quad x \geq 0; \ c, k \geq 1.$$
It is one of a class wherein the distribution function, not the frequency function, is defined in analytical terms.

Busy Period In queueing theory a closed time interval such that all servers in a queueing system are occupied.

Buys-Ballot Table A method of tabular presentation of a time-series used by Buys-Ballot (1847), in his meteorological investigations, for the purpose of investigating periodicities. If, for example, a series is suspected of containing a systematic element with period p then the data are arranged as follows:

u_1	u_2	$u_3 \ . \ . \ . \ . \ . \ . \ u_p$	
u_{p+1}	u_{p+2}	$u_{p+3} \ . \ . \ . \ . \ . \ . \ u_{2p}$	

for as many rows (*m*) as there are terms in the series: any terms at the end being neglected.

The column totals will emphasise the systematic effect of period *p* at the expense of other elements which will tend to cancel out, as their differing periods will get out of step between the rows. Any purely random components will be reduced in effect due to the summing up of the *m* rows.

This form of table is also used for estimating the seasonal pattern of a series after it has been corrected for trend.

Call-back   The inability of an investigator to make contact with a particular designated sample unit at the first attempt raises certain problems of bias due to **Non-Response**. One method of dealing with this is for the investigator to 'call-back' on one or more occasions in order to establish contact.

Campbell's Theorem   An evaluation of the asymptotic distribution of the sum of the effects of random impulses acting with given intensity on a damped system.

The impulses are assumed to occur at times in accordance with a Poisson process with parameter λ. Each impulse has a given intensity α and has an effect $\alpha\psi(t)$ after time t has passed. Let $\theta(t)$ be the sum of the effects of all impulses occurring prior to time t. Then for $t \to \infty$ the mean and variance of $\theta(t)$ are respectively

$$\lambda\alpha \int_0^\infty \psi(t)dt, \quad \lambda\alpha^2 \int_0^\infty \{\psi(t)\}^2 dt.$$

Camp-Meidell Inequality   An inequality of the **Bienaymé-Tchebychev** type in which the limits are more precise, the extra precision being obtained by imposing additional conditions on the probability distribution. For distributions which are continuous and unimodal the inequality states that

$$\Pr\{ \mid x - \mu_0 \mid > \lambda\tau \} \leqslant \frac{4}{9\lambda^2}$$

where μ_0 is the mode and τ is defined by $\tau^2 = \sigma^2 + (\mu - \mu_0)^2$, σ^2 being the variance and μ the mean. It is a particular case of the **Gauss-Winckler Inequality**.

Camp-Paulson Approximation   An approximation, by Camp (1951) based on a general result by Paulson (1942), to the sum of the first $t+1$ terms of the point binomial. The sum is expressed as the Normal integral for a variate value ξ dependent on t and the parameter p of the binomial.

Canonical Matrix   In general, any matrix reduces to a canonical form; e.g. in the analysis of **Canonical Correlations**. It is sometimes given specialised meaning. For instance the eigenvalues of a dispersion matrix are sometimes (regrettably) called a canonical set because a transformation based on the eigenvectors reduces the matrix to a diagonal form. In the same vein if the **Incidence Matrix** N of an experimental design can be diagonalised by a matrix L in the sense that L′NN′L is diagonal, L is called the canonical matrix of the design.

Canonical Variate (Correlations)   In multivariate analysis it can be shown, following Hotelling (1936), that variates $x_1, ..., x_p$ and $x_{p+1}, ..., x_{p+q}$ can be transformed linearly into variates $\lambda_1, ..., \lambda_p$ and $\lambda_{p+1}, ..., \lambda_{p+q}$ so that (*a*) the members of each group are uncorrelated among themselves, (*b*) each member of one group is independent of all but one member of the other and (*c*) the non-vanishing correlations between members of different groups are maximised. These quantities are called canonical correlations and the two sets of transformed variates $\lambda_1, ... \lambda_p$ and $\lambda_{p+1}, ..., \lambda_{p+q}$ are called canonical variates. The process of finding the appropriate transformation involves the reduction of two quadratic forms and an associated bilinear form to their respective canonical forms in the mathematical sense.

Cantelli's Inequality   An inequality based upon moments proposed by Cantelli (1910). It is of the **Tchebychev** type.

Capture/Release Sampling   A method of sampling specially suited to the estimation of the size of total populations of wild animals. It is also known as capture/recapture sampling. The method was practised by Lincoln (1930) and involves capturing, marking and releasing a random sample, say, of animals of a particular kind. Subsequently, a further random sample is taken and the proportion of marked animals in this sample forms the basis of estimates of total population. [See also **Lincoln Index**.]

Carleman's Criterion   Under certain conditions it is possible for two different distributions to have the same set of moments. Criteria are therefore desirable to decide when a set of moments determine a distribution uniquely. One such, advanced by Carleman (1925), states that a set of moments $\mu_1, \mu_2 ...$, not necessarily about the mean, determines a distribution uniquely if:

$\sum_{j=0}^{\infty} \frac{1}{(\mu_{2j})^{1/2j}}$ (for distributions ranging from $-\infty$ to ∞)

or

$\sum_{j=0}^{\infty} \frac{1}{(\mu_j)^{1/2j}}$ (for distributions ranging from 0 to ∞)

diverges. This was generalised by Cramér and Wold (1936) to the multivariate case.

Carli's Index   A simple index number of prices proposed by Carli in 1764. If the prices of a set of commodities in the base and given period are respectively $p_o, p_o', p_o'', ...$, and $p_n, p_n', p_n'', ...$, Carli's index number is given by

$$I_{on} = \frac{1}{k} \Sigma \left(\frac{p_n}{p_o} \right)$$

where k is the number of commodities and the summation extends over all commodities. It is thus an unweighted index of **Price-Relatives**.

Carrier Variable A name given to some numerical quantity for which the different values provide the levels of the corresponding factor in a **Factorial Experiment**. For example, in an agricultural experiment, one carrier variable might be the quantity of fertiliser applied per acre of land and the levels of the factor might be zero and one, two, and three times the standard dressing.

Cartogram A device for displaying statistical information of a descriptive nature by means of a symbol on a map. The symbolism may take various forms according to taste, e.g. dots or circles of varying density, or shading in black and white, or use of a full range of colours. The various forms of cartogram are particularly convenient for portraying data according to geographical distribution.

Cascade Process A general class of stochastic process arising, *inter alia*, in the study of cosmic rays. In general, the collision between a primary electron and some material substance gives rise to a cascade of secondary electrons, which may generate further cascades, and so on. The process is a member of the class known as birth-and-death processes.

Categorical Distribution A distribution is said to be categorical if the data are sorted into categories according to some qualitative description rather than by a numerical variable.

Category A homogeneous class or group of a population of objects or measurements. The category may be styled after one of the finite characteristics of the population or according to the limits of measurement for which observations are to be allocated to that category or frequency group. For example, people may be categorised according to Sex (Male or Female) or Age next birthday (1-5 years; 6-10; 11-15; 16-20; and so on).

Cauchy Distribution The name generally used to denote the continuous distribution

$$dF = \frac{dx}{\pi(1+x^2)}, \quad -\infty \leqslant x \leqslant \infty$$

or some simple transform to another origin or scale. The distribution has no finite moments other, perhaps, than the arithmetic mean, which itself may be considered to exist only by convention.

It is a particular case of 'Student's' distribution with one degree of freedom. [See *t*-distribution.]

Causal Chain Model A macroeconomic model pioneered by Tinbergen (1939) consisting of a (time) chain pattern of relations between **Endogenous Variables**. The structural form of the model is

$$y_t = By_t + \Gamma z_t + e_t$$

where B and Γ are matrices of coefficients and the matrix B is specified to be sub-diagonal.

Causal Distribution An alternative designation for **Deterministic Distribution**. It should be avoided, being neither causal nor a distribution except in a degenerate way.

Cause Variable When a relation such as $y = f(x)$ is interpreted in a causal sense, e.g. y is regarded as 'caused by' x, the latter is sometimes called a cause variable and the former an effect variable. The cause variable is also known as an explanatory variable. [See also **Regression**.]

Cell Frequency When a frequency distribution is classified into categories, univariate or multivariate, the sub-categories are sometimes known as cells; the frequency with which observations fall into a particular cell is the cell frequency.

Censoring A sample is said to be censored when certain values are unknown (or deliberately ignored) although their existence is known. The term is usually employed in those cases where, of a number n of values existing, only the k smallest or k largest are observed or k may be expressed as a proportion of n. This is called Type II censoring; Type I is when the sample is censored by reference to a fixed variate value. Whereas Type I censoring can be 'right' or 'left' censored it is more usual for Type II censoring to be only on the right of the variable. The ignoring of the existence of additional values or where such cannot exist is described as **Truncation** and refers to the population.

Census The complete enumeration of a population or groups at a point in time with respect to well-defined characteristics: for example, Population, Production, Traffic on particular roads. In some connection the term is associated with the data collected rather than the extent of the collection so that the term **Sample Census** has a distinct meaning.

The partial enumeration resulting from a failure to cover the whole population, as distinct from a designed sample enquiry, may be referred to as an 'incomplete census'.

Census Distribution A term proposed by Skellam and Shenton (1957) for two event-counting distributions arising in the analyses of renewal processes. In the discrete case they decompose into the sum of **Pascal Distributions** and **Poisson Distributions** for the continuous case.

Centile An abbreviated form of 'Percentile' not in general use, but frequently found in the statistical literature of psychological and educational testing.

Central Confidence Interval A confidence interval for a parameter θ with lower and upper limits t_1 and t_2 is said to be central if
$$\Pr\{(\theta - t_1) < 0\} = \Pr\{(t_2 - \theta) < 0\}.$$
The values t_1 and t_2 are then, in a sense, symmetrically placed with respect to θ.

Central Factorial Moments Strictly speaking, this expression ought to mean the **Factorial Moments** about the centre or some central value. Actually the term is sometimes applied to the factorial moments calculated about some point near to a central value as distinct from the ends of the range.

Central Limit Theorem This theorem, which, although due to Laplace, was first proved rigorously by Liapounov (1901) is the one which gives the Normal distribution its central place in the theory of probability and in the theory of sampling. In its simplest form the theorem states that if n independent variates have finite variances then their sum will, when expressed in **Standard Measure** tend to be Normally distributed as n tends to infinity. It is a necessary and sufficient condition for the validity of the theorem that the variances obey a condition which may be roughly expressed by saying that no one is large compared with their total. More general theorems relating to variates which do not have finite variances or which are correlated may be proved and are also known as central limit theorems.

Central Moment Strictly speaking this expression ought to mean (and occasionally does mean) a moment taken about the centre of a distribution, i.e. the mid-point of its range. More usually it signifies a moment about the mean. When the distribution is symmetrical the meanings coincide.

Central Tendency The tendency of quantitative data to cluster around some variate value. The position of the central value is usually determined by one of the measures of location such as the **Mean, Median** or **Mode**. The closeness with which values cluster round the central value is measured by one of the measures of dispersion such as the **Mean Deviation** or **Standard Deviation**.

Centre (of a Range) In a specialised sense the phrase 'centre of a sample' is sometimes used to denote the variate value which is midway between the two extreme-variate values; correspondingly 'centre' sometimes refers to the mid-point of the range of a distribution.

Centre of Location When parameters of location and scale are simultaneously under estimate it is possible to choose an origin such that the maximum likelihood estimators are uncorrelated. The origin so defined has been called (R. A. Fisher, 1921) the centre of location.

Centroid Method In factor analysis, a method developed by Burt and Thurstone for the extraction of factors. It relies on the idea that, if the variates (tests) are represented as a set of vectors, a common factor may be represented by a vector which passes through the centroid (centre of gravity) of the terminal points of the set. The method is much easier to apply than that of **Principal Components** but suffers from substantial disadvantages.

Cepstrum A method proposed by Bogert, Healy & Tukey (1963) for analysing time series believed to contain lags due to 'echo' effects. A high-resolution spectrum is converted to logarithms and any trend or slow-moving waves eliminated. The spectrum of this smoothed series is termed the cepstrum (kepstrum).

Certainty Equivalence A principle stated by Simon (1956), subsequently generalised by Theil (1957), concerning prediction and regulation through control rules. If there is no future uncertainty concerning the nature of input series, then in a wide class of cases equivalent control rules are obtained by minimising a statistically or a time averaged criterion function.

Chain A sequence of terms such that each term depends in some defined way upon the previous term or terms in the series; for example, the chain-relative used in the calculation of index numbers upon the chain-base method.

The term chain is also used in connection with stochastic processes where the value at one point is determined by values at previous points apart from a random element; or more exactly, the probability distribution at any point, conditional on certain previous values, is otherwise independent of past history. The most common case is the **Markov Chain**.

Chain Binomial Model A model introduced by Greenwood (1931) to describe the development of an epidemic through several generations of cases. The model was developed by Greenwood (1949) introducing the concept of a variable probability of infection and by Bailey (1953, 1956) in specifying probability distribution of p, the chain of infection of the individual.

Chain Block Design A form of partially balanced incomplete block design introduced by Youden & Connor (1953) to deal with situations where the number of treatments considerably exceeds a limited block size. Comparisons within blocks are of high precision so that only one or two replications are needed.

Chain Index An index number in which the value at any given period is related to a base in the previous period, as distinct from one which is related to a fixed base. The comparison of non-adjacent periods is usually made by multiplying consecutive values of the index numbers, which, as it were, form a chain from one period to another. For example, if the value of the index for period 2 based on period 1 is I_{12} and that for period 1 on period 0 is I_{01} the chain index for period 2 based on period 0 is $I_{01} \times I_{12}$ (divided by 100 if the index numbers are based on 100 as the standard).

In practice chain index numbers are usually formed from weighted averages of link-relatives, namely the values of magnitudes for a given period divided by the corresponding values in the previous period.

Chain-relative See **Chain Index**. The term is synonymous with Link-relative.

Champernowne Distributions A three-parameter system of distributions suggested by Champernowne (1952) in connection with the distribution of incomes:

$$F(x) = 1 - (1/\theta) \tan^{-1} \left[\sin \theta / \left\{ \cos \theta + \left(\frac{x_1}{m} \right)^\alpha \right\} \right]$$

where m is the median of x; θ and α are constants.

Chance Constraint In Mathematical Programming, a constraint expressed in probabilistic terms, e.g. that $\Pr(Ax < b) \geqslant \beta$, where β is a vector of corresponding probability measures. A is an arbitrary $m \times n$ matrix and x and b are column vectors of commensurate order.

Chance Variation Variation in statistical observations due to the action of random, as distinct from systematic, factors.

Changeover Trial An alternative name for an experiment or trial using a **Crossover Design**.

Channel Degrees of Freedom An extension of the concept of **Degrees of Freedom** to the area of Information Theory in connection with the number of independent passages through a message channel. The acronym 'cdf' should be resisted in view of its obvious clash with terms like cumulative, or continuous, distribution function.

Chapman-Kolmogorov Equations A set of equations used in the theory of stochastic processes, giving the state of a system (as a probability distribution) at a certain time in terms of the known states at previous times.

*Characteristic (Carattere) In Italian usage, 'carattere' has much the same connotation as the English 'characteristic' used as a substantive. It may refer to quality or to quantity (*carattere qualitativo o quantitativo*). A periodic characteristic (*carattere ciclico*) is one whose values follow a periodic series. An atypical characteristic (*carattere atipico*) is one with a value differing from the most usual.

Characteristic Function The characteristic function of a variate x is the expected value: $\mathscr{E}(e^{itx})$ where t is a real number. This may also be expressed as

$$\phi(t) = \int_{-\infty}^{\infty} e^{itx} dF(x)$$

or similar formulae for discontinuous variates, $F(x)$ representing the distribution function. The expression is often abbreviated to c.f.

Similarly, the characteristic function of several variates $x_1, \ldots, x_n$ is the expected value of $\exp(it_1 x_1 + \ldots + it_n x_n)$. If $x(t)$ is a stochastic process, the random vector $X(t_1)$, $X(t_2), \ldots, X(t_n)$ has a characteristic function. The form of the characteristic function for general n, or a continuous range of t is the characteristic functional.

Characteristic Root The characteristic root of a square matrix A is a value λ such that $|A - \lambda I| = 0$, where I is the identity matrix. For a $p \times p$ matrix there are, in general, p such roots. They are also known as Latent Roots and Eigenvalues.

The corresponding row-vectors u or column-vectors v for which
$$uA = \lambda u \quad \text{or} \quad Av = \lambda v$$
are called characteristic vectors.

Charlier Distribution A little-used term denoting the family of frequency distributions generated by a Gram-Charlier Series. [See also **Gram-Charlier Series—Type A, B and C**.]

Charlier Polynomials The name given to a class of polynomial derived by Charlier in connection with the **Gram-Charlier Series of Type B**.

If $\gamma(m, x)$ is the Poisson term $\dfrac{e^{-m} m^x}{x!}$ and ∇ is the operator (backward difference) defined by
$$\nabla \gamma(m, x-1) = \gamma(m, x) - \gamma(m, x-1)$$
the polynomial G_r is defined by
$$G_r(m, x) = \frac{(-\nabla)^r \gamma(m, x)}{\gamma(m, x)}$$
or equivalently
$$G_r(m, x) = \frac{\dfrac{d^r}{dm_r} \gamma(m, x)}{\gamma(m, x)} \; .$$

Chi Distribution The distribution of the positive square root of the statistic known as the χ^2 or **Chi-squared**.

Chi-squared Distribution A distribution first given, apparently, by Abbe (1863) and rediscovered by Helmert (1875) and K. Pearson (1900). Its frequency function is:

$$dF = \frac{1}{2^{\frac{1}{2}\nu} \Gamma(\frac{1}{2}\nu)} e^{-\frac{1}{2}\chi^2} (\chi^2)^{\frac{1}{2}\nu - 1} d\chi^2, \quad 0 \leqslant \chi^2 \leqslant \infty$$

and the distribution function is an incomplete gamma-function $\Gamma_{\frac{1}{2}\chi^2}(\frac{1}{2}\nu)$. It is a particular case of the Pearson Type III distribution.

The distribution may be regarded as that of the sum of squares of ν independent Normal variates in standard form. The parameter ν is known as the number of **Degrees of Freedom**.

Chi-squared Statistic Strictly speaking, perhaps, this expression should relate to a statistic which is distributed as chi-squared (χ^2), namely as the sum of squares of independent standard Normal variates. For historical reasons, however, it more usually relates to a statistic of a particular kind which is, as a general rule, distributed more or less approximately in the χ^2 form. If a set of n values is distributed over k classes such that the observed frequency in the jth class is n_j and the theoretical (expected) frequency in that class is v_j, the statistic

$$\sum_{j=1}^{k} \frac{(n_j - v_j)^2}{v_j}$$

is called the χ^2 statistic, or simply the value of χ^2, for the data. It is widely used to test agreement between observation, as represented by the n_j, and hypothesis as represented by the v_j.

Chi-squared Test A test of significance based upon the chi-squared statistic. Such tests occur in many ways, the most prominent being:

(*i*) An overall goodness-of-fit comparison of observed with hypothetical frequencies falling into specified classes;

(*ii*) Comparison of an observed with a hypothetical variance in Normal samples;

(*iii*) Combination of probabilities from a number of tests of significance [see **Combination of Tests**].

Chi-statistic The chi-statistic is the square root of the more familiar **Chi-squared Statistic**.

Chunk Sampling A term introduced by Hauser in connection with the technique of sample surveys and defined as a 'slice of a population' dictated by convenience rather than representativeness. Examples of the 'chunk' are: the first n returns in any postal ballot; a group of people who happen to be handy and amenable to questioning.

Cigarette Card Distribution If there are k different types of cigarette cards being mixed equally frequently and the types are randomly distributed, the expected distribution of the k types in a collection of n cards ($n > k$) is

$$\frac{k!}{(k-r)!\,r!} \frac{\Delta^r(0^n)}{k^n}$$

where the second term uses the leading differences of powers of natural numbers.

Circular Chart A method of diagrammatic representation whereby the components of a single total can be shown as sectors of a circle. The angles of the sectors are proportional to the components of the total. Additional visual aid can be obtained with coloured shading or cross-hatching. Also known as a pie chart.

Circular Distribution A frequency distribution of a variate which ranges from 0 to 2π, so that the frequency may be regarded as distributed round the circumference of a circle. The term is used especially of phenomena which have a period of 2π (by a suitable change of scale

if necessary) so that the probability density at any point α is the same as that at any point $\alpha + 2\pi r$ for integral values of r. It is usually expressed in terms of an angle θ. The analogue of the Normal distribution in this class, the so-called circular Normal distribution, is given by

$$f(\theta) = e^{k \cos (\theta - \theta_0)} / 2\pi I_0(k), \quad 0 \leqslant \theta \leqslant 2\pi,$$

where I_0 is a Bessel function of the first kind of imaginary argument. It was derived by von Mises in 1918.

Circular Formula The application of some operations to the terms of an ordered series may present difficulties owing to the fact that end terms have no preceding or succeeding terms. For example, in a series of six terms there are only five first differences but if, for reasons of analytical convenience, it is desired to have six differences then this can be secured by reproducing the first term as a 'pseudo' seventh term. This is equivalent to regarding the series as 'circular' and, hence, any resultant formula in the analysis may be said to be of circular type. The device is used in **Serial Correlation** analysis and also for proving the arithmetic of the **Moving Average**.

More generally, the same device may be used in a stationary stochastic process by regarding successive elements as arranged in a circle. The process is then called circular.

Circular Serial Correlation Coefficient A form of definition of the serial correlation coefficient which gains simplicity of computation and sampling distribution at the expense of some artificiality which is unimportant for series of moderate length. If a series of values $u_1, ..., u_n$ be observed, the kth serial correlation will depend on the sum $\sum_{i=1}^{n-k} u_i u_{i+k}$ over $n-k$ terms. By putting $u_{n+1} = u_1$, $u_{n+2} = u_2$, etc. the sum may be taken as $\sum_{i=1}^{n} u_i u_{i+k}$ over n terms and a correlation coefficient using this form of covariance in its numerator is said to be circular.

Circular Test In the construction of index numbers a decision has to be made as to the period upon which to base the index. If an index for period A based upon period B is I_{BA} and for period B based upon period C is I_{CB}, the circular test, derived by Irving Fisher, requires that the index for period A based upon period C, i.e. I_{CA}, should be the same as if it were compounded of two stages, the calculation of A on B and that of B on C. That is to say we should have

$$I_{CA} = I_{CB} I_{BA}.$$

A similar argument is applied to comparisons between places. Few index numbers in current use satisfy this test.

Circular Triads In **Paired Comparisons** concerning three objects X, Y and Z, if X is preferred to Y, Y to Z and also Z to X the triad XYZ is said (in the terminology of M. G. Kendall) to be circular. The circular triad shows inconsistent preferences, in the sense that preferences are not being made consistently on a linear scale, and no ranking is possible.

Class Apart from its usage in the customary colloquial sense this word has some mild specialisation in the theory of frequency distributions. The total number of observations made upon a particular variate may be grouped into classes according to convenient divisions of the variate range in order to make subsequent analysis less laborious or for other reasons. A group so determined is called a class. The variate values which determine the upper and lower limits of a class are called class boundaries; the interval between them is the class interval; and the frequency falling into the class is the class frequency.

Class Mark A term sometimes used in elementary statistics, but obsolescent for advanced work, to denote mid-value of the class interval. [See **Class**.]

Class Symbol In the theory of attributes, a letter denoting membership or non-membership of a class, e.g. if A denotes 'male' and α 'female', B represents 'living' and β 'dead', AB would represent 'living males'. In Yule's notation the symbol (AB) would represent the number of living males in the population under discussion.

Sometimes the symbol $\bar{B}$ is used instead of β to denote not-B.

Classification Statistic In general, a statistic calculated from a sample for the purpose of assigning the population from which the sample emanated to one of a number of classes. The term is practically synonymous with discriminant function. [See **Discriminatory Analysis**.]

Clipped Time Series A rather unfortunate expression which does *not* mean that a series is truncated in time or that its values are truncated according to the variate. It has been applied to a series where the observations on a continuous variate are in some way approximated by a discontinuous variate. Loss of efficiency in parameter estimation is usually compensated by computational saving. An infinitely clipped series is one where all positive values are set at $+1$ and all negative values at -1 respectively. [See also **Hard Clipping**.]

Clisy A term introduced by K. Pearson in connection with bivariate frequency arrays. For any fixed value of one variate x the distribution of y's has a third mean-moment $\mu_3(x)$: the way in which $\mu_3(x)$ varies with x, or the corresponding $\mu_3(y)$ with y, expresses the variation in clisy of the distribution.

A curve showing values of **Skewness** for the different frequency arrays of one variate is called a clitic curve. In plotting these curves Pearson apparently used the measure of skewness $\beta_1(= \mu_3^2/\mu_2^3)$. Kendall (1943) used the term in a slightly different sense to denote the graph of the third moment of the arrays against the corresponding variate value.

If all arrays have the same skewness they are said to be homoclitic; if not, hetero-clitic. Pearson also defined a hetero-clitic system as nomic or anomic according as the skewness changes continuously or irregularly with the position of the array, but the terms have not come into general use.

[See also **Kurtosis, Scedasticity**.]

Clitic Curve See **Clisy, Regression**.

Closed-ended Question See **Open-ended Question**.

Closed Sequential Scheme In sequential analysis the sampling usually continues until either an acceptance or a rejection boundary is reached. The sample size is not fixed but in order to avoid having (although perhaps, only rarely) to draw large samples before reaching a boundary it may be desirable to fix an upper limit to the sample size. The scheme is then called 'closed'. In the contrary case it is called 'open'.

Closed Sequential *t*-Tests A system of sequential procedures proposed by Schneiderman & Armitage (1962), and developed in later papers (e.g. 1966 with Myers), which represent an improvement on the **Restricted Sequential Procedures** and the set of sampling plans by Sobel & Wald (1949). This particular type of plan is sometimes referred to as 'wedge' from the plotted shape of the boundaries.

Closeness, in Estimation In a sense defined by Pitman (1937), given two estimators, x and y of a parameter θ, if $\Pr\{\,|\,x-\theta\,|\,<\,|\,y-\theta\,|\,\} > \frac{1}{2}$, x is a 'closer' estimator of θ than y. It has been shown (Geary, 1944) that where joint distribution of x and y is Normal the criterion of 'closeness' is equivalent to that of **'Efficiency'**, in the sense that if x is closer than y, var $x <$ var y.

Cluster A group of contiguous elements of a statistical population, e.g. a group of people living in a single house, a consecutive run of observations in an ordered series, or a set of adjacent plots in one part of a field.

Cluster Analysis A general approach to multivariate problems in which the aim is to see whether the individuals fall into groups or clusters. There are several methods of procedure but they all depend on setting up a metric to define the 'closeness' of individuals.

Cluster Sampling When the basic sampling unit in the population is to be found in groups or clusters, e.g. human beings in households, the sampling is sometimes carried out by selecting a sample of clusters and observing all the members of each selected cluster. This is known as cluster sampling.

If the elements are closely grouped they are said to be compact. If they are almost equivalent to a geographically compact group from the point of view of investigational convenience they are said to be quasi-compact. [See also **Elementary Unit**.]

24

Cochran's Criterion A criterion (Q) proposed by Cochran (1950) for the purpose of comparing percentage results in matched samples. In the general case the data are arranged in an $r \times c$ table with each row a matched group and each column a sample. The test criterion is

$$Q = \frac{c(c-1)\Sigma(T_J - \bar{T})^2}{c(\Sigma u_i) - (\Sigma u_i^2)}$$

where T_J is total successes in jth sample (column) and u_i the total number of successes in ith matched group (row). Q has a limiting distribution of χ^2 with $(c-1)$ degrees of freedom if the true probability of success is the same in all samples.

Cochran's Q-Test An alternative designation for **Cochran's Criterion** in matched samples; not to be confused with the **Coefficient of Association** due to Yule.

Cochran's Rule A rule proposed by Cochran (1941) for rejecting an 'outlier' sample from k samples with m observations in each: the samples can be groups from one set of data. The assumption is that all are from a single Normal population and the rejection criterion is based upon the sample variances. [See also **Grubb's Rule, Thompson's Rule, Dixon's Statistics**.]

Cochran's Test A test due to Cochran (1941), for homogeneity of a set of independent estimates of variance. It is based on the ratio of the largest estimate of variance to the total of all the estimates.

Cochran's Theorem A theorem on quadratic forms stated by Cochran in 1934. If x_i ($i = 1, 2, ..., n$) are independent standardised normal variates and q_J ($j = 1, 2, ..., k$) are quadratic forms in the variates x_i with ranks n_J ($j = 1, 2, ..., k$) and if $\sum_{j=1}^{k} q_J = \sum_{i=1}^{n} x_i^2$, then the necessary and sufficient condition for the q_J to be independent χ^2 variates with n_J degrees of freedom respectively is that $\sum_{i=1}^{k} n_J = n$.

Coefficient Generally this word has the same meaning as in mathematics, but occasionally it is used to denote a dimensionless statistic, e.g. the moment-ratio β_2 as a coefficient of **Kurtosis** or the coefficient of product-moment correlation. In this sense the word 'index' is also used.

For particular coefficients see under the appropriate name, e.g. for 'Coefficient of Agreement', see **'Agreement, Coefficient of'**.

*Cograduation In Italian usage, if two sets of terms, equal in number, are arranged each in order of magnitude so as to be both non-decreasing or non-increasing, the values of terms with the same ordinal number are said to be cograduated, namely to have the same **Grade** or rank. The process is called cograduation.

If one series is non-decreasing and the other non-increasing two values with the same ordinal number are said to be contragraduated (*contragraduati*).

A table of double entry or bivariate frequency table which serves to cograduate the **Marginal** distributions is called a cograduation table (*tabella di cograduazione*); and similarly for a contragraduation table.

*Cograduation, Gini's Index of In Italian usage, a measure of agreement between the ranks of a set of objects when arranged in rank order according to two different criteria. A rank **Correlation Coefficient**. The *indice di cograduazione quadratico* is the same as **Spearman's ρ**. The *indice di cograduazione semplice* (simple index of cograduation) may be represented as

$$\frac{1}{k} \sum_{i=1}^{n} |p_i + q_i - n - 1| - \frac{1}{k} \sum_{i=1}^{n} |p_i - q_i|$$

when there are n objects ranked, the ranks of the ith object according to the two qualities are p_i and q_i and k is $\frac{1}{2}n^2$ or $\frac{1}{2}n^2 - 1$) according to whether n is even or odd. [See also **Spearman's Footrule**.]

Coherency The relationship, in the form of a measure of correlation, between the frequency components of two stochastic processes, or time series. The coefficient of coherence at ω is

$$C(\omega) = \frac{c^2(\omega) + q^2(\omega)}{f_x(\omega) f_y(\omega)}$$

where $c(\omega)$ is the **Co-spectrum** and $q(\omega)$ the **Quadrature Spectrum** and $f_x(\omega), f_y(\omega)$ are the spectral densities of the series x and y. The concept is analogous to the square of the correlation coefficient (r). The functions in the expression for $C(\omega)$ always obey the 'coherence-inequality'

$$c^2(\omega) + q^2(\omega) \leqslant f_x(\omega) f_y(\omega).$$

Coherent Structure A term proposed by Birnbaum, Esany & Saunders (1961) in connection with the reliability of multi-component structures whereby functioning components do not interfere with the functioning of the structure. The components, and the structure, in this model take only one of two states: functioning or failure.

Collapsed Stratum Method A method for estimating the variance of the mean of a stratified sample where the sample is based upon two random selections from each of a number of strata of equal size and the number of non-respondents results in too few completed strata. Two or more strata are amalgamated (collapsed) to form one stratum.

Colligation See **Association, Coefficient of**.

Combination of Tests The combination of a number of probabilities obtained from tests on different groups of data, undertaken so as to assess the probability of the tests as a whole. One such test is based on the fact that if k tests give probabilities $p_1, ..., p_k$ the statistic

$-2 \sum\limits_{i=1}^{k} \log p_i$ is distributed as χ^2 with $2k$ degrees of freedom, provided that the variates giving rise to the p's are continuous. In the case of discontinuity various modifications are required.

*Combinational Power Mean In Italian usage, a power mean (*media potenziata*) of a set of values $x_1, x_2, ..., x_n$ is given by

$$M_k = \left\{ \frac{1}{n} \sum_{i=1}^{n} x_i{}^k \right\}^{\frac{1}{k}}.$$

When distinction is necessary this is called monoplane as compared with the biplane (*bipiana*) from

$$\left\{ \frac{1}{n} \Sigma x^p \middle/ \frac{1}{n} \Sigma x^q \right\}^{\frac{1}{q-p}}$$

If c members are chosen from the n and multiplied together the cth root of the mean of all possible such products

$$\left\{ \sum_{j=1}^{\binom{n}{c}} P_j{}^c(x_i) \middle/ \binom{n}{c} \right\}^{1/c}$$

is called the combinatorial mean. If the x's are raised to power p we have the combinatorial power mean (*media combinatoria potenziata*)

$$\left\{ \frac{1}{\binom{n}{c}} \sum_{j=1}^{\binom{n}{c}} P_j{}^{(c)}(x_i{}^p) \right\}^{1/cp}.$$

This also is called monoplane in distinction to a biplane form of similar type.

Many of the customary means of statistics are particular cases of the combinatorial power mean.

Combinatorial Test A test of significance in which the sampling distribution of the test statistic is obtained by the algebra of combinatorial analysis. For example, the nature of the process governing the partitioning of N units into k groups may be tested by counting the number of zero groups. On the hypothesis of equiprobability of occurrence of a unit in any group the probability of r empty groups is:

$$\Pr\{r \mid k, N\} = \frac{\binom{k}{t}}{k^N} \Delta^{k-t} 0^N$$

where $\Delta^{k-t} 0^N$ is the $(k-t)$th leading difference of the Nth power of the natural numbers.

Common Factor In factor analysis the factor(s) are classified according to the way in which they contribute to the variables under analysis. Any factor which appears in two or more variates is called a common factor. If the factor appears in all the variates it is called a general factor. If it is common to a group of variates it is called a group factor. A factor appearing in only one variate is said to be specific. [See also **General Factor.**]

Common Factor Space In one geometrical representation of a multivariate situation the variation is regarded as taking place in a space of which each factor represents a dimension. When there are, say, m common factors and s specific factors the whole factor space is one of $n = m+s$ dimensions. Of this the m dimensional sub-space is the Common Factor Space.

Common Factor Variance In factor analysis, that part of the variance of a variate which is attributable to the factor or factors which it has in common with other variates, the remainder being due to specific factors or error terms. It is also known as the communality when expressed as a proportion of the total variance.

Communality See **Common Factor Variance.**

Communicate Two states, j and k, in a **Markov Chain** are said to communicate if j is accessible from k and k is accessible from j; that is to say, if the distribution of j, given k, is not independent of k and *vice versa*.

Communicating Class Given a state j of a **Markov Chain,** its communicating class $C(j)$ is defined as the set of all states k in the chain which communicates with j.

Compact (Serial) Cluster See **Cluster, Serial Cluster.**

Comparative Mortality Figure The ratio of the standardised death rate to the crude death rate in a standard population. It may also be regarded as an index number in the **Laspeyres' Form.** [See also **Standardised Mortality Ratio.**]

Comparative Mortality Index This is a variant of the **Comparative Mortality Figure** and is a weighted average death rate, where the weights are the mean of the actual (current) population and the standard population both expressed in proportions on a common basis of absolute size. In this sense it is an index number of the **Marshall-Edgeworth-Bowley** form.

Compensating Error In general, any error which compensates for other errors. More specifically, a class of error with zero mean (unbiased) and subject to the central limit effect, so that the occurrence of several errors will tend to cancel out and their effect become reduced as the errors cumulate. In this sense the term is not to be recommended.

Competition Process A stochastic process, a two-dimensional extension of the **Birth and Death Process,** proposed by Reuter (1961). Two fields of application are competition between species and the spread of an epidemic. A multivariate generalisation was covered by Iglehart (1964).

Complete Class (of Decision Functions) In decision function theory, a class which contains all admissible decision rules.

26

Complete Class (of Tests) A class of tests is complete if for every test outside the class there is one within the class which is uniformly better in the sense of a measurable decision function.

Complete Correlation Matrix See **Correlation Matrix.**

Complete Regression A concept introduced by Tukey (1958) in connection with estimation by order statistics where, if the distribution function $F(z|y)$ satisfies
$$F(z|y'') \leqslant F(z|y') \quad y'' \leqslant y',$$
the regression is complete negative; complete positive regression being defined analogously.

Complete System of Equations A term used mainly in econometrics to denote the equations determining the behaviour of an economic system or part of such a system. The set of equations is said to be complete when it includes all the determining equations governing the system, or a set from which all equations can be deduced. The point of emphasising completeness is that in some cases it is possible to estimate parameters occurring in an incomplete set of equations, but the estimators may then be biassed, a fact to which attention was drawn by Haavelmo in 1943.

Completely Balanced Lattice Square See **Square Lattice.**

Completely Randomised Design A very simple form of experimental design in which the treatments are allocated to the experimental units purely on a chance basis.

Completeness If, for a parametric family of univariate or multivariate distributions $f(x \mid \theta)$, depending on the value of a vector of parameters θ, $h(x)$ be any statistic independent of θ, and, y for all θ
$$\mathscr{E}\{h(x)\} = \int h(x)f(x \mid \theta)dx = 0$$
implies that $h(x) = 0$ identically (save possibly on a set measure zero), then the family $f(x \mid \theta)$ is called complete.

*Complex Abnormal Curve See ***Abnormal Curve.**

Complex Demodulation A variant of the technique of **Demodulation** where the basic series or stochastic process is multiplied by a complex non-random function.

Complex Experiment Generally, an experiment of a complicated kind. More specifically, an experiment in which special devices such as confounding, splitting of plots, and incomplete blocks, are used to reduce the error variance associated with certain comparisons.

Complex Gaussian Distribution A univariate complex Gaussian distribution is based upon a random variable $z = x+iy$ where the real and imaginary parts are each in a p variate Gaussian (Normal) form. This concept, introduced by Draper (1963) is important in the spectrum analysis of multiple time series.

Complex Table A table which shows the classification of a set of data according to more than two different features: as distinct from the one or two features of the simple table. For example, a human population might be tabulated in a complex table according to age, civilian status and sex. The complexity lies, not only in the manifold nature of the classification, but in the difficulty of printing the results in a convenient form. This method of tabulation is sometimes referred to as multiple cross-classification.

Complex Unit A statistical unit of record which is derived from a combination of two or more simple units. For example: national income *per capita* or net-ton-miles-per-train-mile (which is equivalent to the average train load) or shillings-per-week-earned-per-person.

Complex Wishart Distribution A form of the **Wishart Distribution** required in the analysis of the **Complex Gaussian Distribution.**

Component Analysis Component analysis is a branch of multivariate analysis which represents a k-dimensional variation as due to a number of orthogonal components; fewer than k if possible, but if not, in such a way that a few components account for as much of the variation as possible. The components sought in practice are linear functions of the original variates. [See also **Factor Analysis.**]

Component Bar Chart A **Bar Chart** which shows the component parts of the aggregate represented by the total length of the bar. These component parts are shown as sections of the bar with lengths in proportion to their relative size. Visual presentation can be aided by devices of cross-shading or colours.

Component of Interaction See **Interaction.**

Component of Variance See **Variance Component.**

Components of Variance An alternative designation for Model II (or Second Kind) in analysis of variance.

Composite Hypothesis A composite statistical hypothesis is frequently defined as a statistical hypothesis which is not simple. This is not entirely satisfactory and in practice the expression usually refers to a hypothesis which is 'composed' of a group of simple hypotheses. For example, the hypothesis that a frequency function is Normal with unspecified mean and variance is composite since there is a double infinity of values of mean and variance which, when specified, would yield a simple hypothesis.

Composite Index Number A rather vaguely defined term relating to an index number for which the component series are from groups which are different in nature. The definition is somewhat arbitrary in practice since much depends upon the point of view of both the compiler and user of the index. For example, an index number of retail prices would not be regarded as composite from the point of view of a general analysis of the national economy in which 'price' was a single element but it would be regarded as composite by, say, a trade organisation operating in only one retail market. In a slightly different sense a national index of production or of business activity is said to be composite at the national level and also composite geographically at a regional level. To be logical, any index number compiled from more than one homogeneous commodity should be called composite; but the expression has its practical uses.

Composite Sampling Scheme A scheme in which different parts of the sample are drawn by different methods; for example, a sample of a national population might be taken by some form of area sampling in rural districts and by a random or systematic method in urban districts.

Compound Frequency-Distribution This expression occurs in three senses: (1) If several sets of individuals, each with a frequency-distribution, are mingled to form a single set, the frequency-distribution of the latter may be said to be compounded of the separate distributions; e.g. the distribution of heights of a population of human beings may be compounded of the distributions of heights of males and females. (2) A frequency-distribution arising from the **Convolution** of variates may be said to be compounded of the distributions of the individual variates. (3) If a distribution depends on a parameter θ which itself has a distribution, the distribution obtained by summing over θ is said to be compound. This usage appears undesirable, since the word 'compound' then refers to the way in which the distribution was reached, not the properties of the distribution itself. So also, it appears preferable to refer to a distribution arrived at under (2) as 'convoluted'.

Compound Hypergeometric Distribution A distribution that is produced by averaging the hypergeometric distribution for given x over all possible values of X according to a prior distribution. A number of well-known distributions may be considered as special cases of the compound hypergeometric distribution. For example, the Pólya, the binomial and the rectangular.

Compound Negative Multinomial Distribution If, in the context of a **Negative Multinomial Distribution** we consider distinguishable sets of trials with the probabilities fixed within a set but randomly varying between sets, the resultant distribution will be a compound version of the negative multinomial. When the random variation is a **Multivariate Beta-distribution** the resultant compound form was given by Mosiman (1963).

Compound Poisson Distribution A name sometimes given to a distribution resulting from a Poisson distribution of parameter λ where λ itself has a distribution. If the distribution of λ's is represented by $dF(\lambda)$ the probability of observing the number k is

$$\Pr(k) = \frac{1}{k!} \int_0^\infty e^{-\lambda} \lambda^k dF(\lambda).$$

Compressed Limits In quality control, limits which are more stringent than necessary in the sense that items falling outside them may still be within limits acceptable to the consumer. The object of setting compressed limits is to reveal departure from a controlled state sooner, or with smaller sample size, than might be exhibited by wider limits.

*Concentration (Concentrazione) The extent to which a quantity is concentrated in some individuals of an aggregate, in space or in time. Where the word is used without qualification it is understood to relate to individuals; that is to say, a characteristic is more or less concentrated according as the proportion of the total exhibited by a given proportion of individuals is greater or less. For example, the wealth of a country is more concentrated if a greater fraction is possessed by the rich and a correspondingly smaller fraction by the poor.

If the frequency function of a variate is $f(x)$ with distribution function $F(x)$; if the range of the variate lies to the right of the origin; and if the incomplete first moment is defined by

$$\Phi(x) = \frac{1}{\mu_1'} \int_0^x x f(x) dx$$

where μ_1' is the complete first moment, the graph of Φ as ordinate against F as abscissa is called the curve of concentration (*curva di concentrazione*). It is convex to the F axis and ranges from $(0, 0)$ to $(1, 1)$. The area between it and the line $F = \Phi$ is called the area of concentration. Twice this area is called the coefficient of concentration or the concentration ratio (*rapporto di concentrazione*). It is equal to the mean difference divided by twice the arithmetic mean.

The concentration curve is called *curva di concentrazione culminante*, when its maximum distance from the equidistribution line $F = \Phi$ falls on the perpendicular to the middle point of such line.

Concentration, Coefficient of A coefficient advanced by Gini (1912) as a measure of dispersion. It may be defined in terms of the Mean Difference (Δ_1) also due to Gini as,

$$G = \frac{\Delta_1}{2\mu_1'},$$

where μ_1' is the arithmetic mean. [See **Mean Difference, Coefficient of.**]

Concentration, Curve of See ***Concentration.**

Concentration, Ellipse of For a bivariate Normal population with means m_1 and m_2 variances σ_1^2 and σ_2^2 and correlation ρ, the ellipse of concentration is given by

$$\frac{1}{1-\rho^2}\left\{\left(\frac{x-m_1}{\sigma_1}\right)^2 - \frac{2\rho(x-m_1)(y-m_2)}{\sigma_1\sigma_2} + \left(\frac{y-m_2}{\sigma_2}\right)^2\right\} = 4.$$

It is such that a uniform distribution bounded by the ellipse has the same first and second moments as the Normal population. If the ellipse of concentration of one distribution lies wholly inside that of another, the former distribution is said to be more concentrated.

Concentration, Index of A descriptive index proposed by Gini (1909) to measure the extent to which a quantitative characteristic is concentrated in a few units. If a variable X can take values $x_1, x_2, ..., x_n$ (in that order) with frequencies $f_1, f_2, ..., f_n$, the sum of the last m units as compared with the total sum obeys the inequality

$$\frac{\sum\limits_{i=n-m+1}^{n} f_i x_i}{\sum\limits_{i=1}^{n} f_i x_i} > \frac{\sum\limits_{i=n-m+1}^{n} f_i}{\sum\limits_{i=1}^{n} f_i}$$

and the extent to which these two expressions depart from equality is taken as a measure of concentration. For incomes and several other characterics a descriptive index of concentration is the number δ such that

$$\left(\frac{\sum\limits_{i=n-m+1}^{n} f_i x_i}{\sum\limits_{i=1}^{n} f_i x_i}\right)^{\delta} = \frac{\sum\limits_{i=n-m+1}^{n} f_i}{\sum\limits_{i=1}^{n} f_i}$$

*Concomitance (Concomitanza) In Italian usage, the relation between two variates in time; especially of the variation of two time-series in the same direction (positive concomitance) or in opposite directions (negative concomitance).

*Concordance (Concordanza) In Italian usage, a particular form of relationship (*connessione*) between quantitative variables, or qualitative variables when they have comparable **Modalities**. If the variables are such that positive values of one are associated with negative values of another there is said to be discordance (*discordanza*), that is to say, concordance attempts to represent the sign as well as the intensity of the relationship.

Concordance, Coefficient of In ranking theory, a coefficient measuring the agreement among a set of rankings. If m rankings of n objects are arranged one under another and the rankings summed for each of the n objects; and if S is the sum of squares of deviations of these sums from their common mean $\frac{1}{2} m (n+1)$, the coefficient of concordance W is

$$W = \frac{12 S}{m^2(n^3-n)}.$$

Complete agreement between the rankings gives $W = 1$ and lack of agreement results in W being zero or very close to it.

Concordant Sample This concept was introduced by Pitman (1937-38) in connection with his distribution free test of the difference between two samples. Given two sets of observations $a_1, ..., a_n$ and $b_1, ..., n_n$ with means $\bar{a}$ and $\bar{b}$, there are $\binom{m+n}{n}$ equiprobable ways of separating these $m+n = N$ observations into two sets of which the available set is one. If $\bar{a}-\bar{b}$ is defined as the 'spread' of a given separation we choose certain separations (with small spreads) as acceptable in the sense that their occurrence does not lead us to infer a real difference in parent populations. A set of a's and b's which forms one of these separations is 'concordant'. In the contrary case they are 'discordant'.

Concurrent Deviation Let (x_1, y_1), (x_2, y_2), ..., (x_n, y_n) be pairs of observations taken from some convenient origin such as the means of the x's and y's. If x_i, y_i have the same sign they are said to exhibit concurrent deviation. A coefficient of correlation between x and y may be derived as the proportion of p of concurrent deviations to the total of n deviations. If the x's and y's are distributed in a bivariate normal form an estimator of the correlation parameter is given by $\sin \frac{1}{2}\pi p$. [See also **Kendall's Tau.**]

Conditional In some contexts this word may merely have its ordinary meaning and imply that there exist certain conditions obeyed by the quantities under discussion. It occurs most often in a specialised sense relating to variates. If a set of variates $x_1, x_2, ..., x_p, x_{p+1}, ..., x_q$ have a joint frequency distribution the sub-distribution obtained by holding some of them fixed is said to be conditional; thus the distribution of $x_1, x_2, ..., x_p$ for fixed $x_{p+1}, ..., x_q$, usually written

$$F(x_1, ..., x_p \mid x_{p+1}, ..., x_q)$$

is the conditional distribution of $x_1, x_2, ..., x_p$ given $x_{p+1}, ..., x_q$. By extension, if certain functions of the x's, say statistics $t_1, ..., t_k$, are held constant, the distribution of $x_1, x_2, ..., x_q$ is conditional given $t_1, ..., t_k$.

The expectation of any x in a conditional distribution is its conditional expectation. The conditional expectations of a set A of x's when a set B is fixed is a function of the B set and this relationship leads to the concept of **Regression** of A on B.

In a similar manner, if E and F are events occurring according to a probability distribution, the probability of E given the occurrence of F is called the conditional probability of E given F; for example, the product-rule of the probability calculus may be written

$$\Pr(E \text{ and } F) = \Pr(E \mid F)\Pr(F).$$

Conditional Expected Value See **Conditional**.

Conditional Failure Rate An alternative term to **Hazard** used in life analysis of physical systems or components.

29

Conditional Power Function A concept introduced by F. N. David (1947) in connection with the power of tests of randomness in a sequence of alternative events. As in **Conditional Tests**, the actual sample observed is used to define a sample sub-space and the power function considered in this sub-space.

Conditional Regression A regression estimated under certain conditions known *a priori* to apply to some of the parameters concerned; for example, in estimating price and cross elasticities from time-series data the income elasticities involved can sometimes be assumed to be known from cross-section data. It might be better to find an alternative name, owing to the intimate connection between regression and conditional distributions. [See **Conditional**.]

Conditional Statistic A statistic whose distribution is conditional, that is to say, depends upon some quantity which is held constant; the quantity in question is usually itself some function of the variables or variates entering into the statistic.

Conditional Survivor Function The ordinary survivor function is the complement of the distribution function, i.e. $S(x) = 1 - F(x)$. The conditional expression for $x > x_0$ is
$$S(y \mid x) = \Pr(x - x_0 > y \mid x > x_0) = S(y + x_0)/S(x_0).$$
The interpretation of this function, together with the **Hazard**, is of importance in life testing and reliability assessment.

Conditional Test A test of significance is sometimes difficult to apply because the distribution of the test statistic involves unknown parameters of the parent population. This difficulty may sometimes be avoided by introducing restrictions on the sampling distribution, e.g. by considering only samples which have the same mean as that of the observed sample. This is equivalent to making the inference in a sub-population of samples which have a fixed mean. The distribution and the inference based on it are then said to be conditional.

Conditionally Unbiassed Estimator An estimator t of a parameter θ is said to be conditionally unbiassed with respect to statistics $u_1, ..., u_n$ if the expectation of t for constant $u_1, ..., u_n$ is equal to θ. Symbolically
$$\mathscr{E}(t \mid u_1, u_2, ..., u_n) = \theta.$$

Confidence Belt The area between the upper and lower confidence limits.

Confidence Coefficient The measure of probability α associated with a **Confidence Interval** expressing the probability of the truth of a statement that the interval will include the parameter value.

Confidence Curves A unified concept proposed by Birnbaum (1961) to include point, confidence limit and interval estimation.

Confidence Interval If it is possible to define two statistics t_1 and t_2 (functions of sample values only) such that, θ being a parameter under estimate,
$$\Pr(t_1 \leqslant \theta \leqslant t_2) = \alpha$$
where α is some fixed probability, the interval between t_1 and t_2 is called a confidence interval. The assertion that θ lies in this interval will be true, on the average, in a proportion α of the cases when the assertion is made.

Confidence Level An alternative term for **Confidence Coefficient**.

Confidence Limits The values t_1 and t_2 which form the upper and lower limits to the **Confidence Interval**.

Confidence Region When several parameters are under estimate it may be possible to define regions in the parameter space such that there will be assigned confidence α that the parameters lie within them. This is the generalisation of the confidence interval to the case of more than one parameter and the domain so determined is called the confidence region.

Configuration A set of n observations on a variate may be represented as a vector in n-dimensional space. A number k of vectors in the space may be regarded as having geometrical properties of interrelationship independently of the coordinate system. In factor analysis, the arrangement of variate vectors among themselves, and without regard to any frame of reference, is called their configuration.

The expression 'configuration of a sample' also occurs, in a somewhat different sense. The n observations define a point in the n-dimensional space and any set of such points lying in a subspace, e.g. a hyperplane may be said to have the same configuration.

'Configurational sampling' is sometimes used as a synonym for **Grid Sampling**.

Confluence Analysis A method of analysis introduced by Frisch in 1934 in an attempt to overcome certain difficulties in regression analysis when there may be linear relations between the independent (predicated) variables or errors of observation introduce 'nearly' linear relations in the observed independent (predicated) variables.

The technique devised for the purpose is known as **Bunch-Map Analysis**.

A relation between the independent variables which results in the indeterminacy of the coefficients of a regression equation, or approximate indeterminacy where observational errors exist, is called a confluent relation.

Confluent Relation See **Confluence Analysis**.

*****Conformity (Conformità)** In Italian usage, the agreement between experimental results and those expected according to some theoretical scheme.

Confounding A device whereby, in large factorial experiments, the size of blocks is limited by sacrificing some of the independent comparisons relating to the higher order interactions. These particular interactions may be deemed unimportant or of little practical consequence from a policy point of view.

The totality of possible treatment combinations is not replicated in each block but is divided amongst blocks in such a way that the main contrasts can be made within blocks but the others are not distinguishable from contrasts between blocks, with which they are said to be confounded.

More generally, when certain comparisons can be made only for treatments in combination and not for separate treatments, those treatment effects are said to be confounded. Confounding is often a deliberate feature of the design but may arise from inadvertent imperfections.

Congestion Problems Problems concerned with a process which may lead to congestion in a flow of goods or services. An equivalent term more customarily used by British writers is **'Queueing Problems'**.

Conjugate Latin Squares Two Latin squares are conjugate if the rows of one are the columns of the other.

Conjugate Ranking Given two rankings of n objects, if one is arranged in the natural order and the other (correspondingly rearranged) designated by A; and then the latter is arranged in the natural order and the first (correspondingly rearranged) designated by B; then A and B are said to be conjugate. For example two rankings of six objects $0_1, 0_2, ..., 0_6$:

	0_1	0_2	0_3	0_4	0_5	0_6	
	4	1	3	6	5	2	
	2	4	1	5	6	3	
rearranged as	0_2	0_6	0_3	0_1	0_5	0_4	
	1	2	3	4	5	6	
	4	3	1	2	6	5	A
and	0_3	0_1	0_6	0_2	0_4	0_5	
	3	4	2	1	6	5	B
	1	2	3	4	5	6	

give the two rankings indicated by A and B as conjugate.

*Connection (Connessione) 'Connessione' is used in Italian to denote statistical relationship or dependence in the widest sense. Two phenomena are 'connected' if the distribution of the characteristic associated with one depends on the value of the characteristic associated with the other. The dependence is said to be actual (*concreta*) if it is deduced from the available data observed, as distinct from systematic or limiting (*sistematica o limite*) dependence which relates to populations or infinite sets of observations and is not subject to sampling fluctuation. Relationship between the values of one characteristic

and the means of another is called *connessione delle modalità medie*, a term difficult to translate into English.

*Connection, Index of (Indice di Connessione) A measure of **Connection**. Let the **Indices of Dissimilarity** between the total distribution of one variable A and the sub-distributions of A for the possible fixed values of a second variable B be $D_1, D_2, ..., D_s$, based on $n_1, n_2, ..., n_s$ members. A simple (*semplice*) index is given by

$$C_{AB} = \frac{1}{N} \sum_{k=1}^{s} n_k{}^1 D_k, \text{ where } N = \Sigma n_k.$$

This may be standardised by dividing by its maximum value Δ_R the mean difference with **Repetition** of the whole group of N.

A quadratic index may be constructed by using the square root of $\Sigma n_k({}^2D_k)^2/N$ where ${}^2D_k$ is a quadratic index of dissimilarity; and may be standardised in the same manner.

The quadratic index of connection of the mean values of A to the values of B is equivalent to the **Correlation Ratio**.

Conservative Process A stochastic process governing the behaviour of a population which has constant total size but the members of which can assume independently one of a finite number of states, the variation consisting of transfer from one state to another.

Conservative Confidence Interval A confidence interval or region is regarded as 'conservative' when the actual confidence coefficient exceeds the nominal or stated value.

Consistence, Coefficient of In the analysis of **Paired Comparisons** the fundamental inconsistency or consistency of preferences may be expressed in terms of **Circular Triads**. The coefficient of consistence may be defined (Kendall & Babington Smith) as $1 - \dfrac{24d}{n^3 - n}$ for n odd and $1 - \dfrac{24d}{n^3 - 4n}$ for n even, where d is the observed number of circular triads and n is the number of objects being compared.

Consistent Estimator An estimator which converges in probability, as the sample size increases, to the parameter of which it is an estimator. An example of an inconsistent estimator is the sample mean as estimator of θ in the distribution

$$dF = \frac{dx}{\pi\{1 + (x-\theta)^2\}}, \quad -\infty \leqslant x \leqslant \infty.$$

Consistent Test A test of a hypothesis is consistent with respect to a particular alternative hypothesis if the power of the test tends to unity as the sample size tends to infinity; and, similarly, it is consistent with respect to a class of alternatives if it is consistent with respect to each member of the class.

Constraint A constraint in a set of data is a limitation imposed by external conditions, e.g. that a number of variate values shall have zero mean, or that the sum of frequencies in a set of classes shall be a prescribed constant.

There is another sense in which statistical data may be said to be constrained. This is the case of **Subnormal Dispersion** discussed by Lexis.

Consumer Price Index A price index designed to measure changes in the cost of some specified standard of living. [See **Laspeyres' Index, Konyus Conditions.**]

Consumer's Risk In acceptance inspection the risk which a consumer takes that a lot of a certain quality q will be accepted by a sampling plan. It is usually expressed as a probability of acceptance and depends, of course, on q as well as the sampling plan itself. It is equivalent to the probability of an **Error of the Second Kind** in the theory of testing hypothesis in the sense of corresponding to the acceptance of a hypothesis when an alternative is true. [See also **Producer's Risk.**]

Contagious Distribution A class of probability distribution of a compound kind, usually derived from probability distributions dependent on parameters by regarding those parameters as themselves having probability distributions. The name derives from the use of such compound distributions in the study of contagious events such as accidents, occurrences of disease or 'persistence' in weather. [See also **Compound Frequency Distributions.**]

Contaminated Distribution Those distributions containing observations which are of doubtful accuracy or validity.

Contingency The contingency is the difference in the cells of the **Contingency Table** between the actual frequency and the expected frequency on the assumption that the two characteristics are independent in the probabilistic sense. If f_{ij} is the frequency in the ith row and jth column and $f_{i\cdot}$, $f_{\cdot j}$ are the respective row and column totals, and if the total frequency is n, the difference in question is

$$f_{ij} - \frac{f_{i\cdot} f_{\cdot j}}{n}.$$

The square contingency is given by

$$\chi^2 = \sum_{i,j} \frac{n(f_{ij} - f_{i\cdot} f_{\cdot j}/n)^2}{f_{i\cdot} f_{\cdot j}}.$$

The mean-square contingency, usually denoted by ϕ^2, is given by

$$\phi^2 = \chi^2/n.$$

Contingency, Coefficient of A coefficient purporting to measure the strength of dependence between two characteristics on the basis of a contingency table. In the notation of the previous term, K. Pearson's coefficient is defined by

$$C = \left(\frac{\chi^2}{n+\chi^2}\right)^{\frac{1}{2}} = \left(\frac{\phi^2}{1+\phi^2}\right)^{\frac{1}{2}}.$$

Tschuprov's coefficient is defined as

$$T = \left(\frac{\chi^2}{n\sqrt{\{(p-1)(q-1)\}}}\right)^{\frac{1}{2}}$$

where p, q are the number of rows and columns in the contingency table.

Contingency Table The members of an aggregate may be classified according to qualitative or quantitative characteristics. Where the characteristics are qualitative a classification according to two of them may be set out in a two-way table known as a contingency table. For example, if the characteristic A is p-fold and a characteristic B is q-fold then the contingency table will be one of p rows and q columns. The cell corresponding to A_j and B_k contains the number of individuals bearing both of those characteristics. In general, the order of the rows and columns is arbitrary. [See also **Contingency, Coefficient of.**]

Continuity A parameter or a variate is said to be continuous when it may take values in a continuous range. A frequency or probability distribution is sometimes said to be continuous when it relates to a continuous variate, and sometimes when the function itself is continuous. Although in statistical practice the two are often equivalent it is better to keep the usage clear by referring to a distribution of a continuous variate or a continuous distribution of a variate as the case may be.

Continuity Correction See **Yates' correction.**

Continuous Population A population is sometimes called continuous when considered in regard to some variate which is continuous. The usage is not very exact, since a population may be continuous in one variate and discontinuous in another; but is permissible when the meaning is clear from the context.

Continuous Probability Law A **Probability Law** is called continuous if it corresponds to a continuous distribution function. A more stringent definition, as used in advanced probability theory, would prefix the word 'absolutely'.

Continuous Process A name sometimes employed to denote a stochastic process $[x_t]$ which depends on a continuous parameter t. It is apt to lead to confusion with the continuity of x and is, perhaps, better avoided. The same applies to 'discontinuous' or 'discrete' process.

Continuous Sampling Plans A type of sampling inspection plan for use where a production process is continuous in the sense that batching is not a rational procedure. It may also be applicable where immediate process control is required and the rectification of outgoing products. The plans may be single-level (Dodge, 1943) or multi-level.

Contour Level See **Patch.**

Contragraduation See **Cograduation**.

Contrasts A contrast among the parameters (β) in analysis of variance or treatments and interactions in an experiment, is a linear function with known constant coefficients which sum to zero.

Control There are two principal ways in which this term is used in statistics. If a process produces a set of data under what are essentially the same conditions and the internal variations are found to be random, then the process is said to be statistically under control. The separate observations are, in fact, equivalent to random drawings from a population distributed according to some fixed probability law.

The second usage concerns experimentation for the testing of a new method, process or factor against an accepted standard. That part of the test which involves the standard of comparison is known as the control.

Control Chart A graphical device proposed by Shewhart (1924) used to show the results of small scale repeated sampling of a manufacturing process. It usually consists of a central horizontal line corresponding to the average value of the characteristic under investigation, quantitative or qualitative, together with upper and lower control limits between which a stated proportion of the sample statistics should fall. Any marked divergence above or below these control limits will tend to indicate that new causes are at work beyond those responsible for the random variations inherent in large scale production. Points outside the limits will signal the need for special enquiries to identify the new factor(s) at work.

An alternative approach is through using cumulative results for which the **Cumulative Sum Chart** was devised.

Control of Substrata A term used in sampling inquiries to denote the employment of prior knowledge of the population cell values in an n-way table formed according to the n factors which are being used in a scheme of **Multiple Stratification**.

For example, if a population is stratified by age and sex, a knowledge of the number of individuals of each sex in each age group enables the sample to be controlled by these substrata. If only the marginal frequencies were known the sample could be controlled by strata but not by substrata, as, for example, if the numbers of each sex in the population and the age distribution of the population were known, but not the numbers of each sex in each age group.

Control Limits See **Control Chart**.

Controlled Process An industrial process is said to be controlled when the mean and variability of the product remain stable. The variation is then due to random effects or the combination of small factors of a non-cumulative

kind. The expression 'under statistical control' is to be avoided in favour of 'statistically in control' or 'statistically stable'.

Convergence in Measure See **Stochastic Convergence**.

Convergence in Probability See **Stochastic Convergence**.

Convolution Let $F_1(x)$, $F_2(x)$. ..., $F_n(x)$ be a sequence of distribution functions. The distribution

$$F(x) = \int_{-\infty}^{\infty} dF_1(x_1) \ldots$$
$$\int_{-\infty}^{\infty} F_n(x - x_1 - \ldots - x_{n-1}) dF_{n-1}(x_{n-1})$$

is called the convolution of the distributions. The relationship is sometimes written

$$F(x) = F_1(x)^* \, F_2(x)^* \, \ldots \, F_n(x).$$

If the associated variates are independent, $F(x)$ is the distribution function of their sum.

Coordinatograph In Indian usage a simple apparatus to determine random points on a plane by using a pair of random numbers as coordinates.

Corner Test See **Medial Test**.

Cornish-Fisher Expansion A form of the Edgeworth expansion of a frequency function used by E. A. Cornish and R. A. Fisher (1937) to tabulate the significance points of certain probability integrals.

Corrected Moment A moment of a set of observations which has been adjusted for some effect such as the bias arising from its being calculated from a grouped frequency distribution rather than from the original data. [See also **Correction for Grouping; Sheppard's Corrections**.]

Corrected Probit This term is synonymous with **Working Probit** but should be avoided because of its false implication that it puts right a value that was previously wrong.

Correction for Continuity When a statistic is essentially discontinuous, but its distribution function is being represented approximately by a continuous function, the probability levels can sometimes be more accurately ascertained by entering the tables of the continuous function, not with the actual values of the statistic, but with slightly corrected values. These values are then said to be corrected for continuity.

Correction for Grouping When data are grouped into frequency distributions the approximation which becomes necessary by reason of having to regard frequencies as being concentrated at the mid-points of class intervals may impart a bias to the calculations of the moments of the distribution. Under certain conditions it is possible to correct for this effect, the best-known corrections being due to Sheppard.

Other corrections have been advanced for distributions which are **Abrupt** at one, or both, terminals. The problem of **Average Corrections** has been shown to lead to expressions similar to those for **Sheppard's Corrections.**

Corrections for Abruptness A system of corrections to the moments of frequency distributions which do not have high-order contact at the limits of the range. Such corrections were devised by Pairman and K. Pearson in 1919 and were proposed as suitable for use in cases where **Sheppard's Corrections** do not apply.

Correlation In its most general sense correlation denoted the interdependence between quantitative or qualitative data. In this sense it would include the association of dichotomised attributes and the contingency of multiply-classified attributes. The concept is quite general and may be extended to more than two variates.

The word is most frequently used in a somewhat narrower sense to denote the relationship between measurable variates or ranks. In Italian usage the two senses are distinguished by different words, 'connection' for the wider sense and 'concordance' for the narrower sense. Where no ambiguity arises it is used in a still narrower sense to denote **Product-Moment Correlation.**

Correlation, Coefficient of A correlation coefficient is a measure of the interdependence between two variates. It is usually a pure number which varies between -1 and 1 with the intermediate value of zero indicating the absence of correlation, but not necessarily the independence of the variates. The limiting values indicate perfect negative or positive correlation.

If there are two sets of observations $x_1, ..., x_n$ and $y_1, ..., y_n$, and a score is allotted to each pair of individuals, say a_{ij} for the x-group and b_{ij} for the y-group, a generalised coefficient of correlation may be defined as

$$\Gamma = \frac{\Sigma a_{ij} b_{ij}}{\sqrt{(\Sigma a_{ij}{}^2 \Sigma b_{ij}{}^2)}}$$

where Σ is a summation over all values of i and j ($i \neq j$, from 1 to n.

This general coefficient includes the **Kendall** τ, **Spearman** ρ and Pearson **Product-Moment Correlation** r as special cases according to the method of scoring adopted. In the last case, for example, the scoring is based on variate values with $a_{ij} = x_i - x_j$; $b_{ij} = y_i - y_j$.

If positive values of one variate are associated with positive values of the other (measured from their means) the correlation is sometimes said to be direct or positive; as contrasted with the contrary case, when it is said to be inverse or negative.

There are numerous other correlation coefficients of a different character.

Correlation Index An obsolescent term referring to a quantity which purported to measure correlation where regression relationships between the variates concerned were not linear.

Correlation Matrix For a set of variates $x_1, ..., x_n$ with correlations between x_i and x_j denoted by r_{ij}, the correlation matrix is the square matrix of values (r_{ij}). Its determinant is the correlation determinant. The matrix is symmetric since $r_{ij} = r_{ji}$.

Unless otherwise specified the diagonal elements r_{ii} ($i = 1, ..., n$) are unity and in psychological work the matrix with unit diagonals is said to be complete.

Correlation Ratio In a bivariate frequency table with variates x and y the correlation ratio of x on y is defined by

$$\eta^2{}_{xy} = \frac{\Sigma(\bar{x}_i - \bar{x})^2}{\Sigma(x - \bar{x})^2}$$

where the summation in the numerator takes place over y-arrays, $\bar{x}_i$ is the mean of the ith array and $\bar{x}$ is the mean of x in the whole distribution; and the summation in the denominator takes place over all the values. It may be regarded as the ratio of the variance between arrays to the total variance.

There is an analogous expression for the correlation ratio of y on x.

Correlation Surface An almost obsolete synonym for a bivariate frequency surface.

Correlation Table The frequency-table of a bivariate distribution. The difference between a correlation table and a **Contingency Table** is that the former usually denotes a grouped frequency distribution with intervals defined in terms of the variate values and therefore possessing a natural order and clearly defined width.

Correlogram In time series analysis, the graph of the **Serial Correlation** of order k as ordinate against k as abscissa. It is generally presented only for non-negative values of k since $r_k = r_{-k}$ (conventionally).

This term was introduced by Wold in 1938 but the diagram was earlier used by Yule, who called it a 'correlation diagram'.

Cospectrum The covariance between the two cosine components and between the two sine terms in a spectrum analysis of the relationship between two time series. The cospectrum measures the covariance of the components which are in phase. [See also **Quadrature Spectrum, Coherency.**]

Cost Function In sampling theory, a function giving the cost of obtaining the sample as a function of the relevant factors affecting cost. It may relate to only a part of the entire cost, e.g. by providing for the cost of collecting the sample but not for the cost of tabulation.

Counter Model, Type I A stochastic process related to the physical behaviour of Geiger-Müller counters. A model of Type I is one which records a count at the first arrival not covered by the pulse of the previous count.

34

Counter Model, Type II A model of Type II, see entry for Type I for background information, is one in which a count is recorded as the first arrival not covered by the pulse of any previous arrival.

Counting Distribution A probability distribution formed by the number of events occurring in a fixed period of time. If the counting period begins just after an event the distribution may be termed a synchronous distribution (Haight, 1965).

Covariance The first product moment of two variates about their mean values. The term is also used for the estimator from a sample of a parent covariance.

Covariance Analysis This is an extension of the **Analysis of Variance** to cover the case where members falling into the classes bear the values of more than one variate. Interest centres on one of these (chosen as dependent variate) and the question is whether its variation between classes is due to class effects or to its dependence on the other variates which themselves vary among classes. This is discussed by considering the regression of the dependent variate on the other variates and the variation of the regressions (or equivalently of covariances) among classes. The technique is similar to that of variance analysis but considerably more complicated.

Covariance Function A colloquial form of the term auto-covariance function. [See **Autocorrelation Coefficient** and **Function**.]

Covariance Kernel See **Mean Value Function**.

Covariance Matrix For n variates $x_1, ..., x_n$, for which the covariance of x_i and x_j is c_{ij}, the square matrix (c_{ij}) is called the covariance matrix. The diagonal terms are the variances: var x_i.

Alternative names are the Dispersion Matrix and the Variance-Covariance Matrix.

If the variates are standardised so as to have unit variances the covariances become correlations and the matrix becomes the **Correlation Matrix**.

Covariance Stationary Process A stochastic process is stationary in the covariance or covariance stationary if the covariance function
$$R(v) = \mathscr{E}[x(t)\,x(t+v)]$$
exists and is independent of t for all integral v.

Covariation The joint variation of two or more variates.

Covarimin See **Factor Rotation**.

Coverage A term used in sampling in two senses: (1) to denote the scope of the material collected from the sample members (as distinct from the extent of the survey, which refers to the number of units included); (2) to mean the extent or area covered by the sampling as in expressions such as '50 per cent coverage', which means that one-half of the population under discussion have been examined.

Cox's Theorem A theorem due to D. R. Cox (1952) which states specific conditions under which a sequential test of a mean of a Normal distribution against a **Composite Hypothesis** can be constructed. The test is based upon the application of the **Sequential Probability Test** to transformed values of the original observations.

Craig Effect When certain short-cut methods of periodogram analysis are employed on series consisting of a random element superposed upon a scheme of harmonic components there results a bias in the estimate of the periods. The magnitude is dependent upon the variance of the random component. This is known as the Craig effect after J. I. Craig who published it in 1916.

Craig's Theorem A theorem, stated by A. T. Craig (1933), on the independence of quadratic forms in the analysis of variance. If x_i $(i = 1, ..., n)$ be n independent standardised Normal variates, then the quadratic forms $\mathbf{x'Ax}$ and $\mathbf{x'Bx}$ are statistically independent if and only if the product of the symmetric matrices $\mathbf{A}$ and $\mathbf{B}$, namely $\mathbf{AB}$, is zero.

Cramér-Rao Efficiency The efficiency of an estimator $\tilde{\theta}$ which will permit the condition of equality to exist in the **Cramér-Rao Inequality**.

Cramér-Rao Inequality An inequality giving a lower bound to the variance of an estimator of a parameter. If t is an estimator of θ in a distribution with frequency function $f(x, \theta)$: and if the bias $b(\theta)$ is given by
$$b(\theta) = \mathscr{E}(t) - \theta$$
the inequality states that
$$\operatorname{var} t \geqslant \mathscr{E}\,\frac{\left(1+\frac{\partial b}{\partial \theta}\right)^2}{\left(\frac{\partial \log f}{\partial \theta}\right)^2}.$$

Results on these lines have been given by many authors and the question of priority is unsettled. In English writings the inequality is almost invariably known by the names of Cramér and Rao, singly or in conjunction.

Cramér-Tchebychev Inequality An inequality of the **Bienaymé-Tchebychev** type depending on the second and fourth moments, namely
$$\Pr\{|x-a| > t\sigma\} \leqslant \frac{\mu_4 - \sigma^4}{\mu_4 - 2t^2\sigma^4 + t^4\sigma^4}; \; t > 1$$
where σ^2 is the variance and μ_4 the fourth moment of the distribution and a is the mean. Like many inequalities of this type, it has several names. Berge published one version in 1932.

Cramér-von Mises Test A test for the difference between an observed distribution function and a hypothetical distribution function. It was proposed by Cramér in 1928 and, independently, by von Mises in 1931. If $F_n(x)$ is the observed distribution function and $F(x)$ its hypothetical counterpart, the criterion is

$$\omega^2 = \int_{-\infty}^{\infty} \{F_n(x) - F(x)\}^2 dx.$$

The sampling distribution of ω is not known. To meet this difficulty Smirnov (1936) considered an alternative form

$$\omega_n^2 = \int_{-\infty}^{\infty} (F_n - F)^2 dF.$$

ω_n^2 is independent of F and therefore provides a distribution free test. In 1952 the test was modified still further by T. W. Anderson and Darling in the form

$$W_n^2 = n \int_{-\infty}^{\infty} \{F_n(x) - F(x)\}^2 \psi\{F(x)\} dF(x)$$

where $\psi(t)$ is some real non-negative function defined for $0 \leqslant t \leqslant 1$.

Criterion This word is used in statistics in its colloquial sense in a number of contexts, e.g. the likelihood criterion for testing hypotheses.

In earlier literature the phrase '*the* criterion', otherwise unqualified, is found to denote a function which distinguishes between various types of **Pearson Curves**. The criterion is

$$\kappa = \frac{\beta_1(\beta_2 + 3)^2}{4(2\beta_2 - 3\beta_1 - 6)(4\beta_2 - 3\beta_1)}$$

where β_1 and β_2 are the Pearson measures of **Skewness** and **Kurtosis**.

Critical Quotient In the analysis of extreme values from unlimited distributions, Gumbel (1958) defined a critical quotient

$$Q(x) = \frac{-f^2(x)}{f'(x)\{1 - F(x)\}} > 0$$

where $f'(x)$ is the first derivative of the frequency function.

Critical Region A test of a statistical hypothesis is made on the basis of a division of the **Sample Space** into two mutually exclusive regions. If the sample point falls into one (the region of acceptance) the hypothesis is accepted; if in the other region (the region of rejection) it is rejected. Both regions are, in a sense, critical, but it is customary to denote the second by the term critical region.

If, among critical regions of fixed **Size** there is one which minimises the probability of **Error of the Second Kind** it is called the best critical region. If, for a set of alternative hypotheses, the probability of an error of the second kind is less than the probability of an error of the first kind (or equivalently, the power is greater than the size) the region is said to be unbiassed.

Critical Value The value of a statistic corresponding to a given significance level as determined from its sampling distribution; e.g. if $\Pr(t > t_0) = 0.05$, t_0 is the critical value of t at the 5 per cent level.

Cross Amplitude Spectrum A procedure used in spectrum analysis of time series for describing the covariance between the two series which form a bivariate time series. It shows whether and how the amplitude of a component at a particular frequency in one series is associated with a large or small amplitude at the same frequency in the other series.

Cross-correlations Correlations between series ordered in time, or space, with or without a lag between the series. Thus if $u_1, \ldots, u_n, v_1, \ldots, v_n$ are two series, correlations between u_i and v_i, or between u_i and v_{i+j} (for fixed j) are cross-correlations. They are the extension, to more than one series, of **Serial Correlations**.

Cross Intensity Function In a **Renewal Process** events of several types might be occurring along the one dimensional time axis. The cross intensity function is a method proposed by Cox & Lewis (1967) for dealing with the relationship between these different types of events considered jointly against time.

Cross-over Design In its original sense, a design involving two treatments which could be applied more than once to the same set of subjects. The subjects would be divided into pairs and each pair treated first with the treatments A and B and then with the treatments B and A, 'crossed-over'.

More recent usage has extended the meaning to cases where the pairs of subjects are divided into two sets and each pair consists of one where the response is expected to be better and one where it is expected to be worse. In the first set A is applied to the 'better' members and in the second set to the 'worse' members. The method can be extended to cases where there are more than two treatments but if the number is large other designs are usually preferable.

Cross Spectrum In connection with the spectrum analysis of bivariate time series, the product of the **Amplitude Spectrum** and the **Phase Spectrum** is known as the cross spectrum. It is equivalent to the Fourier transform of the cross covariance function.

Crossed Classification A feature of the largest of **Nested Designs**. If, say, there are two factors A and C then where every level of C appears with every level of A this two-way layout is termed completely crossed. Anything less than this is deemed to be partially crossed. [See also **Hierarchical design**.]

Crossed Factors See **Nested Design**.

Crossed Weight Index Number An index number is said to have crossed weights if it results from two subsidiary index numbers, with different weights, after the application of some process of averaging.

The most commonly quoted crossed weight formula is that of Fisher's **'Ideal' Index** which is the result of geometrically crossing (averaging) the index number formulae attributed to **Paasche** and **Laspeyres**. It may be written as a price index:

$$I_{on} = \sqrt{\left(\frac{\Sigma p_n q_o}{\Sigma p_o q_o} \frac{\Sigma p_n q_o}{\Sigma p_n q_n} \right)}$$

where o and n are subscripts relating to the base year and current year respectively. The **Marshall-Edgeworth-Bowley Index** also has crossed weights in this sense.

Crude Moment See **Raw Moment**.

Crypto-Deterministic Process A particular kind of stochastic process, due to Sir Edmund Whittaker (1943), where the initial conditions contain all the uncertainty. Apart from this uncertainty the development of the process in time is of a completely determinate character.

C.S.M. Test A test of significance developed by Barnard (1947) for data, in the form of a 2×2 table, arising from comparative trials; for example, where pre-arranged numbers are taken from each of two sources to compare the proportions of some attribute in the two sets. The name C.S.M. derives from the three conditions of convexity, symmetry and maximum number of outcomes which determine the **Critical Regions** of the test.

Cubic Designs With Three Associate Classes A group of experiment designs investigated by Raghavarao & Chandrasekhararao (1964). The geometric configuration of the association scheme indicated the term 'cubic' and a new method of arranging a p^3 factorial experiment in blocks of size different from p and p^2 was given.

Cubic Lattice An extension of the **Square Lattice** in which the number of treatments is a perfect cube and they are regarded as arranged on the points of a cubic lattice. From the point of view of factorial experiments the treatments are regarded as the combination of three factors each at k levels. [See **Quasifactorial Design**.]

Cuboidal Lattice Design This experimental design is a development of the **Cubic Lattice Design** in much the same way as the rectangular lattice is a development of the square lattice. The cubic lattice design is suitable for numbers of treatments which are perfect cubes (k^3) whereas the cuboidal lattice design is appropriate to a number of treatments of the form $k^2(k+1)$.

Cumulant The cumulants are constants of a frequency distribution defined in terms of the moments by the identity in t

$$\exp\left(\sum_{r=0}^{\infty} \frac{\kappa_r t^r}{r!} \right) = \sum_{r=0}^{\infty} \frac{\mu'_r t^r}{r!}.$$

They are thus given by the coefficients in the expansion of a power series formed from the logarithm of the char-

acteristic function of a variable, if such an expansion exists. The earlier name for the quantities was semivariant or half-invariant, a term introduced by Thiele. The word cumulant is due to Cornish and Fisher (1937).

Cumulant Generating Function A function of a variable t which, when expanded in powers of t, has the cumulants of a distribution (or numerical multiples of them) as the coefficients in the expansion. The only cumulant generating function in common use is the logarithm of the **Characteristic Function**, which results in

$$K(t) = \log \phi(t) = \sum_{r=0}^{\infty} \kappa_r \frac{(it)^r}{r!}$$

where κ_r is the rth cumulant.

Cumulative Distribution (Probability) Function A synonym for the **Distribution Function**.

Cumulative Error An error which, in the course of the cumulation of a set of observations, does not tend to zero. The relative magnitude of the error does not then decrease as the number of observations increases.

Cumulative Frequency (Probability) Function A synonym for the **Distribution Function**.

Cumulative Frequency (Probability) Curve See **Distribution Curve**.

Cumulative Normal Distribution The cumulative frequency function (distribution function) of the **Normal Distribution**.

Cumulative Process A development of the **Regenerative Process**, and a generalisation of the **Renewal Process**, introduced by Smith (1955). As its name implies it is concerned with accumulation of regeneration points, or some attribute occurring at such a point, with the passage of time. An alternative formulation by D. G. Kendall (1948) presented this process as a variant of the standard **Birth and Death Process**.

Cumulative Sum Chart These control charts are intended to replace the standard form, the aim is to make it unlikely that 'lack of control' will be indicated when a process is 'in control' or that a marked change in population mean will remain undetected. In cumulative sum control charts cumulative totals are plotted against the number of observations. If $\bar{x}_j$ denote the mean of the jth sample, and σ the known standard deviation of $\bar{x}_j$ then it is convenient to consider the points on the control chart as having coordinates (m, X_m), where

$X_m = \sigma^{-1}[(\bar{x}_1 - \mu) + (\bar{x}_2 - \mu) + ... + (\bar{x}_m - \mu)]$ and μ is the target mean. In practice this can be effected by calculating the cumulative sum

$$\sum_{j=1}^{m} (\bar{x}_j - \mu)$$

and using an appropriate scale.

Cumulative Sum Distribution Generally, if $x_1, \ldots, x_n$ are a set of variates the distribution of $X_n \equiv \sum_{i=1}^{n} x_i$ is the cumulative sum distribution.

Curtailed Inspection In quality control, inspection is said to be curtailed if it is stopped at some point otherwise than is provided for by the sampling inspection plan. Usage, however, is not uniform and the expression is also found to denote the stoppage of inspection provided by the plan itself, e.g. in 'cutting off' before acceptance or rejection boundaries are reached in **Open Sequential Schemes**. [See also **Cut-off, Truncation.**]

Curtate A word used in vital statistics to denote the integral number of years, as distinct from the nearest number of years, for which a given state has existed. For example, if an assurance matures in 3 years 9 months, the curtate duration is 3 years. An analysis which uses curtate periods will, in most cases, be subject to downward bias.

Curve Fitting An expression used in two rather different senses in statistics: (*a*) to denote the fitting of a mathematically specified frequency curve to a frequency distribution (*cf.* Pearson curves); (*b*) to denote the fitting of a mathematical curve to any statistical data capable of being plotted against a time or space variable, e.g. regression data or time series.

Curvilinear Correlation An expression used to denote correlation in bivariate data when the regressions are not linear. Strictly speaking it is a misnomer, a correlation being a number and not a function admitting curvature, and is perhaps better avoided.

Curvilinear Regression A regression which is not linear. A form often considered is that for which the dependent variate is expressed as polynomial in the independent variables.

Curvilinear Trend A trend which is not linear. It may be expressed as a polynomial, a more complicated mathematical expression such as a logistic curve, or by some smoothing process such as a moving average.

Cut Off The artificial truncation of a sampling process at a point when it becomes apparent that enough data have been collected for the purpose in view. [See also **Sequential Analysis.**]

Cycle Strictly speaking, a periodic movement in a time-series, that is, a component with the property that $f(t+\tau) = f(t)$ where τ is the period of the cycle.

The word is also used in a less exact sense to denote up and down movements which are not strictly periodic. The usage is to be deprecated.

Cyclic Design A class of balanced and partially balanced experimental design in which the blocks are constructed by cyclic permutations of the treatments.

Cyclic Order An arrangement of n permutations of n objects such that if the first is denoted by $1, \ldots, n$ the second is $2, \ldots, n, 1$ and the third $3, \ldots, n, 1, 2$, etc. The process provides some of the basic **Latin Square Designs**; for example, with four treatments

$$\begin{array}{l} \text{ABCD} \\ \text{BCDA} \\ \text{CDAB} \\ \text{DABC.} \end{array}$$

***Cyclic Series (Serie ciclica)** In Italian usage, data classified according to a variable which cyclically repeats itself in time or space, such as the days of the week; thus numbers of marriages by days of the week would be a cyclic series. Such a variate has no natural starting point and may be represented on the circumference of a circle. The period between two successive delimiting points, e.g. two successive midnights, is called an interval (*intervallo*).

The mean for such data can be defined in several ways, e.g. by finding that interval for which the sum of deviations or their squares, weighted by number of observations, is zero; but such a mean may not be unique. If an interval is chosen as origin and the mean falls in that interval it is called *media ordinaria* (ordinary mean); if it falls in some other interval it is a *media fittizia* (fictitious mean).

Cylindrically Rotatable Design A development of the **Rotatable (Response Surface) Design** proposed by Herzberg (1966) in which the variances of the estimated responses at points on the same $(k-1)$-dimensional hyper-sphere centred on a specified axis are equal.

D^2 Statistic A statistic introduced by Mahalanobis (about 1930) as a measure of the **'Distance'** between two populations with differing means but identical dispersion matrices. The distance between the populations

$$\Delta^2 = \Sigma \alpha_{jk}(\mu_{1j} - \mu_{2j})(\mu_{1k} - \mu_{2k})$$

is known as Mahalanobis' generalised distance. The sample value of Δ^2 is

$$D^2 = \sum_{j,\,k=1}^{p} a_{jk}(\bar{x}_{1j} - \bar{x}_{2j})(\bar{x}_{1k} - \bar{x}_{2k})$$

and (a_{jk}) is the inverse of the pooled dispersion matrix. Developments of this concept, principally by the Indian school of statisticians, are sometimes collectively known as the 'distance method'.

D_n^+ Statistic A statistic introduced by Wald & Wolfowitz (1939) in connection with setting confidence limits to continuous distribution functions:

$$\begin{aligned} D_n^+ &= \sup_{-\infty < x < \infty} \{F_n(x) - f(x)\} \\ &= \max_{1 < i < n} (i/n - U_i) \end{aligned}$$

where U_i is an ordered sample from a uniform distribution (0, 1). This statistic may also be used as a goodness-of-fit test (see **Kolmogorov-Smirnov Test**) and may be generalised as $D_n^+(\gamma) = \text{Sup} \{F_n(x) - \gamma F(x)\}$.

δ-index (Gini)   See **Concentration, Index of.**

Damped Oscillation   In an oscillatory time series, if the amplitude from peak to trough progressively decreases along the series it is said to be subject to a damped oscillation. Certain derivative series of a time series, such as the **Correlogram** are also said to be subject to a damped oscillation when they exhibit this effect.

Damping Factor   If, in a damped oscillation, the amplitude (peak to trough) diminishes at a constant rate along the series, the ratio of one amplitude to the preceding amplitude is called the damping factor. For example, the correlogram of a second order autoregressive scheme may be written in the form

$$r_k = \frac{p^k \sin(k\theta + \psi)}{\sin \psi}; k \geqslant 0,$$

and the damping factor is p.

Dandekar's Correction   An adjustment proposed by Dandekar (1955) in the calculation of χ^2 for a 2×2 table. [See also **Yates' Correction.**]

Darmois-Koopman's Distributions   An exponential family of distributions introduced by Darmois (1935), Koopman (1936) and Pitman (1936) of the form

$$f(x) = \exp \{ \sum_{j=1}^{k} A_j(\theta)B_j(x) + C(x) + D(\theta)\}.$$

It is the most general form possessing set of k jointly sufficient statistics for its k parameter. [See **Sufficiency.**]

Darmois-Skitovich Theorem   A theorem due to Darmois (1951) and Skitovich (1954) of importance in minimum variance unbiased estimation. The theorem states that a sufficient condition for independent, but not necessarily identically distributed variables to have Normal distributions is the independence of any two linear functions of the form

$a_1 x_1 + \dots + a_n x_n$
$b_1 x_1 + \dots + b_n x_n$ with $a_i b_i \neq 0 (i = 1, \dots, n)$.

De Finetti's Theorem   A theorem due to de Finetti (1937) which states that a **Simple Sample** that generates a permutable binary sequence can always be regarded as having been obtained by binomial sampling in which the **Type I Probability** P has a Type II initial distribution which is unique.

Death Rate   The number of deaths in a given period divided by the population exposed to risk of death in that period. For human populations the period is usually one year and if the population is changing in size over the year the divisor is taken as the population at the mid-year. The death rate as so defined is called 'crude'. If some refinement is introduced by relating mortality to the age and sex constitution (or other factors) for comparative purposes the rate is said to be standardised.

Decapitated Negative Binomial Distribution   A negative binomial distribution from which the zero class is missing. The term 'decapitated' was employed by Yule to characterise a truncated distribution from which the zero class only had been removed.

Decile   One of the nine variate values which divide the total frequency into ten equal parts. [See also **Quantiles.**]

Decision Function   A decision function is a rule of conduct which, at any stage of a sampling investigation, tells the statistician whether to take further observations or whether enough information has been collected, and in the latter case, what decision to make upon it. At each stage beyond the first the decision function is a function of the preceding observations.

Until the development of sequential methods decision functions were mostly of the simple type, based on a fixed sample size, which enjoined the acceptance or rejection of a hypothesis or set limits to a parameter under estimate. The above definition provides for a sequential situation wherein the investigator may not reach a decision about the hypothesis but proceeds to take further observations.

The class of all decision rules which are admissible in the circumstances of a particular case is called a complete class.

Decision Space   In sequential analysis and the theory of decision functions, the decision space is the set of all possible decisions.

Decomposition   The act of splitting a time series into its constituent parts by the use of statistical methods. A typical time series is often regarded as composed of four parts:

(a) A long-term movement or trend;
(b) Oscillations of more or less regular period and amplitude about this trend;
(c) A seasonal component;
(d) A random, or irregular, component.

Any particular series need not exhibit all three of these but those which are present are presumed to act in an additive fashion, i.e. are superimposed; and the process of determining them separately is one of decomposition.

A more modern approach (Wold, 1938) seeks to decompose the series into deterministic and non-deterministic elements. This is known as Wold's decomposition or predictive decomposition.

Decreasing Hazard Rate   Conventionally, a decreasing hazard rate is one which does not increase; it can, therefore, be constant (Barlow et al., 1963).

Deep Stratification A term which is sometimes used to denote stratification by a substantial number of factors with respect to the marginal distribution of the factors only. [See also **Lattice Sampling, Control of Sub-strata.**]

Defective Probability Distribution A probability distribution normally assigns a probability measure over the appropriate interval in such a way that the sum of all the assigned probabilities is unity. However, probability measures are sometimes encountered which assign a total mass, say p, which is less than unity. Such a measure is called a defective probability measure with defect $1-p$: the probability distribution is a defective distribution.

Defective Sample A sample resulting from an inquiry which has been incompletely carried out, e.g. because certain assigned individuals have not been examined, because records have been lost, or because (in plant or animal experiments) certain members have died.

Defective Unit In quality control, a unit which does not reach some prescribed standard and is therefore to be rejected, in contrast to an effective unit.

Defining Contrast (or Relation) In the analysis of **Half-replicate Factorial Designs** the comparison between those treatment combinations which are used with those that have not been used is called the defining contrast (Finney, 1946). It defines the halves into which the whole replicate is divided.

Degenerate Distribution An alternative designation for **Deterministic Distribution.**

Degree of Belief Some writers on the axiomatics of probability prefer to regard the probability of a proposition as expressing the intensity of belief in its truth. By suitable postulates and axioms this concept may sometimes be made objective and measurable on a linear scale. The resulting (numerical) probability may then be regarded as a 'degree of belief' in the proposition.

Degrees of Freedom This term is used in statistics in slightly different senses. It was introduced by Fisher on the analogy of the idea of degrees of freedom of a dynamical system, that is to say the number of independent co-ordinate values which are necessary to determine it. In this sense the degrees of freedom of a set of observations (which *ex hypothesi* are subject to sampling variation) is the number of values which could be assigned arbitrarily within the specification of the system. For example, in a sample of constant size n grouped into k intervals there are $k-1$ degrees of freedom because, if $k-1$ frequencies are specified, the other is determined by the total size n; and in a contingency table of p rows and q columns with fixed marginal totals there are $(p-1)(q-1)$ degrees of freedom.

A sample of n variate values is said to have n degrees of freedom, whether the variates are dependent or not, and a statistic calculated from it is, by a natural extension, also said to have n degrees of freedom. But if k functions of the sample values are held constant, the number of degrees of freedom is reduced by k. For example, the statistic $\sum_{i=1}^{n}(x_i-\bar{x})^2$ where $\bar{x}$ is the sample mean, is said to have $n-1$ degrees of freedom. The ultimate justifications for this are that (a) the sample mean is regarded as fixed or (b) in normal variation the quantities $\bar{x}_i-\bar{x}$ are distributed independently of $\bar{x}$ and hence may be regarded as $n-1$ independent variates or n variates connected by the linear relation $\Sigma(x_i-\bar{x})=0$.

By a further extension the distribution of a statistic based on n independent variates is said to have n degrees of freedom, particularly in relation to $\chi^2=\Sigma x_i^2$. The statistic $\Sigma(x-\bar{x})^2$ is said similarly to have $n-1$ degrees of freedom. By a further extension still, statistics based on combinations of χ^2 are said to have degrees of freedom; notably the ratio of two independent χ^2's, which is said to have n_1 and n_2 degrees, n_1 and n_2 being the degrees of freedom which enter into the numerator and denominator of the ratio.

From a different viewpoint the expression 'degrees of freedom' is also used to denote the number of independent comparisons which can be made between the members of a sample.

Degrees of Randomness Strictly speaking there are no degrees of randomness, except in the theory developed by M. G. Kendall under which randomness is defined relative to an assigned test domain. The expression has however (rather regrettably) been introduced into the theory of stochastic processes.

A random function $\phi(t)$ has n degrees of randomness if n is the smallest integer such that if n values are known at distinct points of time then it is almost certain that a functional relationship exists between $\phi(t)$ $[t>t_n)$ and these available values.

Demodulatiom A technique used in spectrum analysis of time series for the purpose of detecting and describing the more unusual kinds of non-stationary behaviour. The method consists of multiplying a series by a suitable function and applying a **Filter** to leave only the frequency in a narrow band.

Density Function See **Frequency Function.**

Departure Process The stochastic process formed by the intervals between the departure times from the service counter or station of a queueing situation. This departure process is of importance in the study of tandem queues (Jackson, 1957).

Dependence Quantities are dependent when they are not

independent (see **Independence**). For dependence in regression analysis see **Regression**.

Dependent Variable   See Regression.

Derived Statistics   A derived statistic is one which is obtained by an arithmetical operation from the primary observations. In this sense, almost every statistic is 'derived'. The term is mainly used to denote descriptive statistical quantities obtained from data which are primary in the sense of being mere summaries of observations, e.g. population figures are primary and so are geographical areas, but population-per-square-mile is a derived quantity.

*Descriptive Indices (Indici Descrittivi)   In Italian usage an index (or coefficient) is called descriptive when it exhibits a relationship concerning each value of the variable, as distinct from the global or mean index or coefficient (*indice globale o medio*) which relates only to the aggregate or the mean of the values. Means, standard deviations, mean differences, etc., are global indices; **Pareto's** α and **Gini's** δ for incomes purport to be descriptive indices. An index, which is descriptive for one characteristic, may be only global for another; the α and δ indices are descriptive for incomes, but only global for fortunes.

Descriptive Statistics   A term used to denote statistical data of a descriptive kind or the methods of handling such data, as contrasted with theoretical statistics which, though dealing with practical data, usually involve some process of inference in probability for their interpretation. The distinction is very useful in practice but not, perhaps, entirely logical.

Descriptive Survey   A (sample) survey where the principal objective is to estimate the basic statistical parameters (means, totals, ratios) of the population or its sub-divisions.

Design Equation   A basic equation in connection with **Rotatable Designs** whose roots are the complex values of the various sample points. It was shown by Bose & Carter (1959) that, for a design of order d to be rotatable, the first $2d$ terms after the initial term in the design equation must be zero.

Design Matrix   An expression which might be supposed to mean the exhibition in matrix form of the experimental design, but in fact has been applied to the schedule of observations.

Design Type O : PP   A design of experiment proposed by Hoblyn *et al.* (1954) for three classifications where the rows and columns are orthogonal to each other and the treatments partially balanced with respect to rows and columns.

Destructive Test   Under certain conditions it is possible that the carrying out of an inspection test on a manufactured product will result in destruction of the particular test specimen. This is called a destructive test. For example, a test of the fusing current in an electric fuse is of such a kind and many of the inspection tests for explosive products must necessarily be destructive. Under such conditions there is every incentive to design sampling schemes which minimise the number of items to be tested.

Determination, Coefficient of   The square of the product-moment correlation between two variates, r^2, so-called because it expresses the proportion of the variance of one variate, y, given by the other, x, when y is expressed as a linear regression on x.

More generally, if a dependent variate has multiple correlation R with a set of independent variates R^2 is known as the coefficient of determination.

The quantity is also known as the index of determination.

Determining Variable   See Predicated Variable.

Deterministic Distribution   A degenerate distribution consisting of a mass concentrated at a single point $x = k$ so that its probability function is

$$\Pr(x) = \begin{cases} 1 & x = k \\ 0 & x \neq k \end{cases}$$

Deterministic Model   A deterministic model, as opposed to a stochastic model, is one which contains no random elements and for which, therefore, the future course of the system is determined by its position, velocities, etc., at some fixed point of time.

Deterministic Process   A stochastic process with a zero error of prediction; one in which the past completely determines the future of the system.

Detrimental Variable   In confluence analysis, when a new variable is added to a set of explanatory variables the fit may be made worse. On the **'Bunch Map'** the pencil of beams becomes less compact than before. The new variable is then regarded as detrimental.

Deviance   A term proposed by M. G. Kendall to denote the sum of squares of observations about their mean.

Deviate   The value of a variate measured from some standard point of location, usually the mean. It is often understood that the value is expressed in standard measure, i.e. as a proportion of the parent standard deviation.

Diagonal Regression   When two variables are subject to errors of observation and are connected by a linear relation, a set of observed values will not, in general, lie on a straight line. It has been customary to estimate the linear relation by regression techniques (although, strictly speaking, the problem is quite different from the determination of regression lines); and this method has the embarrassing property of yielding two different regressions. Frisch (1934) proposed compromises between these two in the form of 'Orthogonal Regression' and diagonal regression. The latter is a line lying between the regression lines, passing through their point of intersection, and having the equation

$$\frac{Y-m_y}{\sigma_y} = \frac{X-m_x}{\sigma_x}$$

where m_y, m_x are the means and $\sigma_y{}^2$, $\sigma_x{}^2$ the variances of y and x respectively. If $\sigma_y = \sigma_x$ the diagonal regression coincides with the orthogonal regression. Neither, strictly speaking, is a regression.

Dichotomy A division of the members of a population, or sample, into two groups. The definition of the groups may be in terms of a measurable variable but is more often based on quantitative characteristics or attributes.

Difference Sign Test When considering observations on a phenomenon which is moving through time and generates an ordered set, i.e. a time series, one useful test against trend, etc. is the difference sign test. This consists of counting the number of positive first differences of the series, that is to say, the number of points where the series increases. It is used especially as a test against linear trend and as such is superior to the **Turning Point** test but inferior to tests based on rank order.

Differential Process See **Additive Process.**

Diffusion Index A term proposed by Burns (1950) and Moore (1950) to denote the proportion of a set of time series in a given collection of series which are increasing at a given point of time.

Diffusion Process A type of **Additive Process** describing certain kinds of diffusion. The process is such that the 'displacement' of the variate (its increment) in time dt follows a Normal distribution with variance proportional to dt.

Digital Computer A machine for carrying out mathematical processes by operations based on counting, as distinct from an **Analogue Computer** which simulates the processes and produces results which are measured in terms of physical quantities.

Dilution Series The estimation of bacterial density may sometimes be undertaken by the assessment of the presence or absence of growth colonies in samples at one or more levels of dilution from the original suspension. If observations are taken on a set of samples obtained by diluting the original suspension by varying amounts they are said to form a dilution series.

Direct Correlation See **Correlation Coefficient.**

Direct Probability An expression which is supposed to be antithetical to **Inverse Probability,** although neither expression is very logical. It usually denotes probability when used to proceed from the given probabilities of prior events to the probabilities of contingent events; e.g. if it is given that the probability of throwing each number with an ordinary six faced die is $\frac{1}{6}$, the probability of throwing a score of 15 in three throws is directly ascertainable. The relation is analogous to the deductive relations of logic.

Direct Sampling A term used when the sample units are the actual members of the population and not, for instance, some kind of record relating to such numbers, such as census form, ticket or registration card. The term relates to the directness of the observation of the units which enter into the sample, not to the process by which they are selected. [See also **Indirect Sampling.**]

Dirichlet Distribution The k-variate analogue of the **Beta Distribution.**

Disarray, Coefficient of Two rankings A and B may be transformed one into the other by a successive interchange of pairs of neighbouring items in one of them. There will be a minimum number of moves in which the operation can be made and a coefficient of disarray can be constructed from it. This coefficient is in fact the same as **Kendall's Tau.**

Discontinuous Process See **Continuous Process.**

Discontinuous Variate A variate which can take only a discontinuous set of values.

***Discordance (Discordanza)** See *Concordance.**

Discordant Sample See **Concordant Sample.**

Discounted Least Squares Method An extension by D'Esopo (1961) of the basic least squares analysis incorporating weights to the squared deviations in the form of an exponentially decaying series.

Discrepance In variance-analysis, a term used by some writers to denote the **Error Sum of Squares.**

Discrete Lognormal Distribution A misleading term used to denote a distribution obtained as the compound of a Poisson distribution with the parameter λ distributed lognormally.

Discrete Normal Distribution A discrete analogue of the continuous Normal distribution (Haight, 1957) of the form

$$\Pr(x) = A \exp \left\{ -\frac{(x-m)^2}{2v} \right\}$$

where $x = 0, \pm 1, \pm 2, \ldots$ and $m \approx \mu$ with $v \approx \sigma^2$ and

$$1/A = \Sigma \exp\left\{-\frac{(x-m)^2}{2v}\right\}$$

with v and $m^2/v \geq 10$.

Discrete Pareto Distribution This is the discrete analogue of the continuous **Pareto** law:

$$\Pr(x) = \frac{1}{\zeta(s)}\frac{1}{x^s} \quad x = 1, 2, \ldots, \infty$$

where ζ is the Riemann zeta function.

Discrete Power Series Distribution A general probability distribution (Noack, 1950) of the form

$\Pr(x) = a_x\theta^x/f(\theta)$ with $x = 0, 1, 2, \ldots$ and $f(\theta) = \sum\limits_{x=0}^{\infty} a_x\theta^x$.

By appropriate choice of $f(\theta)$ we may derive several common discrete distributions; for example, the Poisson, negative binomial, Pólya-Eggenberger.

Discrete Probability Law A **Probability Law** is called discrete if it corresponds to a discrete distribution function.

Discrete Rectangular Distribution A distribution (of finite range) in which each of a discrete set of x variables, n in number, has the same probability $1/n$. It is mostly used for equi-spaced variate values, e.g. the numbers 0-9 if equally probable, from a discrete rectangular distribution.

Discrete Process See **Continuous Process**.

Discrete Type III Distribution A discrete analogue of the continuous **Type III Distribution** of the form
$$\Pr(x) = A(x+a)^k e^{-\alpha x},$$
$x = 0, 1, 2, \ldots$ and
$$1/A = \sum\limits_{x=0}^{\infty} (x+a)^k e^{-\alpha x}$$
(Haight, 1957).

Discrete Variate See **Discontinuous Variate**.

Discriminatory Analysis Given a set of multivariate observations on samples, known with certainty to come from two or more populations, the problem is to set up some rule which will allocate further individuals to the correct population of origin with minimal probability of misclassification. This problem and sundry elaborations of it give rise to **Discriminatory Analysis**.

Dishonest Process An alternative name for a pathological process, and the opposite in a mathematical sense, of an '**Honest**' **Process**.

*Disnormality (Disnormalità) In Italian usage, a measure of the dissimiliarity between an observed curve and a Normal curve of the same median and dispersion about the median, taking account not only, as in **Abnormality**, of the intensity but also of the sign of the differences. It is then analogous to a measure of skewness. As in the case of abnormality it is obtained by comparing the observed distribution with a Normal distribution of the same median and the same dispersion about the median but taking account of the sign of the differences.

The absolute index of disnormality is given by $\mu_1 - m_e$ where μ_1 is the arithmetic mean and m_e is the median; and the relative index by

$$D = \frac{\mu_1 - m_e}{'S_M}$$

where $'S_M$ is the mean deviation about the median. If D is zero the distribution is said to have *disnormalità nulla*; if $D > 0$ *disnormalità positiva*; if $D < 0$ *disnormalità negativa*.

Dispersion The degree of scatter shown by observations. It is usually measured as an average deviation about some central value (e.g. mean deviation, standard deviation) or by an order statistic (e.g. quartile deviation, range) but may also be a mean of deviations of values among themselves (e.g. Gini's mean difference and also standard deviation).

Dispersion Index This is not, as the name might suggest, a measure of the dispersion of a set of values expressed as an index number. It is the name given to certain statistics which are used to test the homogeneity of a set of samples, i.e. it refers to the dispersion of the samples. [See **Binomial Index of Dispersion, Poisson Index of Dispersion, Lexis Ratio**.]

Dispersion Matrix See **Covariance Matrix**.

Dispersion Stabilising Transformation A term proposed by Ruben (1966) for a generalisation of the variance stabilising transformation. The generalisation covers the stabilisation of covariance matrices.

Displaced Poisson Distribution A situation in which the number of events in excess of a threshold value r, when it is assumed that at least the r events do occur, has a displaced Poisson distribution with parameters λ and r

$$\Pr(x) = \frac{e^{-\lambda}\lambda^{(x+r)}}{I(r, \lambda)(x+r)!}$$

where $\quad I(r, \lambda) = \sum\limits_{\gamma=r}^{\infty} (e^{-\lambda}\lambda^\gamma)/\gamma!$

Disproportionate Sub-Class Numbers In the analysis of variance, except in the case of analysis by reference to a single classification, the arithmetic and estimation of class effects are greatly simplified when there are equal numbers of observations in the sub-classes or when the numbers are proportionate. In the contrary case, the sub-class numbers are said to be disproportionate and the analysis, though theoretically straightforward, is much more complicated.

Dissection (of Heterogeneous Distributions) The dissection into its components of a distribution which is composed of two or more distributions, e.g. a distribution of the lengths of petals from two separate species of primrose might require analysis into the distribution for each species.

*Dissimilarity, Index of (Indice di Dissomiglianza) In Italian usage, a measure of dissimilarity of a particular kind between two distributions. If the distributions are cograduated (see **Cograduation**) and x_i, y_i, refer to their respective cograduated variables the simple index (*indice semplice*) of dissimilarity is given by

$$D' = \frac{\Sigma \, |\, x_i - y_i \,|}{\max \Sigma \, |\, x_i - y_i \,|} = \frac{\Sigma \, |\, x_i - y_i \,|}{n\mu_1' + (n-2)\mu_2'}$$

where there are n values, μ_1' and μ_2' are the respective means of x and y and $\mu_1' \geqslant \mu_2'$.

A corresponding quadratic index can be constructed on the basis of $\Sigma(x_i - y_i)^2$.

When the two distributions have not the same number of values the index of dissimilarity is calculated for two distributions similar to those given and having the same number of values.

*Dissymmetry (Dissimmetria) See ***Symmetry**.

Distance This word is used in many statistical contexts in its ordinary sense, e.g. the 'distance' of a value x from some origin a is $x - a$. A specialised use occurs in the notion of 'distance' between two variates, x and y, which may be defined as the expected value of $x - y$; or the 'distance' between two populations, which may be defined as the difference of their means (though other definitions are possible). [See also **Bhattacharyya's Distance**, D^2 **Statistic**.]

Distributed Lag A term introduced by Irving Fisher (1925) in connection with the analysis of correlation between time series. It is based on the assumption that a given cause occurring at one point of time will exert its effect at various future points and will thus be 'distributed' over terms which lag behind the original cause.

Distribution Curve The graph of cumulated frequency as ordinate against the variate value as abscissa, namely the graph of the **Distribution Function**.

The curve is sometimes known as an ogive, a name introduced by Galton, because the distribution curve of a Normal function is of the ogive shape; but not all distribution curves have this form and the term ogive is better avoided or confined to the Normal or nearly-normal case.

Distribution Free Method A method, e.g. of testing a hypothesis or of setting up a confidence interval, which does not depend on the form of the underlying distribution; for example, confidence intervals may be obtained for the median, based on binomial variation, which are valid for any continuous distribution. In the case of testing hypotheses, the expression is to be understood as meaning that the test is independent of the distribution on the **Null Hypothesis**. Distribution free inference or distribution free tests are sometimes known as non-parametric but this usage is confusing and should be avoided. It is better to confine the word 'non-parametric' to the description of hypotheses which do not explicitly make an assertion about a parameter.

Distribution Free Sufficiency A concept introduced by Godambe (1966) and related to **Linear Sufficiency**. An estimation is said to be distribution free sufficient if any other estimator is independent, in terms of the accumulated prior knowledge, of the original estimator and the population value under estimation.

Distribution Function The distribution function $F(x)$ of a variate x is the total frequency of members with variate values less than or equal to x. As a general rule the total frequency is taken to be unity, in which case the distribution function is the *proportion* of members bearing values $\leqslant x$. Similarly, for p variates $x_1, x_2, ..., x_p$ the distribution function $F(x_1, x_2, ..., x_p)$ is the frequency of values less than or equal to x_1 for the first variate, x_2 for the second and so on.

Distribution of Run Lengths The distribution of lengths of runs of attributes in a series. A run is defined as a sequence of items of the same kind terminated by one or more items of another kind.

Disturbancy, Coefficient of An obsolete coefficient introduced by Charlier to measure departure from **Bernoulli Variation** such as occurs in **Lexis Variation** or **Poisson Variation**. [See also **Lexis Ratio**.]

Disturbed Harmonic Process This form of stochastic process was proposed by Yule (1927) to explain the continual change of amplitude and shift of phase which appears to be typical of time series from the economic and meteorological sciences. If a series is observed at points $t = 1, 2, ...$ an ordinary harmonic movement may be expressed by the equation

$$u_{t+2} - 2u_{t+1} + u_t = 0.$$

If we replace the zero by a random term ϵ_{t+2} on the right-hand side the motion is said to be that of a disturbed harmonic. It is a limiting case of an **Auto-regressive Process** but is not stationary.

Disturbed Oscillation A time series which exhibits a continual shift of period and amplitude in its oscillation is said to possess a disturbed oscillation.

Divergence, Coefficient of A coefficient introduced by Lexis to measure departure from simple or Bernoullian sampling in the sampling of attributes. [See **Lexis Ratio**.]

*Dividing Value (Valore Divisorio) In Italian usage, given a non-decreasing series $a_1, a_2, \ldots a_k, \ldots, a_n$, the value a_k is called a dividing value if it can be regarded as the sum of two parts a_k' and a_k'' such that

$$\sum_{i=1}^{k-1} a_i + a_k' = \sum_{i=k+1}^{n} a_i + a_k''.$$

If $\sum_{i=1}^{k} a_i = \sum_{i=k+1}^{n} a_i$ any value between a_k and a_{k+1} is regarded as a dividing value. [See also **Median**.]

Divisia's Index An index number due to F. Divisia (1925). It is in the form of a chain index. If prices p_t and quantities q_t are regarded as functions of the time, the price index is defined as

$$I_p = \exp \left(\int_c \frac{\Sigma q_t \, dp_t}{\Sigma q_t p_t} \right)$$

where c denotes the path of the prices. There is a similar form for a quantity index, I_q. The indices have the property that changes in total expenditure are proportional to the product of I_p and I_q.

Divisia-Roy Index An index number of Divisia's type for prices, constructed as a **Konyus Index** by taking the quantities to be optimal for a fixed nominal income of the consumer.

Dixon's Statistics A class of statistics for the rejection of the largest observation proposed by Dixon (1950, 1951) of the form

$$r_{ij} = \frac{x_{(n)} - x_{(n-i)}}{x_{(n)} - x_{(j+1)}} \qquad \begin{array}{l} i = 1, 2 \\ j = 0, 1, 2. \end{array}$$

where $x_{(k)}$ means the kth value of a sample of n when ranked by size.

Dodge Continuous Sampling Plan The first **Continuous Sampling Plan** was proposed by Dodge (1943). It assumed a knowledge of the production process. This restriction was relaxed by Derman *et al.* (1959) who made other developments.

Domain of Study In sample surveys a certain sub-group of a population may be of particular interest. Such a sub-group of interest is called a 'domain'. Domains frequently cut across the strata and the various stage units of a sample. Formulae for domains cutting across strata have been devised.

Dominating Strategy A strategy which is in no respect worse than all others and in some respect is better.

Doolittle Technique A method of solving the normal equations in least squares proposed by Doolittle (1878) which proceeds by systematic backward elimination of the variables. It is a more compact version of an approach originally proposed by Gauss (1873). It has been modified and extended by others.

Dose Metameter See **Metameter**.

Double Binomial Distribution This distribution is a weighted average of two binomials with parameters p_1 and p_2; $p_1 < p_2$ and weights w_1 and w_2; $w_1 + w_2 = 1$.

Double Confounding **Confounding** of two different groups of treatment contrasts with two different sources of variation in an experimental design. Thus in the row and column layout of a **Quasi-Latin Square** certain interactions are confounded with rows, and certain others with columns.

Double Dichotomy The division of a set of members by two dichotomies, usually according to attributes; thus a set may be divided into A's and not-A's and each of these two into two subsets according to whether they bear a second attribute B or not.

Double Exponential Distribution The distribution with frequency function $f(x) = a \, e^{b|x-c|}$, $-\infty \leqslant x \leqslant \infty$, a, b and c being constants, $b < 0$. The distribution may be regarded as an ordinary exponential together with its reflexion about the point c. An example of the double exponential distribution is that of the distribution of the mid-range from a rectangular population.

Double Exponential Regression An expression sometimes used for a regression equation in which the dependent variable (y) is a linear combination of two simple exponential terms in (x). [See also **Mixed Exponential Response Law**.]

Double Hypergeometric Distribution If, from a finite population containing two kinds of item, a random sample is drawn without replacement and then a second random sample without replacing the first, the distribution of the (x_1, x_2) successes takes the form of the double hypergeometric. (K. Pearson, 1924.)

Double Logarithmic Chart A chart in which both the horizontal and the vertical axis are scaled in logarithms usually to base 10.

Double Pareto Curve A continuous frequency function whose ordinate is the sum of two functions of the **Pareto** type, e.g.

$$f(x) = \frac{A}{x^{1+\alpha}} + \frac{B}{x^{1+\beta}}, \qquad a, \beta > 0; \ 0 \leqslant x \leqslant \infty.$$

Double Poisson Distribution A distribution in which the parameter λ of the Poisson series is itself regarded as distributed in the Poisson form. The distribution arises particularly in ecology where the numbers of offspring of parents which are themselves distributed over space in the Poisson form are also distributed in that form in the neighbourhood of the parent.

Double-ratio Estimator An estimator built up from four variates, x_1, x_2, y_1, y_2 by using the ratio of the ratios y_1/x_1 and y_2/x_2. [See **Ratio Estimator**.]

Double Reversal Design An extension of the **Switch-Back** or reversal design to the case of two treatments over four periods: A_1, B_2, A_3 B_4 for one-half of the experimental units and B_1 A_2 B_3 A_4 for the other half; the suffixes are the time periods. This design allows for trends in the responses which are independent of the treatment effects.

Double Sampling A standard form of sample design for industrial inspection purposes. In accordance with the characteristics of a particular plan, two samples are drawn, n_1 and n_2, and the first sample inspected. The batch can then be accepted or rejected upon the results of this inspection or the second sample be inspected and the decision made upon the combined result. The term has also been used somewhat loosely for what is called **Multi-phase Sampling** and the two-stage version of **Multi-stage Sampling**. There is a further usage whereby a first sample provides a preliminary estimate of design parameters which govern the size of the second sample to achieve a desired overall result. [See **Stein's Two-sample Procedure**.]

Doubly Stochastic Matrix See **Stochastic Matrix**.

Doubly Stochastic Poisson Process A point stochastic process, suggested by Bartlett (1963), where the rate of occurrence in a Poisson process is itself a stochastic process.

Double-Tailed Test A test for which the **Region of Rejection** comprises areas at both extremes of the sampling distribution of the test function. It is usual but not essential to allot one-half of the probability of rejection to each extreme, giving a symmetrical test.

Down cross A down cross is the point where a time series, measured about its mean, changes in sign from positive to negative. Correspondingly a point where it changes in sign from negative to positive is called an up cross.

Downward Bias Bias which tends to reduce a magnitude below its true value. In index number theory the expression frequently occurs but has a rather obscure meaning owing to the somewhat arbitrary nature of the definition of bias in index numbers. In the theory of estimation an estimator t is biassed downwards if $\mathscr{E}(t) < \theta$, the parameter under estimate.

Similarly, upward bias is such as to tend to raise a magnitude above its true value.

Dragstedt-Behrens' Method See **Behrens' Method**.

Dual Process For every queueing process it is possible to create an associated process by interchanging the distributions of inter-arrival and service times respectively: this associated process was termed the dual process by Takacs (1965).

Dual Theorem A fundamental theorem of linear programming. The 'primal' problem in the variables x_J $(j = 1, ..., n)$ is to maximise $z = c_1x_1 + c_2x_2 + ... + c_nx_n$ subject to the linear programming constraints

$$a_{i1}x_1 + a_{i2}x_2 + ... + a_{in}x_n \leqslant b_i \ (i = 1, ..., m),$$

$x_J \geqslant 0$ $(j = 1, ..., n)$. The 'dual' problem in variables u_i $(i = 1, ... m)$ is to minimise $v = b_1u_1 + b_2u_2 + ... + b_mu_m$ subject to $a_{1j}u_1 + a_{2j}u_2 + ... + a_{mj}u_m \geqslant c_J$ $(j = 1, ..., n)$; $u_i \geqslant 0$ $(i = 1, ..., m)$.

The theorem states that for values of x_J $(j = 1, ..., n)$, u_i $(i = 1, ..., m)$ satisfying the above inequalities then $z \leqslant v$, and that if z^*, v^* are the solutions of the optimisation problems $z^* = v^*$.

Dummy Observation In variance-analysis with **Disproportionate Sub-class Numbers** it is sometimes possible to obtain good approximations to the correct analysis by adding 'dummy' observations. These observations are generally inserted with values equal to the means of the particular cells concerned. An analogous use of class means is made to replace missing values in the '**Missing Plot' Technique**. The word 'dummy' in these connections is, perhaps, better avoided.

Dummy Treatment In order to preserve the symmetry or other features of an experimental design it is sometimes useful to pretend that treatments are applied to certain units when, in fact, no treatments are applied; they are then called dummies. For example, a factorial experiment on the effect of two fertilisers may provide for the application of each in two different concentrations and also for a control (nil) application. One possible design would provide for two factors at three levels, but the third level of each, involving no application of fertiliser, would be a dummy treatment.

Dummy Variable A quantity written in a mathematical expression in the form of a variable although it represents a constant; for example, in a regression equation

$$y = \beta_0 + \beta_1X_1 + \beta_2X_2 + ... + \beta_pX_p$$

it may be more convenient to attach to the coefficient β_0 a dummy X_0 which is always unity so that the expression may be written

$$y = \sum_{i=0}^{p} \beta_iX_i.$$

The term is also used, rather laxly, to denote an artificial variable expressing qualitative characteristics; for example, the presence or absence of an attribute may be indicated by attaching the values 0 or 1 to the individuals concerned. In this sense the word 'dummy' should be avoided.

Duncan's Test A modification (1952, 1957) to the **Newman-Keuls Test** with the object of redistributing the probabilities of error among the components of the multiple comparison procedure.

Duo-Trio Test A test in which three objects, two of which are alike, are presented to a judge who attempts to select the dissimilar object after being given the identity of two of them. [See also **Triangular Test**.]

Duplicate Sample A sample collected concurrently, i.e. in the course of the same sample survey under comparable conditions, with a first sample. It acts in much the same way as a **Replication** except that in some cases the only thing which can be replicated is the act of taking a second independent sample. For example, a second independent sample in a survey will afford additional information on the sampling error but nothing has been replicated in the sense of a repeated experiment beyond the act of selecting a duplicate sample. The act of taking several such samples may be called replicated sampling. [See **Interpenetrating Sample**.]

Duplicated Sample A sample which is taken up twice for enquiry by two different parties of investigators. Two **Interpenetrating Samples** assigned to two investigating parties are sometimes so arranged, for purposes of control of field operations, as to have some common sample units these common units constitute a duplicated (sub-)sample. This should be distinguished from **Duplicate Sample**.

Durbin-Watson Statistic A statistic testing the independence of errors in least squares regression (1950) against the alternative of serial correlation. The statistic is a simple linear transformation of the first serial correlation of residuals and, although its distribution is unknown, it is tested by two bounding statistics which follow R. L. Anderson's distribution.

Dvoretsky's Stochastic Approximation Theorem A general theorem on the convergence of transformations with superimposed random errors (Dvoretsky, 1956). It applies to the stochastic approximation procedures of **Robbins-Munro** and **Kiefer-Wolfowitz**. [See also **Sach's Theorem**.]

Dynamic Model In econometrics a model is said to be dynamic if it possesses either or both of these properties: (1) at least one variable occurs in the structural equations with values taken at different points of time or in the form of time-derivatives, etc.; (2) at least one equation contains a function of time.

If the first property is present the system is sometimes called multitemporal and, if neither is present, unitemporal. These terms are not ideal; they do not mean that several 'time' variables or one 'time' variable are involved.

Dynamic Programming A method for securing a sequence of consistently optimum solutions to multistage decision problems; sequential decision taking, in fact.

Dynamic Stochastic Process A class of second order process introduced by Stigum (1963) which combine a wide sense stationary process and a deterministic non-stationary process in a single representation.

Edgeworth Index See **Marshall-Edgeworth-Bowley Index**.

Edgeworth's Series This series was introduced by Edgeworth (1904) as a method for representing certain skew frequency distributions. It is based upon an expansion in terms of the Normal distribution and its derivatives and is very similar to the **Gram-Charlier Series**.

If the cumulants of the distribution are prepresented by κ_1, κ_2, etc., the Normal function $\frac{1}{\sqrt{(2\pi)}} e^{-\frac{1}{2}x^2}$ by $\alpha(x)$ and the process of differentiating $\frac{d}{dx}$ by D, the Edgeworth expansion of a frequency function $f(x)$ with zero mean and unit variance may be written

$$f(x) = \exp\left\{ \sum_{j=3}^{\infty} \frac{(-D)^j}{j!} \kappa_j \right\} \alpha(x)$$

where the exponential is expanded as a power series before the operations with D are carried out.

Effect Variable See **Cause Variable**.

Effective Range A group of observations may contain a limited number of outlying observations at either or both ends of the range. A very rough measure of dispersion may be obtained by taking the 'effective' range, i.e. the range after these outlying values have been removed. The removal may have to be a matter of subjective judgment and inferences based on effective range are of somewhat doubtful value; in fact the term itself is not a good one.

Effective Unit See **Defective Unit**.

Efficiency The concept of efficiency in statistical estimation is due to Fisher (1921) and is an attempt to measure objectively the relative merits of several possible estimators.

The criterion adopted by Fisher was that of variance, an estimator being regarded as more 'efficient' than another if it has smaller variance; and if there exists an estimator with minimum variance v the efficiency of another estimator of variance v_1 is defined as the ratio of v/v_1. It was implicit in this development that the estimator should obey certain other criteria such as **Consistency**. For small samples, where other considerations such as bias enter, the concept of efficiency may require extension or modification.

The word is also used to denote the properties of experimental designs, one design being more efficient than another if it secures the same precision with less expenditure of time or money.

Efficiency Equivalence A broad extension of the concept of **Asymptotic Relative Efficiency** to the case of regression parameters in second order processes (Striebel, 1961).

Efficiency Factor The efficiency factor of an experimental design is usually expressed in terms of a ratio of **Error Variances**, or, from a slightly different viewpoint, in terms of the size or number of replications needed to attain a given precision. Thus in an agricultural field trial laid out in incomplete blocks, the efficiency factor is the ratio by which the variance of a given treatment estimate would be reduced by ignoring blocks, if this did not alter the variance per plot. In general the efficiency factor will vary according to the standard of comparison.

Efficiency Index A concept proposed by Armitage (1959) in connection with the comparison of survival curves. If the difference between the two death rates (λ and λ') is $\delta\lambda$, the efficiency index is
$$\psi(\lambda T) = \chi^2 (\lambda/\delta\lambda)^2/n.$$

Ehrenfest Model A diffusion model in Markov chain form, proposed by Ehrenfest (1906), without absorbing states and with the probabilities converging to zero with increasing time.

Eigenvalue See **Characteristic Root**.

Eisenhart Models In variance analysis it is customary to draw a distinction between type 1, in which the classification variables are fixed, and type 2 in which they are themselves random variables. The distinction has been pointed out by Eisenhart whose name is frequently attached to the models, especially in balanced incomplete block experiments. [See also **Variance Components**.]

Elementary Unit One of the individuals which, in the aggregate, compose a population: the smallest unit yielding information which, by suitable aggregation, leads to the population property under investigation. Cases occur where the term may be ambiguous; e.g. if an age distribution is to be estimated from a sample of households then the person is the elementary unit; but if, at the same time, the size of household is to be estimated, the household is the elementary unit. [See **Basic Cell**.]

Elfving Distribution An approach to the distribution of the range in samples from a Normal population proposed by Elfving (1947) making use of the probability integral transformation.

Empirical Bayes' Estimator An estimator derived according to an **Empirical Bayes' Procedure**. The non-parametric form was given by Johns (1957) and a 'smooth' form for discrete distributions by Maritz (1966).

Empirical Bayes' Procedure A term proposed by Robbins (1956) to denote the approximation to an optimum Bayes' decision procedure which may be derived from the use of data previously obtained by the same selection procedure as is now proposed for use in a new experiment on the same population. It is essentially an adaptive procedure (Robbins, 1964).

Empirical Distribution Function Given an ordered sample of n independent observations
$$x_{(1)} \leqslant x_{(2)} \leqslant x_{(3)} \leqslant \ldots \leqslant x_{(n)},$$
the function $S_n(x)$ defined as:
$$S_n(x) = \begin{cases} 0, & x \leqslant x_{(1)} \\ \dfrac{k}{n}, & x_{(1)} < x < x_{(k+1)} \\ 1, & x_{(n)} < x \end{cases}$$
is called an empirical distribution function.

Empirical Probit The analysis of experimental results concerned with the relationship between levels of a stimulus and the responses thereto is usually made in terms of the percentage of test subjects reacting to the stimulus. The analysis is often facilitated by transforming the percentages into **Probits**. The probits corresponding to these observed percentages are known as empirical probits, in contradistinction to **Expected Probits** and **Working Probits**, which relate to a fitted regression line.

Empty Cell Test A rather inefficient test of randomness in sampling, based on the number of cells in a uni- or multi-variate classification, in which a sample member does not fall.

End Corrections An expression used in several statistical contexts, to denote corrections made to extreme values. For example, in taking a systematic sample from material which varies continuously it may be possible to obtain some increase in accuracy of estimation by employing corrections to the term at each end of the sample. Again, in fitting a curve to a time series, corrections may be made to end values to improve the fit. The expression also occurs in the calculation of serial correlations where certain alterations are necessary to formulae relating to infinite series to allow for the fact that the observations have finite length.

In many such cases the word 'adjustment' is better than 'correction', which carries an implication that the values concerned are in error. [See **Finite Multiplier**.]

Endogenous Variate The statistical representation and analysis of multivariate systems generally involves a primary division of variates into those which are endogenous and those which are exogenous. Endogenous variates are those which form an inherent part of the system, as for instance price and demand in an economic model. Exogenous variates are those which impinge on the system from outside, e.g. rainfall or epidemics of disease. It is possible for a variate to be endogenous in one model and exogenous in another, for example, rainfall might be regarded as exogenous to an economic model of the automobile industry but endogenous to a meteorological model describing climatic states.

Engset Distribution A distribution occurring in congestion theory, it is equivalent to the binomial truncated at the higher values:

$$\Pr(x) = \frac{\binom{n}{x} p^x q^{n-x}}{\sum_{j=0}^{N} \binom{n}{j} p^j q^{n-j}}. \quad N < n.$$

Ensemble The infinite set of **Realisations** of a single **Stochastic Process**. The concept is frequently extended to apply to a sample set of realisations.

Ensemble Average Given a number of independent sample **Realisations** from the same generating process, the mean of the process can be estimated by the overall mean of the samples. The estimator is called an 'ensemble average'. If the process is stationary, its mean can be estimated by a simple average over time instead of by the 'ensemble average'.

Entry Plot A plot through which a **Serial Cluster** is entered or determined.

Envelope Power Function A power function based on a critical region which itself is obtained as the envelope of a number of critical regions. Where no best critical region exists for all alternative hypotheses, such an envelope provides a 'good' test in the sense that it picks out and amalgamates parts of critical regions which are best for particular alternatives.

Envelope Risk Function A concept analogous to the **Envelope Power Function** but applied to the risk functions associated with decision functions.

'Epsem' Sampling See **Equal Probability of Selection Method**.

Equal Ignorance, Principle of In connection with the use of Bayes' formula, the absence of definitive information on prior probabilities generally leads to an assumption of equal ignorance, or a uniform distribution of prior probabilities.

Equal Probability of Selection Method Any method of sampling in which the population elements have an equal probability of selection. 'EPSEM' sampling can result from either equal probability selection throughout, or, from variable probabilities that compensate each other in multi-stage selection.

Equal-tails Test A symmetrical **Double-Tailed Test**.

*Equalising Value (Adeguato numerico) In Italian usage, a quantity q, dependent on observations x_1, x_2, ... x_n, which obeys the relation

$$f(x_1, x_2, ..., x_n) = f(q, q, ..., q)$$

where f is some given function. As q may lie outside the range of the x's it should be distinguished from a mean value properly so-called.

Equally Correlated Distribution A multivariate distribution is equally correlated if the correlation between each pair of variates is equal to ρ.

Equidetectability, Curve of An expression used by Neyman and Pearson in the testing of simple hypotheses concerning two parameters. If the parameters are represented on a pair of rectangular axes, the power function in the neighbourhood of the values specified by the null hypothesis is given approximately by a quadratic term in the parameters. The curves for which the function is constant, namely the curves of constant power, are called curves of equidetectability.

More generally, for n parameters, there are $(n-1)$-dimensional hypersurfaces of equidetectability.

*Equidistribution, Line of (Retta di Equidistribuzione) In Italian usage, the concentration curve in the case of zero **Concentration**.

Equilibrium In queueing theory, systems in 'statistical equilibrium' are those in which the number of customers or items waiting in the queue oscillates in such a way that mean and distribution remain constant over a long period.

Equilibrium Distribution A physical, economic or social system which has settled down to a stable statistical behaviour has a stationary, or equilibrium, distribution which specifies the limiting proportions spent in the designated states of the system.

Equitable Game An alternative name for **'Fair' Game**.

Equivalence Class A term used in the construction of cyclic designs whereby certain permutations of treatments emerge as rearrangements of others (David & Wolock, 1965).

Equivalent Deviate If P is a probability or proportion and $f(x)dx$ is the frequency element of a distribution, almost always continuous, and usually in a standard form free of parameters, the equivalent deviate of P relative to the distributions is Y where

$$P = \int_{-\infty}^{Y} f(x)dx.$$

Particular examples are the **Normal Equivalent Deviate** and the **Logit**.

Equivalent Dose The dose of a standard preparation having the same effect, i.e. the same expected mean response, as a specified dose of a test preparation. It is of particular importance under conditions of **Similar Action** when the ratio of any dose of the test preparation and its equivalent is a constant, the **Relative Potency**.

Equivalent Samples Any two samples selected by the same sampling scheme which contain the same set of $d (\leqslant n)$ population members, yield the same information about the parameters, irrespective of whether these d members appear with different frequencies in the two samples.

Ergodicity Generally, this word denotes a property of certain systems which develop through time according to probabilistic laws. Under certain circumstances a system will tend in probability to a limiting form which is independent of the initial position from which it started. This is the ergodic property.

A stationary stochastic process $\{x_t\}$ may be regarded as the set of all realisations possible under the process. Each such realisation may have a mean m_r. If the process itself has a mean $\mathscr{E}(x_t) = \mu$, the ergodic theorem of Birkhoff & Khintchine states that m_r exists for almost all realisations. If, in addition, $m_r = \mu$ for almost all realisations the process is said to be ergodic. In this sense ergodicity may be regarded as a form of the law of large numbers applied to stationary processes.

Erlang Distribution A family of Gamma-type distributions of the form

$$f(t) = \frac{\lambda}{\Gamma(k)}(\lambda t)^{u-1}e^{-\lambda t}$$

proposed by Erlang for the inter-arrival times and service times in queueing problems. It is equivalent to a **Chi-squared Distribution** with an even number $(2k)$ of degrees of freedom.

Erlang's Formula An early result in congestion theory given by Erlang (1917) concerning the degree of hindrance experienced by a telephone subscriber who is unable to effect a call because all lines or channels are in use. The assumption is that the call is then lost and the formula is frequently known as Erlang's Loss Formula. The probability that, of N channels, all are in use is obtained by placing $x = N$ in

$$\Pr(x) = \frac{\lambda^x}{x!} \Big/ \left(\sum_{j=0}^{N} \frac{\lambda^j}{j!} \right) \quad x = 0, 1, 2, ..., N.$$

Error In general, a mistake or error in the colloquial sense. There may, for example, be a gross error or avoidable mistake; an error of reference, when data concerning one phenomenon are attributed to another; copying errors; an error of interpretation.

In a more limited sense the word 'error' is used in statistics to denote the difference between an occurring value and its 'true' or 'expected' value. There is here no imputation of mistake on the part of a human agent; the deviation is a chance effect. In this sense we have, for example, **Errors of Observations, Errors in Equations, Errors of the First and Second Kinds** in testing hypotheses, and the **Error Band** surrounding an estimate; and also the Normal curve of errors itself.

Error Band In estimation or prediction the estimated or predicted value is bracketed by a range of values determined by standard errors, confidence-intervals or similar methods within which the value may be supposed to lie with a certain probability. This is called the Error Band.

Error in Equations An equation in variables or variates may be inexact either because the equation is not a complete representation of the situation, as in a demand-supply equation which omits other factors such as income or employment, or because it is disturbed by extraneous sources of variation as in an autoregression equation. These departures from the relationship expressed by the equation are known as errors in the equation; as distinct from effects such as observational errors in the variables themselves.

Error Mean-Square The residual or **Error Sum of Squares** divided by the number of degrees of freedom on which the sum is based. It provides an estimator of the residual or error variance.

Error of Estimation In general, the difference between an estimated value and the true value.

More specifically, in regression analysis where the regression equation is used to estimate the 'dependent' from given values of the 'independent' variates, the difference between the estimated and the observed value of the dependent variate. The standard deviation of these differences in repeated samples is sometimes known as the 'error of estimate' and, far preferably, as the 'standard error of estimation'.

Error of First Kind If, as the result of a statistical test, a statistical hypothesis is rejected when it should be accepted, i.e. when it is true, then an error is committed. This class of error is termed an error of the first kind and is fundamental to the theory of testing statistical hypotheses associated with the names of Neyman and E. S. Pearson.

The frequency of errors of the first kind can be controlled by an appropriate selection of the regions of acceptance and rejection; that is to say, by choice of appropriate **Critical Regions** it is possible to ensure that the probability of committing an error of the first kind is an assignable constant.

Error of Observation An error arising from imperfections in the method of observing a quantity, whether due to instrumental or to human factors.

Error of Second Kind If, as the result of a test, a statistical hypothesis is accepted when it is false, i.e. when it should have been rejected, then an error is made. This class of error is termed an error of the second kind and, like the **Error of First Kind**, it is fundamental to the Neyman-Pearson theory of testing statistical hypothesis.

Unlike the error of the first kind, however, it is not, in general, controlled by the simple process of selecting regions of acceptance and rejection. The customary procedure in choosing tests of hypotheses is to fix the frequency of the first kind of error and, with this restriction, to minimise the frequency of the second kind of error. [See also **Power Function**.]

Error of Third Kind In 1947 F. N. David, perhaps not entirely seriously, suggested that there was a third kind of error which might be committed in testing statistical hypotheses: that of selecting the test falsely to suit the significance of the particular sample data available.

A somewhat different type of error of the third kind was suggested by Mosteller (1948) in proposing a nonparametric test for deciding whether one population, out of k populations characterised by a location parameter, has shifted too far to the right of the others. He defines it as 'the error of correctly rejecting the **Null Hypothesis** for the wrong reason'.

Error Reducing Power A term used in connection with the smoothing of time series. Each observation is regarded as composed of a true value and an error of observation which is independent of the errors in other observations. The smoothing process is an attempt to approximate to the true values and to reduce the errors. The success of any process is measured by its error reducing power, one common measure being the extent to which the variance of a random series is reduced if the process is applied to it.

Error Sum of Squares In variance analysis it is customary to regard the data as generated by a model (usually linear) consisting of certain class effects plus a stochastic component. When estimates are made of the class effects and subtracted from the observations, the residuals are estimates of the contribution from the stochastic component and the sum of squares of these residuals is known as the 'error sum of squares', though 'residual sum of squares' is preferable from many points of view. [See also **Pooling of Error, Variance Components**.]

Errors in Variables In contradistinction to **Errors in Equations**, errors in the values of the variables concerned; usually errors of observation.

Error Variance The variance of an error component. Thus, if the generating model of a set of data consists of certain systematic components together with a stochastic component, the variance of the latter is the error variance.

The expression can also be understood in a wider sense, as the variance of error in repetitions of an experimental situation, whether the 'error' is due to sampling effects or not. It makes for clarity if expressions such as 'error variance' are eschewed in favour of 'residual variance' but the use of the former type of wording is very widespread.

Errors in Surveys The errors in a sample survey arise both from sampling effects and from other sources not connected with sampling, i.e. they would also be present for a complete survey. It has become customary to use the word 'error' to cover all these types of departures from representativeness, whereas in some statistical contexts 'error' denotes 'sampling error' and the other effects are called 'biasses'.

Esseen-Type Approximation In connection with the Normal approximation to the common discrete distributions, Esseen (1945) proposed an improvement to the Lyapounov results concerning conditions for, and speed of, convergence.

Estimable This usage has departed completely from the standard interpretation of 'capable of being esteemed'. What word should have been used for the concept of 'capable of being estimated in a particular sense' is a matter of debate; one suggestion is 'estimative'.

One sense of a parameter being of this kind is that it possesses an **Unbiassed Estimator**.

Estimate In the strict sense an estimate is the particular value yielded by an **Estimator** in a given set of circumstances. The expression is, however, widely used to denote the rule by which such particular values are calculated. It seems preferable, following Pitman, to use the word **Estimator** for the rule of procedure and 'estimate' for the values to which it leads in particular cases.

Estimating Equation An equation involving observed quantities and an unknown which serves to estimate the latter; one of a set of such equations involving several unknowns.

Estimation Estimation is concerned with inference about the numerical value of unknown population values from incomplete data such as a sample. If a single figure is calculated for each unknown parameter the process is called point estimation. If an interval is calculated within which the parameter is likely, in some sense, to lie, the process is called interval estimation.

Estimator An estimator is a rule or method of estimating a constant of a parent population. It is usually expressed as a function of sample values and hence is a variate whose distribution is of great importance in assessing the reliability of the estimate to which it leads.

Even Summation In the smoothing of time series, a moving average taken over an even number of terms. The method produces a complication when applied to equally spaced data because it yields a term which is midway between the two central observations of the portion of the series being summed and averaged. A second summation of the first results, again based upon an even number of terms, will bring the results back into line with the original data.

Event Space See **Sample Space**.

*Evolution, Index of (Indice di Evoluzione)** In Italian usage, an index purporting to represent the tendency of a series to increase or decrease. If the first and last terms of the series of n terms are u_1 and u_n, the index is given by $(u_n - u_1)/(n-1)$.

Evolutionary Operation A technique devised by Box for the optimisation of established full-scale processes. It is used for experimenting on production processes. The basic elements are (i) provision to introduce a routine of systematic small changes in the levels at which process variables are held; (ii) provision to feed back the results derived from making these small changes to the operating supervisor; and (iii) an organisation which continually reviews the results and suggests new action to be introduced later.

Evolutionary Process Any non-stationary stochastic process. The probability distributions associated with the process are not independent of the time.

Evolutionary Spectrum In a non-stationary process or time series, a spectrum will apply strictly only to a limited period of time. Hence, for the whole realisation the spectral function will be time dependent or evolutionary.

Exact Chi-Squared Test See **Fisher-Yates' Test**.

'Exact' Statistical Method Used in interval estimation to imply that the probability distributions involved are completely known and that probability levels quoted are exactly achieved.

Exceedance Life Test This test is mathematically equivalent to the **Precedence Test**. [See also **Exceedances, Distribution of**.]

Exceedances, Distribution of In two random samples S_1 and S_2 of size n_1, and n_2, the number of observations in S_2 which exceed in magnitude at least $n_2 - m + 1$; $m = 1$, 2, ..., n_2 observations in S_1 has the distribution of exceedances (Gumbel & von Schelling, 1950).

Excess, Coefficient of A name given to a measure of **Kurtosis**. It is defined as

$$\gamma_2 = \frac{\mu_4}{\mu_2^2} - 3 = \frac{\kappa_4}{\kappa_2^2}$$

where μ's represent moments and κ's cumulants.

The coefficient is simply the excess of the value of κ_4/κ_2^2 over the value it takes in the case of a Normal distribution.

Exhaustive Sampling A term occasionally encountered which indicates that, in order to achieve the sample size necessary for the required precision, the sampling has included all of the population under study. That is to say that a complete enumeration rather than a sample has been taken.

Exogenous Variate The opposite to an **Endogenous Variate**.

Expectation The expected value of a function of variate values is its mean value in repeated sampling. Thus, if $t(x_1, x_2, ..., x_n)$ is some statistic dependent on variates $x_1, x_2, ..., x_n$ with a joint distribution $dF(x_1, x_2, ..., x_n)$ the expected value of t, if it exists, is

$$\int t \, dF(x_1, x_2, ..., x_n).$$

The 'expected' value is not necessarily the most frequently occurring value or even a possible value; e.g. if a variate can take each of the values 0 and 1 with probability $\frac{1}{2}$ and no other value is possible, the expected value is $\frac{1}{2}$.

Expected Probit The probit at some experimental dose calculated from the **Probit Regression Equation** fitted to the data. The expression is also used for the corresponding value obtained from the provisional line at any stage of iteration.

Experimental Error In general, any error in an experiment whether due to stochastic variation or bias. More specifically, the expression is used to denote the essential probabilistic variation to be expected under repetition of the experiment, not actual mistakes in design or avoidable imperfections in technique. It is the aim of good experimental design to provide valid measures of the experimental error in the more restricted sense.

Experimental Unit See **Plot**.

Explanatory Variable See **Cause Variable**.

Exploratory Survey See **Pilot Survey**.

Explosive Process A rather too vivid term to describe a stochastic process which has no bound to the expectation of the mean square. A process the values of which may increase without limit (in absolute magnitude, so that oscillations are possible) as time goes on. The term is not to be recommended.

Explosive Stochastic Difference Equation The difference equation $x_t + \alpha_1 x_{t-1} + \alpha_2 x_{t-2} + ... + \alpha_k x_{t-k} = u_t$ derived from an **Explosive Process**. The one or more roots of the individual equation $\sum\limits_0^k a_j z^{t-j}$ lie outside the unit circle and the values of x_t increase without limit as t increases. The same expression would be applied to more general difference equations with a variable possessing the same property of not remaining within finite bounds.

Exponential Curve A series of observations ordered in time which has a constant, or approximately constant, rate of increase can be represented over a long period by the curve:

$$y = ae^{bt}$$

where *a* and *b* are constants and *t* is time. This, or some simple transformation, is called the exponential curve. The fitting of an exponential trend of this form by the **Method of Least Squares** is facilitated by transforming into the logarithmic form:

$$\log_e y = \log_e a + bt.$$

Exponential Distribution A distribution of the form

$$dF = \frac{1}{\sigma} \exp\left\{-\frac{x-m}{\sigma}\right\} dx, \quad m \leqslant x \leqslant \infty.$$

The parameter σ is the standard deviation of the distribution and is also equal to the distance of the mean from the start.

Exponential Regression A term used to denote a relationship of the form

$$y_t = a + \beta e^{\lambda x_i} + \epsilon$$

where ϵ is a random residual. The term 'asymptotic regression' would be preferable.

Exponential Smoothing A method used in time series to smooth or to predict a series. There are various forms, but all are based on the supposition that more remote history has less importance than more recent history. For instance, in time series analysis a predictor y_t is often expressed as an average of some variable observed at previous points of time, with weights which diminish in importance as the period becomes more remote

$$y_t = \beta x_t + \beta^2 x_{t-1} + \beta^3 x_{t-2} + \dots.$$

Formulae of this kind lie at the basis of prediction methods developed by Holt (1957), Brown (1959), Harrison (1964) and later writers. Originally designed to deal with a stationary series, they have been extended to cover series containing linear trend and seasonal variation.

Extended Hypergeometric Distribution The distribution arises in testing the hypothesis that two proportions p_1 and p_2, derived from binomial random variables x_1 and x_2, are equal. The distribution is that of x_1 conditional on the fixed sum $x_1 + x_2$ and is of the form

$$f(x; t) = g(x)t^x / P(t)$$

where $t = p_1 q_2 / p_2 q_1$ and $P(t)$ is the factorial generating function of the **Hypergeometric Distribution**.

Extended Group Divisible Design The extension of the ordinary group divisible partially balanced incomplete design was for an *m*-associate class and proposed by Hinkelmann & Kempthorne (1963). The definition is lengthy; see Hinkelmann (1964).

*Extensive Magnitudes In Italian usage, the terms *grandezze estensive* and *grandezze intensive* (extensive and intensive magnitudes) are used to denote variables and attributes respectively, that is to say, quantities arising from measurement or from counting. Equivalent terms (which also occur in other European languages) are *heterograde* (*eterograde* = extensive) quantities and *homograde* (*omograde* = intensive) quantities; measurable (*misurabile* = extensive) quantities and enumerable (*enumerabili* = intensive) quantities. [See **Variable, Attribute**.]

Extensive Sampling A term used to denote sampling where the subject matter, or geographical coverage, of a sample is diffuse or widespread as opposed to intensive, where it is narrowed to a small field. Extensive sampling may refer either to a case where a wide variety of topics are covered superficially, rather than a few topics in detail or a large area is surveyed broadly, rather than a small area studied in detail. The term could also be used with reference to time, that is to say, of sampling covering a long period.

It would be convenient to distinguish the cases as space-extensive, item-extensive and time-extensive respectively.

External Variance An obsolete term introduced by Schultz (1930) to denote the variance of predictions of the future movements of a particular time series based upon a given form for its trend. The components contributing to the variance are errors in the estimation of the parameters, not those arising from an incorrect choice of trend line or from superposed random variation.

This term is also used by Deming to denote variance between primary units in a two stage sampling scheme, in contradistinction to internal variance between second stage units in a single primary unit. This appears much the same as variance between and within groups.

Extra Period Change Over Design The extension of a change over design for a further period not contemplated in the original design. A feature of these extra period designs is that, in the analysis, direct and residual effects are orthogonal and hence the new data can be incorporated without difficulty in the analysis.

Extremal Intensity The particular values of the **Intensity Function**, or **Hazard**, which appear as parameters in asymptotic distributions of extreme values.

Extremal Process A class of stochastic process proposed by Dwass (1964), derived from three extreme value distribution functions.

Extremal Quotient The ratio of the absolute value of the largest observation to the smallest observation in a sample. For continuous variates which are symmetrical and unlimited at both ends of the range the logarithm of the extremal quotient is symmetrically distributed.

Extremal Statistic Any function which depends upon observations at both extremes of the sample. For example, the **Range** and the **Extremal Quotient**.

Extreme Mean One of a set of mean values, e.g. in the analysis of variance which lies at the extreme of those values, i.e. the largest or the smallest.

Extreme Rank Sum Test A non-parametric test proposed by Youden (1963); if I objects are independently ranked by J judges, the individual ranks r_{ij} can be summed for each object. An extreme rank sum derived by this method can be tested using tables in Thompson & Willke (1963).

Extreme Studentised Deviate A statistic, given by
$$t_n = (x_{(n)} - \bar{x})/s \quad \text{or} \quad (\bar{x} - x_{(1)})/s$$
where s is the sample standard deviation based on $n-1$ degrees of freedom, n is the sample number, $x_{(1)}$ and $x_{(n)}$ are the smallest and largest members of the sample and $\bar{x}$ is the sample mean.

Extreme Value Distribution The distribution of the largest (smallest) observations in a sample. The distribution may be exact or asymptotic in form (Gumbel, 1958) and the form differs in accordance with the population from which the sample is drawn. [See also **Range, *m*th Values**.]

Extreme Values The largest or smallest variate values borne by the members of a set. Slightly more generally, the expression signifies values neighbouring the end-values.

F-Distribution See **Variance Ratio Distribution**.

F-test An alternative name for the **Variance Ratio Test**. [See also *z*-test.]

Factor This word occurs in statistical contexts in several senses:
 (*a*) in the ordinary mathematical sense, e.g. a factor of an algebraic expression;
 (*b*) to denote a quantity under examination in an experiment as a possible cause of variation, e.g. in a 'factorial' experiment;
 (*c*) (adapted from psychology) in multivariate analysis, to denote a function of the observed variates, usually linear, which may be regarded as part of those variates; and hence as a 'factor' of the variation;
 (*d*) to denote a constituent item in an average or index number.

Factor Analysis A branch of multivariate analysis in which the observed variates x_i; $i = 1, 2, ..., p$ are supposed to be expressible in terms of a number $m < p$ factors f_i together with residual elements. One such model is expressed by
$$x_i = \sum_{j=1}^{m} a_{ij}f_j + b_i s_i + c_i \epsilon_i$$
where s_i is a factor specific to the ith variate, ϵ_i is an error variate and the a's, b's and c's are structural constants of the model which it is the object of the analysis to estimate. The coefficients a_{ij} are known as factor loadings. That part of the variance of x_i which is attributable to the f's is

called the communality; that attributable to s_i is called the specificity; and that attributable to the ϵ_i is called the unreliability. The complement of the last is called the **Reliability**.

The expressions 'component analysis' and, to a much smaller extent, 'factor analysis' occur in variance analysis with a different meaning, namely in relation to the allocation of variance to different causal factors or components of variation.

Factor Antithesis A term of doubtful utility sometimes employed in index number theory. If, for example, in the **Laspeyres' Index**, namely
$$I_{on} = \frac{\Sigma(p_n q_o)}{\Sigma(p_o q_o)}$$
the roles of p and q are interchanged so as to give
$$I'_{on} = \frac{\Sigma(q_n p_o)}{\Sigma(q_o p_o)}$$
and the result divided into the true value ratio $\Sigma(p_n q_n)/\Sigma(p_o q_o)$ we obtain
$$I''_{on} = \frac{\Sigma(p_n q_n)}{\Sigma(p_o q_n)}$$
which is the factor antithesis of the Laspeyres' index and is, in fact, the **Paasche Index**.

Factor Loading See **Factor Analysis**. The use of the word 'loading' rather than 'weighting' in this context is due to the fact that the terminology of factor analysis developed from psychology. The word 'saturation' is used in a similar sense to denote the extent to which a variate is 'saturated' with a common factor.

Factor Matrix The matrix of coefficients (a_{ij}) appearing in the relations between variates and factors in **Factor Analysis**.

Factor Pattern In the **Factor Matrix** certain items may be known or assumed on prior grounds to be zero; for example, if the jth factor f_j does not appear in the ith variate, $a_{ij} = 0$. The pattern of non-zero coefficients as distinct from their actual values is called the factor pattern. It may be regarded as defining the model of structure, in terms of factors, which is under investigation.

In oblique factor analysis, i.e. factor analysis where the factors are correlated, it may be necessary to distinguish between the factor pattern, as so defined, and the factor structure which expresses the way in which the factors are dependent among themselves.

Factor Reversal Test A test proposed for index numbers by Irving Fisher (1927). The idea was that, in an index number of price, if the symbols for price and quantity are interchanged, there should result an index of quantity; and that this, multiplied by the original price-index, should give an index of changes in total value. The test is obeyed by Fisher's **'Ideal' Index Number** but not by most of those in current use, e.g. those of Laspeyres and Paasche. [See also **Time Reversal Test**.]

Factor Rotation The final stage of **Factor Analysis** is, in geometric terms, a rotation of the factor axes in order to achieve some correspondence between the numerically derived factors and explanatory entities; that is to say to seek a **Structure**. Analytic methods may be divided broadly into those dealing with orthogonal factors, where the loadings are uncorrelated, and oblique factors involving correlated loadings. One group of methods pursues the maximisation of the scatter among factor loadings; for example, Quartimax and Varimax for orthogonal rotation and Oblimax for oblique rotation. An alternative approach, using the minimisation of cross-products, is termed Quartimin and Covarimin, which latter method is the inverse of Varimax.

Factorial Cumulant The factorial cumulant of order r, $\kappa_{[r]}$, is defined as the coefficient of $t^r/r!$ in the expansion of the **Factorial Cumulant Generating Function** as a power series in t.

Factorial Cumulant Generating Function The logarithm of the **Factorial Moment Generating Function**, formed by analogy with the cumulant generating function, which is the logarithm of a moment generating function.

Factorial Distribution A discrete distribution for which the successive frequencies are factorial qualities. Irwin (1963) defined a whole class of such distributions of form

$$f(r) = \frac{ka_r}{\theta(\theta+1)...(\theta+r)}, \quad \theta > 0, r = 0, 1, 2, ...$$

The family is sometimes known as the inverse factorial series distribution. A particular case is the Waring distribution, with $a_0 = 1$, $a_r = a(a+1)...(a+r-1)$ so that it becomes

$$f(r) = (\theta-a) \frac{a^{(r)}}{(\theta+r)^{(r+1)}}.$$

The name was given in recognition of Waring's discovery in the 18th century of the expansion of $(\theta-a)^{-1}$ in reciprocals of factorials. In particular, if $a = 1$, $\theta-a = p$ the distribution becomes $f(r) = \dfrac{pp!}{r(r+1)...(r+p)}$ and it is known as the Yule distribution.

Factorial Experiment An experiment designed to examine the effect of one or more factors, each factor being applied at two levels at least so that differential effects can be observed. The term is frequently used in a slightly narrower sense, as describing an experiment investigating all possible treatment combinations which may be formed from the factors under investigation. The 'level' of a factor denotes the intensity with which it is brought to bear. It may be measured quantitatively, as when fertiliser is applied to plots in a given weight per unit area, or qualitatively, as when patients are considered at two levels 'inoculated' and 'not inoculated'.

Factorial Moment A type of moment used for discontinuous distributions defined at equal variate intervals. If f_r is the frequency at x_r the jth factorial moment about arbitrary origin a is

$$\mu'_{[j]} = \sum_{r=-\infty}^{\infty} (x_r-a)^{[j]}f_r,$$

where

$$(x_r-a)^{[j]} = (x_r-a)(x_r-a-1) ... (x_r-a-j+1).$$

In most cases the variate intervals are taken as units, the variate values as 0, 1, 2, ... and a as zero, in which case

$$\mu'_{[j]} = \sum_{r=0}^{\infty} r(r-1) ... (r-j+1)f_r.$$

Factorial Moment Generating Function A function of a variable t which, when expanded formally as a power series in t, yields the factorial moments as coefficients of the respective powers. It is used almost entirely for discontinuous distributions defined at equal distances of the variate, say at $x = 0, 1, ...$ If f_r is the frequency at x_r a factorial moment generating function is given by $\omega(t)$, say, where

$$\omega(t) = \sum_{r=0}^{\infty} f_r(1+t)^r = \sum_{j=0}^{\infty} \mu'_{[j]} \frac{t^j}{j!}$$

where $\mu'_{[j]}$ is the jth factorial moment about zero.

Factorial Multinomial Distribution See **Multivariate Hypergeometric Distribution**.

Factorial Sum The sums entering into the calculation of **Factorial Moments**. If the frequency of a variate value r $(r = 0, 1, ..., k)$ is f_r the factorial sum of order j is

$$\sum_{r=0}^{k} \{f_r r(r-1) ... (r-j+1)\}.$$

Fair Game In the theory of games, a game consisting of a sequence of trials is deemed to be a 'fair' game if the cost of each trial is equal to the expected value of the gain from each trial. A 'fair' game in this sense may not be fair as between a pair of adversaries with unequal resources: it is well known that at a 'fair' game the player with the larger sum to stake has the better chance of ruining his opponent.

Fellegi's Method Fellegi (1963) proposed a method of sampling without replacement with probability proportional to size, which allows for rotation of the sample simultaneously with the exact calculation of joint probabilities of selection of sets of units.

Fermi-Dirac Statistics See **Bose-Einstein Statistics**.

Fertility Gradient If a field, or other stretch of land which is to be used for agricultural experimentation, is divided into plots ready for the experimental treatments, some part of the differences in yield between contiguous plots may be due to inherent variation in the fertility of the soil. If this inherent variation is slowly decreasing, or increasing, from one side to the other there is said to be a fertility gradient. It is one aim of randomisation and other devices of experimental design to eliminate bias due to the existence of such gradients.

Fertility Rate The number of live births in a unit period expressed as a proportion (usually per thousand) of potentially fertile women in the population concerned. 'Potentially fertile' is usually defined by reference to age, e.g. by taking women from 15-50 years old. This crude cr general fertility rate can be refined or standardised in respect of age, social class, etc.

Fiducial Distribution A distribution of a parameter required for **Fiducial Inference** about that parameter. It is not a probability distribution in the customary sense, but is derived from the distribution of estimators containing all the relevant information in the sample. In earlier literature it is sometimes referred to as a 'fiducial probability distribution'.

Fiducial Inference A type of statistical inference based on **Fiducial Distribution**, introduced by R. A. Fisher (1930). The object of the inference is to make probabilistic statements about the values of unknown parameters and to that extent it resembles the theory of **Confidence Intervals**.

In simple cases, results from fiducial theory agree with those from the theory of confidence intervals, but this is not so in general. When this type of inference and the related confidence interval type were first propounded they were often confused. The confusion survives to the extent that 'fiducial' is sometimes applied to inference of the confidence interval type.

Fiducial Limits In **Fiducial Inference**, limits between which a parameter is considered to lie. The term also occurs as a synonym of **Confidence Limits**. The two are often the same in commonly occurring cases, but their conceptual genesis is different.

Fiducial Probability See **Fiducial Distribution**.

Fieller's Theorem A theorem giving limits of the confidence interval type for a ratio, stated in its general form by Fieller (1940).

Filter Any method of isolating harmonic constituents in a time series; a mathematical analogy of the 'filtering' of a ray of light or sound by removing unsystematic effects and bringing out the constituent harmonics.

Finite Arc Sine Distribution A combinatorial distribution occurring in stochastic process analysis; it expresses the probability of return to zero of the winnings in a series of trials, with probability one-half of success at each trial.

For $n \to \infty$ this tends to the ordinary **Arc Sine Distribution**.

Finite Markov Chain A Markov chain $\{x_n\}$ is said to be a finite chain with k states if the number of possible values of the random variables $\{x_n\}$ is finite and equal to k.

Finite Multiplier If a sample of n values is drawn without replacement from a population of limited size N, the sampling variances of the statistics derivable from it depend, in general, on N as well as n. For example the variance of the sample mean $\bar{x}$ may be written

$$\text{var } \bar{x} = \frac{\sigma^2}{n}\left(1 - \frac{n}{N}\right)$$

where $\sigma^2 = \frac{1}{N-1} \sum_{i=1}^{N} (x_i - \mu)^2$ and $\mu = \frac{1}{N} \sum_{i=1}^{N} x_i$.

The factor $1 - n/N$ is sometimes called the finite multiplier or the finite sampling correction.

The latter usage is bad, since the formula for var $\bar{x}$ is exact and needs no correction. Even the former is not free from objection. Formulae for sampling from a finite population are not, in general, expressible as the product of formulae for the infinite case and terms dependent solely on n and N.

Finite Population A population of individuals which are finite in number.

Finite Sampling Correction An alternative term for the **Finite Multiplier**. It is objectionable on the grounds that it is not a 'correction' to an inexact result.

First Limit Theorem If a sequence of distribution functions tends to a single distribution function F then the corresponding characteristic functions tend uniformly in any finite interval to the characteristic function of F. This is generally known as the First Limit Theorem. Although known to earlier writers, it seems to have been proved rigorously for the first time independently by P. Lévy and Cramér about 1925.

First Passage Time An important concept in the theory of stochastic processes where, if $x_N = a$ and $a \neq 0$, the achievement of $x_N = a$ gives N as the first passage time from the origin to state a. For the various realisations of a process, N is a random variable.

First Stage Unit See **Multi Stage Sampling**.

Fisher's 'B' Distribution The distribution of the square root of a **Non-central Chi-squared Variate**. It is a hypergeometric function and was derived by Fisher in a study of the multiple correlation coefficient.

Fisher Model A term proposed by Ogawa (1963) to denote the general class of experiment designs without technical errors but based upon Fisher's randomisation procedure. [See also **Neyman Model**.]

Fisher's Distribution The distribution of the ratio of the variances of two independent samples from a Normal population or, a little more generally, of the ratio of two independent quantities, each of which is distributed as χ^2. [See also **Variance Ratio Distribution**.]

Fisher's Transformation (of the Correlation Coefficient)
A transformation of the correlation coefficient r according to the formula

$$z = \tanh^{-1} r.$$

The distribution of z for samples from a bivariate Normal population approaches normality of form much more rapidly than does that of r; and its variance, unlike that of r, is little affected by the population correlation coefficient even for samples of moderate size. It is therefore useful in many problems involving estimation and tests of significance.

Fisher-Behrens Test   See **Behrens-Fisher Test**.

Fisher-Yates Test   The use of chi-squared as a test of independence in a double dichotomy has limitations if the cell frequencies are small. Yates (1934) proposed a correction for **Continuity** in these circumstances and, following a suggestion by R. A. Fisher, also gave an 'exact' test in the form of a computation for the probability of any observed set of cell frequencies. If the four cell frequencies in a two-by-two table are denoted by a, b, c and d with a and d diagonal, then the probability of this set of frequencies on the hypothesis of independence is:

$$\frac{(a+b)!\,(c+d)!\,(a+c)!\,(b+d)!}{(a+b+c+d)!\,a!\,b!\,c!\,d!}$$

The test proceeds by calculating the exact probabilities of the frequencies observed and of those deviating more than the observed from the situations of independence, and cumulating the results. It is also known as the Exact Chi-squared Test.

Five Point Assay   The five-point design for a biological assay is one of the general class of symmetrical $(2k+1)$ point designs for **Slope Ratio Assays**. One fifth of the test subjects are allocated to each of two doses of both a standard and a test preparation, the remaining fifth receiving no treatment.

Fixed Base Index   An index number for which the **Base Period** for the calculations is selected and remains unchanged during the lifetime of the index. This is in contradistinction to a **Chain Base Index**.

Fixed Effects (Constants) Model   An alternative designation for **Model I** (or **First Kind**) in analysis of variance.

Fixed Sample   When a survey is repeated on several occasions, but observations are taken on the same sample instead of a new sample for each occasion, the sample is said to be fixed.

Fixed Variate   In regression theory the model under investigation is of the type

$$y = \beta_0 + \beta_1 X_1 + \ldots + \beta_p X_p + \epsilon$$

where the β's are constants, the X's are variables in the mathematical sense and ϵ (and hence y) are random variables or variates. Methods of estimating the β's are the same whether the X's are selected arbitrarily or are themselves the values of variates, provided that in the latter case the conditional distribution of ϵ is the same for all values of the X's. The X's are, in such an event, sometimes known as 'fixed variates'. This is a contradictory and therefore regrettable phrase. The more usual 'independent variable' is better, but still far short of perfection. Other equivalents are 'predictor', 'predicated variable', 'explanatory variable' and 'regressor'. The dependent variate is also known as the 'predictand' or 'regressand'.

*Flexibility, Curve of (Curva di Flessibilità)   In Italian usage, a type of curve, deduced from the curve of **Concentration** showing the variation of $\Phi dF/Fd\Phi$ as ordinate against F as abscissa, F and Φ being respectively the abscissa and ordinate of the concentration curve.

Fluctuation   A movement up or down between consecutive items of a series of numbers or numerical observations.

In a different sense the variation of a statistic from sample to sample is also referred to as a sampling fluctuation.

Fokker-Planck Equation   An equation originally occurring in the theory of diffusion when drift is taken into account. It may be written in the form

$$\frac{\partial v(x, t)}{\partial t} = -2c\,\frac{\partial v(x, t)}{\partial x} + D\,\frac{\partial_2 v(x, t)}{\partial x_2},$$

where $v(x, t)$ is the probability density for displacement x at time t, D is the diffusion coefficient and c represents drift. The equation occurs in the theory of stochastic processes as a limiting case of random walk or additive processes.

Folded Contingency Table   An unnecessary expression to denote a symmetric (square) contingency table, i.e. one in which the frequency with ith row and jth column is equal to that in the jth row and ith column.

Folded Normal Distribution   A measurement may be recorded without its algebraic sign: as a result the underlying distribution of a measurement is replaced by a distribution of absolute measurements. When the underlying distribution is Normal the resulting distribution is termed 'the folded Normal distribution'.

Folding   In empirical data, observations are sometimes recorded without being given a sign, and therefore have to be regarded as all having the same sign (conventionally 'plus'). This affects their distribution since the negative point is 'folded' over and added to the positive part. This situation should be distinguished from that in which a variable is essentially positive.

Follow-up A further attempt to obtain information from an individual in a survey or field experiment because the initial attempt has failed or later information is available.

Force of Mortality A term used in actuarial or other analysis of human life to indicate the **Age Specific Death Rate**.

Forecasting 'Forecasting' and 'prediction' are often used synonymously in the customary sense of assessing the magnitude which a quantity will assume at some future point of time: as distinct from 'estimation' which attempts to assess the magnitude of an already existent quantity. For example, the final yield of a crop is 'forecast' during the growing period but 'estimated' at harvest.

The errors of estimation involved in prediction from a regression equation are sometimes referred to as 'forecasting errors' but this expression is better avoided in such a restricted sense. Likewise terms such as 'index numbers of forecasting efficiency', in the sense of residual error variances in regression analysis, are to be avoided.

Forward Equations See **Kolmogorov Equations**.

Fourfold Table An alternative name for the **Two-by-two Frequency Table**, the name being derived from the four cells into which the frequency is divided by a double dichotomy. The expressions '2×2' or 'two-by-two' seem better.

Fourier Analysis The theory of representing functions of a variable t as the sum of a series of sine and cosine terms of type $a_j \cos(2\pi j/\lambda_j)$, $j = 0, 1, \ldots$. The λ's are not necessarily commensurable and hence the analysis is more general than harmonic analysis which considers series of terms such as $\cos(2\pi j/\lambda)$ where λ is some constant.

Fractile A term introduced by Hald (1948) to denote the variate value below which lies a given fraction of the cumulative frequency. This term is synonymous with the more generally used term **Quantile** and the necessity for its coining is not clear.

Fractile Graphical Analysis The name suggested by Mahalanobis (1961) for a distribution free regression analysis based upon order statistics.

Fraction Defective In quality control, that proportion of a number of units which are defective.

Fractional Replication Where there are a large number of treatment combinations resulting from a large number of factors to be tested, it is sometimes impracticable to test all the combinations with one experimental layout. In such cases resort may be made to a fractional, i.e. a partial replication. This device is likely to be useful only where certain high order interactions can be regarded as negligible.

Frame A list, map or other specification of the units which constitute the available information relating to the population designated for a particular sampling scheme. There is a frame corresponding to each stage of sampling in a multi-stage sampling scheme. The frame may or may not contain information about the **Size** or other supplementary information of the units, but should have enough details so that a unit, if included in the sample, may be located and taken up for inquiry. The nature of the frame exerts a considerable influence over the structure of a sample survey. It is rarely perfect, and may be inaccurate, incomplete, inadequately described, out of date or subject to some degree of duplication. Reasonable reliability in the frame is a desirable condition for the reliability of a sample survey based on it.

In multi-stage sampling it is sometimes possible to construct the frame at higher stages during the progress of the sample survey itself. For example, certain first stage units may be selected in the first instance; and then more detailed lists or maps may be constructed by compilation of available information or by direct observation only of the first-stage units actually selected.

Freehand Method A method of describing the relationship in a series of data, ordered in time or space, whereby the general trend is estimated by drawing a line freehand through or near the series of plotted observations.

Frequency The number of occurrences of a given type of event, or the number of members of a population falling into a specified class.

Where the frequency is expressed as a proportion of the total number of occurrences or the total number of members, it is called the relative or proportional frequency; but where no ambiguity can arise these ratios may simply be called frequencies.

Frequency Curve The graphical representation of a continuous frequency distribution, the variate being the abscissa and the frequency the ordinate. The frequency curve may be viewed as the limiting form of the frequency polygon as the number of observations tends to become infinitely large, and the class intervals indefinitely small.

Frequency Distribution A specification of the way in which the frequencies of members of a population are distributed according to the values of the variates which they exhibit. For observed data the distribution is usually specified in tabular form, with some grouping for continuous variates. A conceptual distribution is usually specified by a **Frequency Function** or a **Distribution Function**.

Frequency Function An expression giving the frequency of a variate value x as a function of x; or, for continuous variates, the frequency in an elemental range dx. Unless the contrary is specified the total frequency is taken to be unity, so that the frequency function represents the proportion of variate values x. From a more sophisticated standpoint the frequency function is most conveniently regarded as the derivative of the **Distribution Function**. The generalisation to more than one variate is immediate.

Frequency Moment If a frequency or probability distribution is given by $dF = y \, dx$ the rth probability-moment is defined by

$$\Omega_r = \int_{-\infty}^{\infty} y^r \, dx.$$

It may be regarded as the moment in the sense of rigid dynamics of the frequency curve about the variate axis, whereas ordinary moments relate to the frequency or y-axis.

In the above definition the total frequency is, as usual by convention, taken as unity. For a population of total frequency N the rth frequency-moment is defined by

$$J_r = \int_{-\infty}^{\infty} y^r \, dx$$

where $N = J_1 = \int_{-\infty}^{\infty} y \, dx$, y now relating to actual and not to relative frequency. Analogous definitions involving sums apply to discontinuous distributions.

Frequency Polygon A diagram showing the form of a frequency distribution; the frequencies are graphed as ordinates against the variate values as abscissae and the tops of the ordinates joined one to the next. The diagram may be used to exhibit the frequencies of a continuous distribution if the frequencies are grouped in variate intervals; it is then customary to erect ordinates at the middle of the intervals.

Frequency Response Function In the analysis of time series a linear weighted average of $u(t)$ may be expressed as:

$$v(t) = \int_{0}^{\infty} a(\tau) \, u(t - \tau) \, d\tau$$

where $a(\tau)$ is a system of weights. The frequency response function, or transfer function, is the **Characteristic Function** or Fourier transform of $a(\tau)$ namely

$$\int_{0}^{\infty} e^{i\tau\alpha} \, a(\tau) \, d\tau.$$

Frequency Surface The bivariate analogue of the **Frequency Curve**.

Frequency Table A table drawn up to show the distribution of the frequency of occurrence of a given characteristic according to some specified set of class intervals. It may be univariate or multivariate but there are difficulties in presenting data tabulated according to more than two variables.

Frequency Theory of Probability The frequency theory of probability regards the probability of an event as the limit of the frequency of occurrence of that event in a series of n trials as n tends to infinity. The existence of this limit is an axiom of the theory as proposed by von Mises (1919), but later axiomatisations (e.g. by Kolmogorov, 1933) avoid the difficulties associated with it by taking the probability as a measure associated with a set of points (events) and proceeding on the basis of measure theory. This avoids the difficulty only for a mathematician. For the statistician the problem of relating probability to frequency of occurrence remains.

Friedman's Test A test based on chi-squared introduced by M. Friedman (1937) to test the dependence of a set of rankings.

Full Information Method In econometrics, a method of deriving estimates of parameters in a stochastic model which are subject to all the *a priori* restrictions of that model. [See also **Limited Information Method**.]

Fundamental Probability Set A set of objects or events which are basic to a probabilistic situation, in the sense that all other objects or events under consideration are derived from them by compounding. It follows that all probabilities are expressible by the rules of addition, multiplication, etc. of probabilities in terms of the probabilities of the fundamental set. Failure to specify the fundamental set explicitly leads to confusion, occasionally to error, and in particular to a number of paradoxes. The fundamental probability set is sometimes called the Reference Set.

By a slight extension the actual probabilities of the fundamental set are also referred to as the fundamental probability set.

Fundamental Random Process See **Brownian Motion Process**.

Furry Process An early variety of a **Birth and Death Process** studied by Furry in 1937.

g-statistics The sample values of **Gamma Coefficients**.

g-Test An application of the log likelihood ratio test for the hypothesis of independence in an $r \times c$ contingency table. If

$$g = \sum_{i=1}^{r} \sum_{j=1}^{s} \left(n_{ij} - \frac{n_i. \, n._j}{n} \right)^2 \Big/ \left(\frac{n_i. \, n._j}{n} \right)$$

where $n_i. = \sum_j n_{ij}$ and $n._j = \sum_i n_{ij}$ then, when the hypothesis is true, the statistic g and $-2 \log \lambda$, where λ is the likelihood ratio criterion converge in distribution to a χ^2 with $(r-1)(s-1)$ degrees of freedom.

Gabriel's Test A simultaneous comparison procedure for mean values (1964) in analysis of variance. It is based upon the 'between-groups' sum of squares for the $(2^k - k - 1)$ subsets of two or more mean values of the k

groups and is tested against the fixed critical value of the F-distribution with $(k-1,\ n-k)$ degrees of freedom. The sample sizes (n_i) need not be equal.

Galton's Individual Difference Problem A problem discussed by Galton in the latter half of the 19th century. In modern terms it is equivalent to determining the differences of variate values or their expectations of certain individuals, based on their ranks. For example, Galton considered the problem: how should a prize be divided between the winner and the second and third members in a contest, assuming that the underlying distribution of abilities is Normal?

Galton Ogive Under certain conditions, e.g. when a frequency function is unimodal, the **Distribution Curve** resembles a letter 'S'—the ogive form, a term borrowed by Galton (1875) from architecture and used particularly for the distribution curve of the Normal distribution.

Galton-McAllister Distribution A frequency distribution in which the logarithm of the variate, or some simple linear function of it is Normally distributed. The distribution was suggested by McAllister (1879). It is now more generally known as the lognormal distribution. [See **Logarithmic-Normal Distribution**.]

Galton's Rank Order Test A simple application of rank order statistics for testing the difference between two treatments which occurred in some work for Darwin (1876); summarised and developed by Hodges (1955) using results of Chung & Feller (1949).

Galton-Watson Process A Markov type **Branching Process** where the application is the survival of family names. It was first stated by Galton (1873) and Watson (1874) gave a partial solution: the first complete solution was by Steffensen (1930).

Gambler's Ruin The name given to one of the classical topics in probability theory. A game of chance can be related to a series of Bernoulli trials at which a gambler wins a certain predetermined sum of money for every success and loses a second sum of money for every failure. The play may proceed until his initial capital is exhausted and he is ruined. The statistical problems involved are concerned with the probability of the ruin of a player, given the stakes, initial capital and chances of success, and with such matters as the distribution of the length of play.

There are many variations to this classical problem, which is closely associated with problems of the **Random Walk**; in particular, of **Sequential Sampling**.

Game Theory Generally, that branch of mathematics which deals with the theory of contests between two or more players under specified sets of rules. The subject assumes a statistical aspect when part of the game proceeds under a chance scheme, e.g. by the throw of a die or when strategies are selected at random. [See also **Strategy, Zero-sum Game, Minimax Principle, Fair Game**.]

Gamma Coefficients These are ratios, analogous to **Moment-ratios** or the **Beta Coefficients**, which are based upon the cumulants:
$$\gamma_1 = \kappa_3/\kappa_2^{3/2}, \ \gamma_2 = \kappa_4/\kappa_2^2$$
and, in general
$$\gamma_r = \kappa_r/\kappa_2^{\frac{1}{2}r}.$$

Gamma Distribution A frequency distribution of the form
$$dF(x) = \frac{e^{-x} x^{\lambda-1}}{\Gamma(\lambda)}\,dx,\ 0 \leqslant x \leqslant \infty.$$
It is also known as Pearson's Type III or simply as the **Type III Distribution**. The distribution function $F(x)$ is an incomplete gamma function: hence the name. Its importance in statistics derives largely from the fact that $c\chi^2$, where c is a numerical constant, is actually or approximately distributed in the gamma form under certain conditions.

Gantt Progress Chart An application of the **Bar Chart** due to Gantt, of use in industrial statistics. An actual performance or output is expressed as a percentage of a quota or planned performance per unit of time. Account may also be taken of the cumulative performance by plotting it, with the planned cumulative performance, as ordinate against time as abscissa.

Garwood Distribution A distribution proposed by Garwood (1940-41) for the waiting time of vehicles arriving randomly at vehicle controlled traffic lights.

Gauss Distribution An alternative name for the **Normal Distribution**.

Gauss-Markov Theorem A fundamental theorem dealing with an unbiassed estimator of a population characteristic, based upon a linear combination of sample observations drawn from that population. The theorem is to the general effect that an unbiassed linear estimator of a parameter is 'best', i.e. has minimum variance, when the estimator is obtained by least squares. The theorem can be extended in many directions, e.g. by considering the simultaneous estimation of several parameters or linear functions of them.

Gauss-Seidel Method A classical method for the iterative solution of a set of linear equations, particularly those arising from least squares solutions: an extension by Seidel (1874) of the fundamental method due to Gauss (1823).

Gauss-Winckler Inequality An inequality concerning the moments of a continuous distribution about the mode.

If ν_r is the absolute moment of order r about the mode, assumed unique,

$$\{(r+1)\nu_r\}^{1/r} \leqslant \{(n+1)\nu_n\}^{1/n} \quad r < n.$$

This was given by Gauss for $r = 2$, $n = 4$ in the form $\mu_4/\mu_2^2 \geqslant 1.8$ and generalised by Winckler in 1866.

The expression is also used to denote an inequality of the **Bienaymé-Tchebychev** type covering limits to the probability of deviations from the mode. There are various forms of the inequality, which is also associated with the names of Camp, Meidell and Narumi. [See **Camp-Meidell Inequality**.]

Geary's Ratio As a test of normality the moment ratio for **Kurtosis** has the drawback that its sampling distribution is skew even for quite high values of n, the sample size. In order to overcome this, Geary (1935) proposed a test in the form of a ratio

$$\frac{\text{mean deviation}}{\text{standard deviation}},$$

which, in samples from a Normal distribution, tends to $\sqrt{(2/\pi)}$ as n tends to infinity. The distribution of Geary's Ratio tends to the Normal form fairly rapidly and, as a test, the ratio aims at detecting departures from meso-kurtosis in the parent population.

General Factor In component analysis, a component which is common to all the observed variates; a factor which is involved in the variances of all the tests in a **Battery of Tests** subjected to a **Factor Analysis**. In the theory due to Spearman the common factor variance reduces to the variance of a single general factor and a psychological significance is attributed to this General Factor or g. [See **Common Factor**.]

General Interdependent System An interdependent model where each residual is assumed to be uncorrelated only with explanatory variables of the same equation. The classic **Simultaneous Equations Model** assumes that each residual is uncorrelated with all explanatory variables irrespective of the equation in which they occur *vis-à-vis* the residuals.

Generalised Binomial Distribution An alternative name for the **Poisson Binomial Distribution**.

Generalised Bivariate Exponential Distribution A generalisation of the bivariate form obtained directly from the **Multivariate Exponential Distribution**. This generalisation (Marshall & Olkin, 1967), is important in connection with the form of the underlying model used in life assessment problems.

Generalised Classical Linear Estimators An alternative name given by Basmann (1957) for the **Two Stage Least Squares Method**.

Generalised Contagious Distribution A wider generalisation of the **Neyman Type A, B or C Distribution** than the **Beall-Rescia Generalisation** proposed by Gurland (1958).

Its probability generating function is

$$H(z) = \exp(-m_1) \exp\{m_1, {}_1F_1[\alpha, \alpha+\beta, m_2(z-1)]\}$$

where ${}_1F_1$ is the confluent hypergeometric function. When $\alpha = 1$ this distribution is the Beall-Rescia distribution.

Generalised Distribution This expression occurs in two senses, (*i*) as a more complicated form similar to some known distribution; (*ii*) as obtained from a known distribution when its parameters are themselves random variables. The latter usage should be discarded in favour of **Compound Distribution**.

Generalised Gamma Distributions While there can be many forms of generalisation (Stacy & Mihram, 1965), the most common uses a positive parameter in the exponential factor:

$$f(x; \alpha, \beta, \gamma) = (\gamma/\alpha^\beta)x^{\beta-1} e^{-(x/\alpha)^\gamma}/\Gamma(\beta/\gamma).$$

The standard gamma, chi, chi-squared, exponential and Weibull distributions all appear as special cases (Stacy, 1962).

Generalised Inverse The generalised inverse of an $m \times n$ matrix A, is an $n \times m$ matrix A^- such that for any y where $Ax = y$ is consistent, $x = A^-y$ is a solution. This concept occurs in least squares when the matrix of normal equations is singular.

Generalised Least Squares Estimator A method proposed by Aitken (1934) for estimating the k parameters of the vector π in a linear equation when the disturbances v are not independent but their variance-covariance matrix is known.

Generalised Maximum Likelihood Estimator An approach to parameter estimation involving the concept of asymptotic efficiency proposed by Weiss & Wolfowitz (1966) to remove some of the problems associated with classical maximum likelihood estimators.

Generalised Multinomial Distribution An n-dimensional distribution of discrete variates proposed by Tullis (1962) which includes an additional parameter ρ as the correlation coefficient between pairs of variates with common mean and marginal distributions which are multinomial.

Generalised Normal Distribution A little used alternative name for **Kapteyn's Univariate Distribution** not to be confused with the **Multivariate Normal Distribution**.

Generalised Polykays An extension of the concept of **Polykays** and **Bipolykays** by Dayhoff (1964) to 'n-way' polykays: i.e. statistics of an n-variate situation.

Generalised Power Series Distribution A random variable X taking non-negative integral values with probability

$$P_x = \Pr\{X = x\} = \frac{a_x \theta^x}{f(\theta)}, \quad x = k, k+1, \ldots, \infty$$

61

is a generalised power series distribution. When x ranges from 0 to ∞ the generalised power series distribution reduces to a power series distribution.

Generalised Right Angular Designs A generalisation proposed by Tharthare (1965) to produce a further class of four associate **Right Angular Designs**.

Generalised Sequential Probability Ratio Test An extension by Weiss (1953) of the basic Wald **Sequential Probability Ratio Test** where constant acceptance and rejection limits (A, B) are not necessarily used at each stage of sampling: at the ith stage pre-determined numbers $A_i B_i$ may be used.

Generalised STER Distribution A generalisation of the **STER Distribution** corresponding to a non-negative integer random variable Y has a frequency distribution of the form

$$\Pr(x) = \sum_{\gamma=x+k}^{\infty} \{1/(\gamma-k+1)\}\Pr_Y(\gamma)/\sum_{\gamma=k}^{\infty} \Pr_Y(\gamma)$$

where $x = 0, 1, 2, ...,$ and $k = 1, 2, 3, ...$ The STER distribution is a special case of this where $k = 1$.

Generalised T^2 Distribution A generalisation to several samples by Lawley (1939) and Hotelling (1951) of the two-sample T^2 test for the equality of mean vectors. The exact form of this distribution is not available but an asymptotic formula was given by Ito (1956, 1960).

Generating Function A function of a parameter t which, when expanded as a power series in t, yields as the coefficients the values of some quantity of statistical interest such as the probability of events or the moments of a frequency distribution. The **Characteristic Function** is an important case of a moment generating function, a phrase which is often abbreviated to m.g.f. The theory of probability has made use of the generating function approach since the time of De Moivre in the first half of the 18th century.

Geometric Distribution A distribution in which the frequencies fall off in geometric progression as the variate values increase. The expression is usually confined to discontinuous distributions and does not, for example, include the **Exponential Distribution**.

Geometric Mean A measure of location which is one of the general class of **Combinatorial Power Mean**. The geometric mean (G) of n positive quantities is the positive nth root of the product of these quantities:

$$G = \left(\prod_{j=1}^{n} x_i \right)^{\frac{1}{n}}.$$

Where it exists the geometric mean lies between the harmonic mean and the arithmetic mean. The geometric mean of a frequency distribution may be written in terms of relative frequencies in the groups:

$$G = \prod_{j=1}^{k} (x_j{}^{f_j})$$

where f_i is the frequency at x_j. For a continuous distribution with frequency function $f(x)$ it may be defined by the equation

$$\log G = \int_{-\infty}^{\infty} \log x f(x) dx.$$

Geometric Moving Average An expression to be avoided, as it is not based on geometric means. [See **Exponential Weights**.]

Geometric Range This expression occurs very infrequently in two senses: (*i*) as the ratio of the extremes of a sample, (*ii*) as the geometric mean of the extremes of a sample. Neither usage has much to recommend it.

Gibrat Distribution A form of the logarithmic Normal or **Lognormal Distribution**. The normal distribution $\frac{1}{\sqrt{(2\pi)}} e^{-\frac{1}{2}z^2} dx$ is transformed by $z = a \log(x-x_0)+b$ to the Gibrat distribution, a, b and x_0 being constants at choice.

***Gini's Hypothesis (Ipotesi di Gini)** In Italian usage, a distribution of prior probability in the form of a **Beta-distribution**.

Given Period In the construction of index numbers it is necessary to relate data at two points of time, a **Base Period** and some other period which may be earlier or later in time, the 'given' period.

Glivenko-Cantelli Lemma This lemma states that for random samples of n from populations with any distribution function $F(x)$, $\sup_{-\infty < x < +\infty} | F_n(x)-F(x)| \to 0$ with probability unity as n increases; where $F_n(x)$ is the empirical distribution function.

Glivenko's Theorem A theorem stated by Glivenko (1943) which shows that the **Empirical Distribution Function** of a sample number of size n tends to the **Distribution Function** of the population as n becomes infinite. [See **Glivenko-Cantelli Lemma**.]

Gnedenko's Theorem A theorem showing the convergence of probability functions of sums of independent random variables, which take on only values of the form $a+kh$ ($h>0$, k an integer), to the density function of the Normal distribution. This theorem contains the de Moivre-Laplace theorem as a particular case.

Gompertz Curve See **Growth Curve**.

Goodman-Kruskal Tau A measure of association for cross classification in contingency tables. It is asymmetric between the two qualities concerned, being based on a probability interpretation involving the notion of predicting one given the other.

Goodness of Fit In general, the goodness of agreement between an observed set of values and a second set which are derived wholly or partly on a hypothetical basis, that is to say, derive from the 'fitting' of a model to the data. The term is used especially in relation to the fitting of theoretical distributions to observation and the fitting of regression lines. The excellence of the fit is often measured by some criterion depending on the squares of differences between observed and theoretical value, and if the criterion has a minimum value the corresponding fit is said to be 'best'. [See also **Kolmogorov-Smirnov Test**.]

Goutereau's Constant The expected value in a random series of a quantity defined for a time series as

$$\Sigma |x_{t+1} - x_t| / \Sigma |x_t - \bar{x}|.$$

Comparison with this value indicates the extent to which the series approximates to randomness. Nowadays the **von Neumann Ratio** or serial correlations would be used.

Grade For a continuous population, the grade of an individual variate value is the proportion of the total frequency with values less than or equal to that value; it is thus equivalent to the (cumulated) distribution function of that value. For discontinuous distributions the grade is similarly defined except that, by convention, an individual bearing the specified variate value counts as half an individual for the purpose of calculating proportional frequencies, the other half being regarded as lying in the remaining part of the range to the right of the specified value. The concept of grade was introduced by Galton for a continuous population to replace that of rank.

Grade Correlation In a continuous bivariate population, the correlation between the grades of its members. In a bivariate Normal population with correlation parameter ρ the grade correlation ρ' is given by a formula due to Karl Pearson:

$$\rho' = 2 \sin \frac{\pi \rho}{6}.$$

Graduation Curve (Curva di Graduazione) The curve which has for ordinate the intensity of a quantitative characteristic, against the number of the individuals with a value of the characteristic inferior to that intensity on abscissa. Apart from a change of the axes, it is equivalent to the English term **Distribution Curve**.

Graeco-Latin Square An extension of the **Latin Square**. Formally, it is an arrangement in a square of two sets of letters (say A, B, ..., and α, β, ...), one of each in each cell of the square, such that no Roman letter occurs more than once in the same row or column, no Greek letter occurs more than once in the same row or column, and no combination of the two occurs more than once anywhere. For example, a 4×4 square of this kind is

$A\alpha$	$B\beta$	$C\gamma$	$D\delta$
$B\gamma$	$A\delta$	$D\alpha$	$C\beta$
$C\delta$	$D\gamma$	$A\beta$	$B\alpha$
$D\beta$	$C\alpha$	$B\delta$	$A\gamma$

The arrangement is used in experimental designs to allocate treatments of three factors so that all comparisons are orthogonal. The arrangement also provides four orthogonal classifications of the 16 cells, by rows, columns, Roman and Greek letters.

Gram-Charlier Series Type A An expression of a frequency function in terms of derivatives of the Normal curve. If $H_r(x)$ is the Tchebychev-Hermite polynomial of order r the series with zero mean and unit standard deviation is

$$\frac{1}{\sqrt{2\pi}} e^{-\frac{1}{2}x^2} \{1 + \tfrac{1}{2}(\mu_2 - 1)H_2 + \tfrac{1}{6}\mu_3 H_3 + \tfrac{1}{24}(\mu_4 - 6\mu_2 + 3)H_4 + ...\}.$$

The name derives from the work of Gram (1883) and Charlier (1905) who used the series to approximate to frequency functions. A possibly better form is due to Edgeworth. [See also **Edgeworth Series**.]

Gram-Charlier Series Type B A series proposed by Charlier in 1905 to represent a discontinuous function in terms of differences of a Poisson variate. There are many difficulties in the use of the series and it is rarely employed at the present time.

Gram-Charlier Series Type C A further series proposed by Charlier in 1928 to avoid difficulties due to negative frequencies which can arise with Type A. The series expands a frequency function in the form

$$f(x) = \exp \{\Sigma \gamma_r H_r\}$$

where the H_r are polynomials. This series has not come into use.

Gram's Criterion A criterion which states that for n continuous functions $f_i(x)$ to be linearly independent in the interval $a \leqslant x \leqslant b$ it is necessary and sufficient that $|d_{ik}| = 0$, where $|d_{ik}|$ is the determinant defined by

$$d_{ik} = \int_a^b f_i(x)f_k(x)dx; \quad i, k = 1, ..., n.$$

[See **Singular Distribution**.]

Graphical Estimator A constant chosen by trial and error from a geometrical representation generally to secure a linear plot of observations on a specially ruled paper.

Grenander's Uncertainty Principle The product of a measure of 'resolvability' in the estimation of spectral densities (Δ_1) and a measure of the reliability, in the sense of variability, of the estimator (Δ_2) was shown by Grenander (1951) always to have a positive lower bound. The product $\Delta_1 \Delta_2$ is a measure of 'uncertainty' in the estimator.

Grid A rectangular mesh on a plane formed by two sets of lines orthogonal to each other, each line of each set being at a constant interval from the adjacent lines. It is used in some forms of area sampling.

Grid Sampling A form of cluster sampling, the clusters being individual areas of a grid and hence consisting of groups of basic cells arranged in some standard geometrical pattern. The term 'configurational sampling' is also used in the same sense.

Group A set of elements, individuals or observations all of which possess one, or more, characteristics in common. The word also occurs occasionally in statistics in its mathematical sense. [See also **Class**.]

Group Comparison A comparison between groups of individuals, usually on the basis of a representative value (such as a mean) from each.

Group Divisible Design A class of experimental design studied extensively by R. C. Bose. The parameters for treatments (v), blocks (b), replicates (r), groups (m), group size (n) and λ_1, λ_2 allocation factors are related
$$v = mn; \quad bk = vr$$
$$\lambda_1(n-1) + \lambda_2 n(m-1) = r(k-1)$$
$$Q = r - \lambda_1 \geqslant 0; \quad P = rk - v\lambda_2 \geqslant 0.$$
When $Q = 0$ the group divisible (GD) design is singular. For $Q > 0$ and $P = 0$ the GD design is semi-regular; regular GD designs are characterised by Q, $P > 0$.

Group Divisible Incomplete Block Design An **Incomplete Block Design** with v treatments replicated r times in blocks (b) of size k is group divisible if the treatments can be divided into m groups of n treatments each so that treatments belonging to the same group occur together in λ_1 blocks and for different groups in λ_2 blocks. If $\lambda_1 = \lambda_2$ the design reduces to a **Balanced Incomplete Block**.

Group Divisible Rotatable Designs A class of response surface design proposed by Das & Dey (1967) in which the factors are divided into two groups. The design is rotatable for each group separately when the factors in the other groups are held constant at some set of levels.

Group Factor See **Common Factor**.

Grouping Lattice A lattice or mesh of equal variate intervals which, when superimposed on a variate scale, defines the intervals in which the frequencies are grouped.

Group Screening Methods Screening designs are directed towards finding the few effective factors out of a large list of possibles which can effect the response in the design of experiments.

Group screening methods consist of putting the factors in groups, testing these groups and then testing the factors in the significant groups.

Grouped Poisson Distribution In the pure birth stochastic process if we assume that the time between two events has the gamma distribution then the population size at a given time has the grouped Poisson distribution with parameters k and λ with probability function

$$\Pr(x) = \sum_{j=1}^{k} e^{-\lambda} \frac{\lambda^{xk+j-1}}{(xk+j-1)!} \text{ where } x = 0, 1, 2, ..., k \text{ is a}$$

positive integer and λ is positive.

Growth Curve In general, an expression giving the size of a population y as a function of a time-variable t, and hence describing the course of its growth. The expression may also be used to denote the growth of an individual.

If the relative growth-rate declines at a constant rate, i.e.
$$\frac{1}{y}\frac{dy}{dt} = -b, \quad b > 0,$$
the curve is known as a Gompertz curve. Explicitly it may be written
$$y = ae^{-b^t}.$$
If the asymptotic value of y as t tends to infinity is a positive constant c, so that
$$y = c + ae^{-b^t}$$
the curve is sometimes known as the modified exponential curve. A growth curve for which
$$\frac{dy}{dt} = by(k-y)$$
is called logistic or autocatalytic. Its explicit form is
$$y = \frac{k}{1 + e^{-kbt}}$$
A rather more general form of type
$$y = \frac{k}{1 + e^{c\phi(t)}}$$
where $\phi(t)$ is some function of time, is also called logistic.

Grubbs' Rule A criterion for the rejection of outlying large observations proposed by Grubbs (1950) and based upon a studentised maximum residual $R_{(n)}$. It may also be used in the form of $R_{(1)}$ to cover the case of an unduly small observation.

Gurland's Generalisation of Neyman's Distribution The mixture of Neyman's type A distribution with two parameters λ_1 and $\lambda_2 p$, where p has the **Beta Distribution** with parameters α and β, is Gurland's generalisation of Neyman's distribution with parameters λ_1, λ_2, α and β where all four parameters are positive. The **Beall-Rescias Generalisation** is a special case of this Gurland generalisation.

Half-drill Strip One of the older systematic experimental designs which was commonly used for comparisons between two agricultural factors, e.g. two cereal strains. Owing to the various disadvantages of this kind of design, compared with the more modern designs based upon randomisation, it has lost popularity.

Half Invariant See **Cumulant**.

Half Normal Distribution A special case of the **Folded Normal Distribution**. If the distribution of x is $N(0, \sigma)$ then the distribution of $z = |x|$ is 'folded' in half and has the density function

$$f(z) = \frac{1}{\sigma}\sqrt{\frac{2}{\pi}}\, e^{-z^2/2\sigma^2}.$$

Half Normal Plots A graphical method for interpreting the contrasts in a two level factorial experiment. By ranking these contrasts they may be shown as a cumulative distribution function on a modified form of arithmetic probability paper. The modification consists of suppressing the lower half of the probability scale and replacing each value on the upper half by $P' = 2P - 100$. The ordinary plotting convention of $P' = (i - \frac{1}{2})/n$ is then used. [See also **Half Normal Distribution**.]

Half-plaid Square An experimental design introduced by Yates (1937) which is related to the **Split Plot Design** and to the **Quasi-Latin Square Design**.

If certain treatment combinations are laid out in the form of a Latin square and an additional treatment at two levels is applied to the rows of the square, one row receiving one level only, the resulting square is said to be half-plaid. For example, if factors B and C are each at two levels and A is the additional factor, such a layout might be

b	c	o	bc
c	b	bc	o
o	bc	b	c
bc	o	c	b

with A applied to each member of the last two rows but not applied or applied at a different level to the first two.

Whereas in quasi-Latin squares the principle of **Confounding** is applied only to interactions, in the case of the half-plaid square design the main effects are also confounded. The half-plaid square may also be regarded as a split plot design with the sub-units laid out in the form of Latin squares.

Half-replicate Design An experimental design based upon the principle of **Fractional Replication**, which employs only one half of the complete number of treatment combinations in a basic design.

Half-width This expression is sometimes used in relation to **Central Confidence Intervals** to denote the upper or lower half of an interval. In fixed interval prediction, such as may be used in a **Control Chart**, the half-width refers to the distance on the scale of the variate between, say, the process average and the upper or lower control limit.

An objectionable use of the term sometimes occurs in elementary texts. If a frequency function is unimodal with a frequency f_0 at the mode, and if there are two points x_1 and x_2 where

$$f(x_2) = f(x_1) = \tfrac{1}{2} f(x_0)$$

the distance $\frac{1}{2}(x_2 - x_1)$ is sometimes called the half-width.

Hamming A procedure for smoothing the spectrum of a time series using weights of $\frac{1}{4}$, $\frac{1}{2}$ and $\frac{1}{4}$. It was proposed by Hamming in 1949 (see Tukey, 1950) and may have had some connection with the earlier work on meteorological series by von Hann. [See also **Hanning**.]

Hanning A procedure for smoothing the spectrum of a time series associated with the Austrian meteorologist von Hann. He used weights of $\frac{1}{4}$, $\frac{1}{2}$ and $\frac{1}{4}$ for smoothing meteorological data (not the spectrum) many years before the period of modern spectrum analysis. [See also **Hamming**.]

Hard Clipping (Limiting) Certain non-linear transformations of stochastic signal processes are used in electronic systems which work in real time. This form of processing, called hard clipping, is used in order to reduce the number of information bits which the system has to process. With hard clipping only one binary digit per sample value of an input signal is used. Information concerning the input signal is lost and the spectrum calculated from the clipped signal is a distorted form of the original signal spectrum. By a sine transformation of the sample covariance of the clipped process a consistent estimator is derived for the spectral density of the input signal.

Hardy Summation Method A method of determining the moments of a frequency function defined at equidistant points, or grouped in equal intervals, by repeated summation of the frequencies. It was suggested by Sir George Hardy in 1903. The summations give **Factorial Moments** from which ordinary moments may easily be derived.

Harley Approximation An approximation to the t- **distribution** proposed by Harley (1957) based upon a transformation of the correlation coefficient.

Harmonic Analysis The analysis of a series of values into constituent periodic terms. [See also **Fourier Analysis, Periodogram, Spectral Function**.]

Harmonic Dial A method of representing harmonic constituents of a time series introduced in a geophysical context by Bartels (1935). A harmonic component is represented by a vector with length proportional to its intensity and angular orientation proportional to its phase. A set of components then appear like a number of hands on a clock.

Harmonic Distribution A term to be avoided. [See **Zipf's Law**.]

Harmonic Mean The harmonic mean of a set of observations is the reciprocal of the arithmetic mean of their reciprocals. It may be written in the discrete case for n quantities $x, x_2, ..., x_n$, as

$$\frac{1}{H} = \frac{1}{n}\sum_{i=1}^{n}\left(\frac{1}{x_i}\right),$$

or, in the continuous case, as

$$\frac{1}{H} = \int_{-\infty}^{\infty} \frac{f(x)}{x} dx,$$

where $f(x)$ is the frequency function, provided of course that the integral exists. For frequency distributions where the variate values are non-negative it may be shown that the harmonic mean is less than either the geometric mean or the arithmetic mean.

Harmonic Process A stationary random process which satisfies the relationship:

$$x_t + b_1 x_{t-1} + b_2 x_{t-2} + \ldots + b_h x_{t-h} = 0$$

is a harmonic process. It is deterministic, the general solution being a sum of harmonic terms which may be damped or explosive.

Harrison's Method See **Exponential Weights.**

Hazard In general, a word implying the existence of chance or risk (from Arabic 'al zhar', meaning a die). Specialised usage occurs principally in connection with life analysis of physical systems or components. The hazard is the probability that an article, functioning at time t_0 will fail in the interval $(t, t+\delta t)$ and this is given by $h(x) = f(x)/[1-F(x)]$. If this **Instantaneous (Death or Failure) Rate** is calculated at a number of survival times, the resulting hazard rates may be regarded as a hazard function.

Helly's First Theorem A theorem which states that every sequence of distribution functions contains a subsequence which tends to some non-decreasing function, not necessarily a distribution function, at all continuity points of this non-decreasing function. This first theorem is sometimes referred to as Helly's Lemma.

Helly-Bray Theorem This theorem, sometimes referred to as Helly's Second Theorem, states that, if $f(x)$ is a continuous function and $\{F_k(x)\}$ a sequence of uniformly bounded non-decreasing functions that converge weakly to $F(x)$ in the interval $[a, b]$, then

$$\lim_{k \to \infty} \int_a^b f(x) dF_k(x) = \int_a^b f(x) dF(x).$$

Helmert Criterion See **Abbe-Helmert Criterion.**

Helmert Distribution The distribution of the sample standard deviation or, equivalently, of the sample variance in samples from a normal population. It may be written

$$dF = \frac{n^{\frac{1}{2}(n-1)}}{\Gamma\{\frac{1}{2}(n-1)\}2^{\frac{1}{2}(n-3)}} \left(\frac{s}{\sigma}\right)^{n-2} e^{-ns^2/2\sigma^2} \frac{ds}{\sigma}, \ 0 \leqslant s \leqslant \infty$$

where σ^2 is the parent variance and s^2 the sample variance. If ns^2/σ^2 is put equal to χ^2 the distribution becomes that of χ^2 with $n-1$ degrees of freedom. It is a form of the **Type III Distribution.**

Helmert Transformation An orthogonal linear variate transformation due to Helmert. If $x_1, x_2, \ldots, x_n$, have zero mean and unit variances the transformation is given by

$$y_1 = (x_1 - x_2)\frac{1}{\sqrt{2}}$$

$$y_2 = (x_1 + x_2 - 2x_3)\frac{1}{\sqrt{6}}$$

$$y_3 = (x_1 + x_2 + x_3 - 3x_4)\frac{1}{\sqrt{12}}$$

$$\cdot \quad \cdot \quad \cdot \quad \cdot \quad \cdot \quad \cdot \quad \cdot \quad \cdot$$

$$y_{n-1} = \{x_1 + x_2 + \ldots + x_{n-1} - (n-1)x_n\}\frac{1}{\sqrt{\{n(n-1)\}}}$$

$$y_n = (x_1 + x_2 + \ldots + x_n)\frac{1}{\sqrt{n}}.$$

Hermite Distribution The probability generating function of a generalised Poisson distribution can be expressed as $g(z) = \exp\{a_1(z-1) + a_2(z^2-1) + \ldots\}$ where $\sum_{i=1}^{\infty} a_i = \lambda$, the Poisson parameter. This generalised distribution tends to the Poisson distribution with parameter a_1 if a_2, $a_3, \ldots$ become negligible compared to a_1 in some limiting process. If a_2 does not become negligible compared to a_1, then limiting distribution is the Hermite distribution with parameters a_1 and a_2.

Heteroclitic See **Clisy.**

***Heterograde** See ***Extensive Magnitudes.**

Heterograde A term used by some German and Scandinavian writers to denote a variable which is quantitative, i.e. is a variable as distinct from a qualitative characteristic or attribute. [See also ***Extensive Magnitudes.**]

Heterokurtic See **Kurtosis.**

Heteroscedastic See **Scedasticity.**

Heterotypic A term used in relation to **Pearson Distributions.** For certain values of the moment ratios β_1 and β_2 the differential equation defining the family of distributions has infinite moments of order 8 or more. In this region the standard error of the sample estimate of β_2 would be infinite and hence the fitting of the distributions by the method of moments would be inappropriate. Distributions of the Pearson family with such values of β_1 and β_2 were called heterotypic; but, as is now realised, certain of the distributions may be valuable for many purposes.

$Hh_n(x)$ Function A function derived by integration and differentiation of the 'normal' function $e^{-\frac{1}{2}x^2}$. The function of zero order is defined as

$$Hh_o(x) = \int_x^{\infty} e^{-\frac{1}{2}t^2} dt$$

and for positive n the function is defined by recurrence:

$$Hh_n(x) = \int_x^{\infty} Hh_{n-1}(t) dt.$$

Similarly

$$Hh_{-n}(x) = \left(-\frac{d}{dx}\right)^n Hh_0(x)$$
$$= \left(-\frac{d}{dx}\right) Hh_{-n+1}(x).$$

The Hh_{-n} functions are Hermite functions. [See **Tchebychev-Hermite Polynomial**.]

Hidden Periodicity, Scheme of A term advanced by Schuster (1898), and later used extensively by other writers, e.g. Wold (1938), to denote a time-series, or more generally a stochastic process, which is generated by the addition of a finite number of harmonic terms and a random residual component. One of the objects of analysing such a series is to determine the amplitude, period and phase of each 'hidden' component.

Hierarchical Birth and Death Process A type of birth and death process proposed by Blom (1960) in which the states of the process are divided into sub-states according to given rules. Although only processes with constant transition probabilities were considered the analysis was generalised to include an infinitely denumerable number of states.

Hierarchical Classification [See **Nested Classification, Nested Design**.]

Hierarchical Group Divisible Design The designation proposed by Roy (1953) for group divisible designs with m associate classes. These designs were developed by Raghavarao under the more general title.

Hierarchy If, in a matrix of **Intercorrelations** of a set of variates, the rows and columns can be so arranged to give the highest correlations in the upper left-hand corner and the lowest correlations in the lower right-hand corner and when this is done there is a constant proportional relationship between adjacent columns, except for diagonal terms, the table is called a hierarchy (Spearman, 1904) and the intercorrelations are said to be an hierarchical order. Thus, for two rows denoted by a and b and two columns by c and d, the correlations obey the so-called tetrad relations $r_{ac}r_{bd} = r_{ad}r_{bc}$. Under certain conditions the fact that the correlation matrix is hierarchical is a necessary and sufficient condition that the variation can be accounted for by a single factor common to the variates.

The quantities $r_{ac}r_{bd} - r_{ad}r_{bc}$ are called tetrad differences.

High Contact In relation to a frequency function $f(x)$, the order of contact of the function with the variate axis, or at infinity if the range is infinite, is said to be high if $x^r f(x)$ or its limit vanishes at the terminals for some high value of r. What constitutes a 'high' value for this purpose is somewhat arbitrary. This property of high contact is one of the conditions for the application of the **Corrections for Grouping** known as **Sheppard's Corrections**. [See also **Abrupt Distribution**.]

High-low Graph A form of graph used to depict ranges of variation in successive intervals of time. For example, daily price variation might be represented by taking time intervals of one month on the abscissa and, at each monthly point, showing the maximum and minimum price attained during the previous month. The maxima can be joined by a line to provide a graph of high points; and similarly for the minima; or, for each month, the high and low points may be joined by a vertical bar.

Histogram A univariate frequency diagram in which rectangles proportional in area to the **Class Frequencies** are erected on sections of the horizontal axis, the width of each section representing the corresponding class interval of the variate. [See also **Block Diagram, Frequency Polygon**.]

Historigram A term used to denote a graph of a time series with the value of the series as ordinate against time as abscissa. Owing to possible confusion with the word 'histogram' the term is not to be recommended.

Hitting Point See **Waiting Time**.

Hodges Bivariate Sign Test This analogue, proposed by Hodges (1955), of the classical **Sign Test**, can be shown to have the same null distribution as a regression test for median $\{Y \mid x\}$ proposed by Daniels (1954).

Hoeffding 'C$_1$' Statistic A test criterion proposed by Hoeffding (1951) for a most powerful rank order test. It was extended by Terry (1957) for specific parametric alternative hypothesis.

Holt's Method See **Exponential Weights**.

Homoclitic See **Clisy**.

Homogeneity This term is used in statistics in its ordinary sense, but most frequently occurs in connection with samples from different populations which may or may not be identical. If the populations are identical they are said to be homogeneous, and by extension, the sample data are also said to be homogeneous. In a more restricted sense populations may be said to be homogeneous in respect of some of their constants, e.g. k populations with identical means but different dispersions are homogeneous in their means.

Homogeneous Process A stochastic process is said to be homogeneous in space if the transition probability between any two state values at two given times depends only on the difference between those state values.

The process is homogeneous in time if the transition probability between two given state values at any two times depends only on the difference between those times.

*Homograde See ***Extensive Magnitudes**.

Homograde A term used by some German and Scandinavian authors to denote a qualitative variate, i.e. an **Attribute**. The word is not in common current use among English writers.

Homokurtic See **Kurtosis**.

***Homophily (Omofilia), Index of** In Italian usage, a measure of **Concordance** which takes into account the influence exerted on the association by dissimilarity in the marginal distributions.

Homoscedastic See **Scedasticity**.

Honest Process For a generalised Markov birth process, the solution of the differential equations is such that for all t, $\Sigma p_i(t) = 1$ if $\Sigma 1/\lambda_i$ is divergent; the process is then designated an 'honest' process. If, however, $\Sigma 1/\lambda_i$ is convergent then, for some t, $\Sigma p_i(t) < 1$ and the process is termed 'dishonest' or pathological. The λ_i are the average times spent in states i.

Horvitz & Thompson Estimator A method of estimating the population total when sampling is without replacement from a finite population and when unequal probabilities of selection are used. The estimator, proposed in 1952, is unbiassed, linear and can be used with a variety of basic sample designs.

Hotelling's T A generalisation by Hotelling (1931) of **'Student's' Distribution** to the multivariate case, and like 'Student's' t, available to test the significance of a broad class of statistics including means and differences of means, regression coefficients and their differences. If, for example, the measurements of p variates on a random sample of n individuals from a multivariate Normal distribution of unknown covariance matrix are to be used to test the hypothesis that the respective population means are ξ_1, ξ_2, ..., ξ_p, T is defined as the non-negative square root of

$$T^2 = n\Sigma l_{ij}(\bar{x}_i - \xi_i)(\bar{x}_j - \xi_j)$$

where $\bar{x}_i$ is the sample mean of the ith variate, the summations are from 1 to p, and l_{ij} is a typical element of the p-rowed matrix inverse to that of sample covariances. Tests of significance involving T^2 can be carried out by the **Variance Ratio Distribution**.

Hunt-Stein Theorem A theorem concerning critical regions associated with a particular significance test. For example, in the set of **Critical Regions** of size α which are also invariant, there may be one, say W_0, which is **Uniformly Most Powerful** and **Most Stringent**.

Hypercube The original term for **Orthogonal Array** as introduced by Rao (1946). To be avoided in this context.

Hyperexponential Distribution A term proposed by Morse (1958) to a mixture of exponential distributions in the general form:

$$\gamma e^{-2\gamma bt} + (1-\gamma)e^{-2(1-\gamma)bt}.$$

Hypergeometric Distribution A distribution of a discrete variate generally associated with sampling from a finite population without replacement. The frequency of r 'successes' and $n-r$ 'failures' in a sample of n so drawn from a population of N in which there are Np 'successes' and Nq 'failures' ($p+q=1$) is

$$\frac{1}{N^n}\binom{n}{r}(Np)^{[r]}(Nq)^{[n-r]}$$

where $N^{[r]} = N(N-1)...(N-r+1)$. As N tends to infinity the distribution tends to the ordinary binomial form. The distribution derives its name from the fact that the probability generating function may be put in the form of a hypergeometric series.

Hypergeometric Waiting Time Distribution An alternative name for the **Inverse Hypergeometric Distribution**.

Hyper-Graeco-Latin Square A generalisation of the **Latin** and **Graeco-Latin Square** in the form of a $p \times p$ square in which each cell contains one of the characters of each of k types ($k > 2$), the characters of each type being p in number and constituting a Latin square; and the types being mutually orthogonal so that no combination of the characters of different types occurs more than once anywhere in the design. The maximum value of k never exceeds $p-1$.

For example, the following 4×4 square:

$A\alpha 1$	$B\beta 2$	$C\gamma 3$	$D\delta 4$
$B\gamma 4$	$A\delta 3$	$D\alpha 2$	$C\beta 1$
$C\delta 2$	$D\gamma 1$	$A\beta 4$	$B\alpha 3$
$D\beta 3$	$C\alpha 4$	$B\delta 1$	$A\gamma 2$

shows how comparisons between 16 observations may be considered in 5 independent sets, corresponding to rows, columns, Roman letters, Greek letters and numerals.

Hyper-Poisson Distribution This discrete distribution has a complicated probability function in two parameters, λ and θ. If $\lambda = 1$ this distribution becomes the **Poisson Distribution** with parameter θ. If $\lambda < 1$ it is sometimes referred to as a sub-Poisson distribution and if $\lambda > 1$ as a super-Poisson distribution.

Hypernormal Dispersion See **Lexis Variation**.

***Hypernormality (Ipernormalità)** See ***Abnormality**.

Hypothesis, Statistical A statistical hypothesis is a hypothesis concerning the parameters or form of the probability distribution for a designated population or populations, or, more generally, of a probabilistic mechanism which is supposed to generate the observations.

Hypothetical Population A statistical population which has no real existence but is imagined to be generated by repetitions of events of a certain type; e.g. the binomial distribution as generated by the throws of a die, or crop-yields on a set of plots imagined as all the possible ways in which a set of yields might occur under the conditions of an experiment.

'Ideal' Index Number In 1927 Irving Fisher advanced certain criteria which should be obeyed by 'good' index numbers. Of the large collection of formulae investigated only a few obeyed his tests. One of these was termed the 'ideal' index. It may be written

$$\left(\frac{\Sigma p_n q_o}{\Sigma p_o q_o} \times \frac{\Sigma p_n q_n}{\Sigma p_o q_n} \right)^{\frac{1}{2}}$$

where p_o, q_o represent prices and quantities in the base period and p_n, q_n those of the period for which the index is being calculated. The ideal index is the geometric mean of the **Laspeyres'** and **Paasche Index Numbers.** [See also **Crossed-Weight Index Number, Factor Reversal Test, Time Reversal Test.**]

Identical Categorisations In the investigation of the relationships between two or more categorised variables, the variables investigated may simply be the same variable observed either on different occasions or on related samples. Such interactions are said to have identical categorisation.

Identical Errors A regression model of type $y = f(x_1, ..., x_p) + \epsilon$ where ϵ has a unique frequency distribution, is said to have identical errors. 'Identity' refers not to individual errors but to their frequency distribution.

Identifiability In certain systems of stochastic equations it may happen that some or all parameters cannot be separately estimated without bias, however extensive the data, even if the number of equations is equal to the number of unknown endogenous variates. For example, if x_1 and x_2 are Normal variates and ϵ_1, ϵ_2 are (unobservable) components, the system

$$a_1 x_1 + x_2 = \epsilon_1 \equiv A, \text{ say}$$
$$a_2 x_1 + x_2 = \epsilon_2 \equiv B, \text{ say}$$

does not permit of the separate estimation of α_1 and α_2, for it is observationally indistinguishable from any system of type $A + kB$, $A + lB$. The system is then said to be unidentifiable.

If no parameters can be estimated on the information available the system is said to be completely unidentifiable. If some parameters, but not others, can be estimated the system is partially identifiable. If all can be estimated it is completely identifiable. If there are more relations given than are necessary for complete identification the system is over-identified.

Illusory Association An association between attributes may be statistically significant without necessarily involving any direct causal connection between them. The association is then sometimes said to be illusory. Such associations may arise between attributes A and B if both are associated with some other attribute C or with a time-variable. For example, if, in Europe, the possession of blonde hair was found to be positively associated with ability to skate, the association would be said to be illu-sory in the sense that one attribute does not 'cause' the other; such association as exists being due to the accidental circumstance that the blonde races inhabit the more northern regions where skating is more often possible. The word 'illusory' in this kind of context has to be used with some caution. The association is not illusory in the sense that it does not really exist; only in the sense that it does not bear an obvious interpretation.

Illusory Correlation As in the case of **Illusory Association,** a correlation may be significant without implying causal connection between two variates. For example, over a period of years there is in Europe a negative correlation between the birth rate and the number of deaths from road accidents, but it is not arguable that, for example, one way of depressing the birth rate is to arrange for more accidents on the road. Both effects can be assigned to a general movement in economic and sociological condi-tions and their co-relationship is due to the relation of each with the time variable. Illusory correlation has been termed by Yule 'nonsense correlation'.

Imbedded Process A stochastic process in continuous time is observed only at the time points where a change of state occurs. These points of discontinuity can be thought of as forming a new discrete time variable.

A new stochastic process can be derived by defining the state of the process at time n to be that immediately following the nth transition in the old process. The new process in discrete time is said to be 'imbedded' in the old process in continuous time. In this sense, the imbedded process can, for example, be Markov derived from a non-Markov process or a renewal process derived from a point process.

Implicit Strata The n groups, zones or sub-classes into which, in systematic sampling, the population is divided by the sampling interval $k = N/n$, where N is the population size and n the sample size.

Importance Sampling This method of sampling, used in Monte Carlo analysis, is to concentrate the distribution of sample points in those parts of the variate interval which are of greatest interest. It is a form of stratified sampling with variable sampling fraction.

Incidence Matrix of Design A **Block** experiment may be described by a treatment matrix of order $(k \times t)$ for each block. If we need only allocation of treatments to blocks without reference to allocation within blocks then the b treatment matrices t_j can be condensed into an inci-dence matrix **n** of order $(t \times b)$. The matrix merely records whether or not a particular plot or treatment appears in the experiment, not the outcome of the observations.

Incidental Parameters See **Partially Consistent Observa-tions.**

Incomplete Beta Function This function is defined as

$$B_t(s, r) = \int_0^t y^{s-1}(1-y)^{r-1}\,dy, \quad s, r > 0;\ 0 \leqslant t \leqslant 1.$$

The ratio of the incomplete beta function to the complete beta function is generally written:

$$I_t(s, r) = \frac{B_t(s, r)}{B(s, r)} = \frac{\displaystyle\int_0^t y^{s-1}(1-y)^{r-1}\,dy}{\displaystyle\int_0^\infty y^{s-1}(1-y)^{r-1}\,dy}.$$

Incomplete Block A basic form of experimental design introduced by Yates (1936). If material is divided into blocks and it is desired to allocate certain treatments to the units of a block, the treatments may be too numerous for them all to appear in each block. When a block contains fewer than a complete replication of the treatments it is called incomplete. Differences between blocks are then discussed by a more elaborate analysis than is required for complete blocks (see **Recovery of Information**).

If each block contains the same number of treatments and they are arranged so that every pair of treatments occurs together in the same number of blocks, the design is said to be balanced.

Incomplete Census See **Census**.

Incomplete Gamma Function A function defined as

$$\Gamma_t(\lambda) = \int_0^t e^{-x} x^{\lambda-1}\,dx \quad \lambda > 0;\ \ 0 \leqslant t \leqslant \infty.$$

It is the distribution function of the **Gamma Distribution** multiplied by $\Gamma(\lambda)$.

Incomplete Latin Square An alternative name for the **Youden Square**.

Incomplete Moment The ordinary moment of a distribution about an arbitrary origin a is given by

$$\mu'_r = \int_{-\infty}^\infty (x-a)^r\,dF(x).$$

If this be modified to make the limits of integration extend only from $-\infty$ to t the expression

$$\int_{-\infty}^t (x-a)^r\,dF(x)$$

is called the incomplete moment of order r, provided that it exists. By contrast the form when $t = \infty$ is sometimes called the complete moment.

Incomplete Multiresponse Design An experiment design introduced by Srivastava (1964) to deal with a situation where not all the variates are measured for each experimental unit. This position can arise from a state of physical impossibility, e.g. irreparable damage, from inconvenience or from other considerations such as expense.

Inconsistent Estimator See **Consistent Estimator**.

Increasing Hazard Rate Conventionally, an increasing hazard rate is one which does not decrease; in other words it may be constant. For the **Hazard Rate** to be increasing the complement of the distribution function must be a **Pólya Frequency Function of Order Two**.

Independence In the calculus of probabilities, independence is usually defined by reference to the principle of compound probabilities. Two events are independent if the probability of one is the same whether the other is given or not, i.e. $\Pr(A) = \Pr(A \mid B)$ and $\Pr(B) = \Pr(B \mid A)$.

From this it follows that the probability of the compound event $\Pr(AB) = \Pr(A)\Pr(B)$ if the events are independent. As a matter of axiomatisation it may be preferable to use relations of the type $\Pr(AB) = \Pr(A)\Pr(B)$ as definitions to avoid difficulties arising when $\Pr(A)$ or $\Pr(B)$ is zero.

In statistics two variates x_1 and x_2 are independent if their distribution functions are related by

$$F(x_1, x_2) = F(x_1, \infty)F(\infty, x_2)$$

or equivalently if their frequency functions, should they exist, are related by $f(x_1, x_2) = f_1(x_1)f_2(x_2)$. Generally, n variates $x_1, x_2, ..., x_n$ are independent if

$$F(x_1, x_2, ..., x_n) = F(x_1, \infty, \infty, ..., \infty)$$
$$F(\infty, x_2, \infty ..., \infty)... F(\infty, \infty, \infty, ..., x_n).$$

It is not enough that they should be independent pair and pair. The word is also applied in the ordinary mathematical sense to describe the independence of two or more variables.

*In Italian usage, if, of a set of variates $h+k$ in number, the set of h are independent of the set of k there is said to be independence of class (h, k). Of n variates there is said to be independence of order (*ordine*) h if $n-h$ of them are mutually independent when the values of the other h are fixed.

Independence Frequency In a contingency table, the frequency which would be found in a particular cell if the attributes defining it were independent; e.g. if the r rows of the table have total frequencies $A_1, A_2 ..., A_r$, and the s columns have frequencies $B_1, B_2, ..., B_s$, and

$$\Sigma A = \Sigma B = N,$$

the independence frequency in the ith row and jth column is $A_i B_j/N$.

Independent Action Suppose that doses x_1 and x_2 of two stimuli have expected quantal response rates of $P_1(x_1)$ and $P_2(x_2)$. The two stimuli are said to display independent action if, for all x_1 and x_2, the expected response rate to a simultaneous application of both doses is

$$P = P_1 + P_2 - P_1 P_2$$
$$= 1 - (1 - P_1)(1 - P_2).$$

The extension to three or more stimuli is obvious.

Independent Increments, Process with See **Additive Process**.

Independent Trials The successive trials of an event are said to be independent if the probability of outcome of any trial is independent of the outcome of the others. The expression is usually confined to cases where the probability is the same for all trials. In the sampling of attributes, such a series of trials is often referred to as 'Bernoullian Trials'. It includes all the classical cases of drawing coloured balls from urns with replacement after each draw, coin tossing, dice rolling and the events associated with other games of chance.

Independent Variable This term is regularly used in contradistinction to 'dependent variable' in regression analysis. When a variate y is expressed as a function of variables x_1, x_2, ..., plus a stochastic term the x's are known as 'independent variables'. The terminology is rather unfortunate since the concept has no connection with either mathematical or statistical dependence. Modern usage prefers 'explanatory variable', 'predicated variable' or, best of all, 'regressor'.

Index Number An index number is a quantity which shows by its variations the changes over time or space of a magnitude which is not susceptible of direct measurement in itself or of direct observation in practice. Examples of these magnitudes are: Business Activity, Physical Volume of Production, Wholesale Prices (General Level of). Important features in the construction of an index number are its coverage, base period, weighting system and method of averaging observations.

The above definition relates to the usual meaning of the expression 'index number'. In full generality, however, the term can also be applied to a series of values which are standardised by being referred to a basic period or area e.g. if the price of a fixed commodity in a basic year is 40 units and those in succeeding years are 60 and 68 units, the index number for those years would be, on the basis of 100 for the first year, 150 and 170. Such simple cases are, however, usually referred to as 'relatives' and the index number is constructed as an average of a number of relatives.

It is also somewhat tendentious to define an index number as the measure of a magnitude when that magnitude relates to an ill-defined concept such as 'business activity'. It is perhaps preferable to regard the index number as not relating to a specific quality but as a measure of location in a complex of concomitant variation.

Index of Dispersion See **Dispersion Index.**

Index of Response An alternative, and simpler, method than the method of **Concomitant Observations** for estimating the effects of treatments in the analysis of experimental data. The method consists of constructing an initial index of response by combining two or more variables and treating this construction as a new variable.

Indicator Function Given a class of events A, on a sample S, the Indicator Function is defined for all points s in S as $I(A; s)$ equal to 1 or zero depending on whether s does or does not belong to A.

*Indifference (Indifferenza Statistica) In Italian usage there is said to be statistical indifference between two variates if their correlation is zero, or more generally, if some index of **Concordance** (*concordanza*) vanishes.

Indifference-Level Index Number A synonym for **Konyus Index Number.**

Indirect Least Squares A term of doubtful usefulness denoting a method of parameter estimation in econometric models. If the system is specified by a set of equations, instead of applying least squares to those equations, the method may be applied to some derived equations; for example the **Reduced Equations.**

Indirect Sampling Sampling from documents, or some record of the characteristics of a population, rather than the recording of information obtained at first hand from units of the population themselves. For example, it is becoming customary to obtain preliminary information on the results of, say, a national census by analysing a sample of the census forms before the full analysis is undertaken; the population is then subject to indirect sampling. [See also **Direct Sampling.**]

*Individuality, Coefficient of (Coefficiente di Individualità) A coefficient introduced by Gini (1908). Its object is to assess the systematic component of a variate as distinct from its sampling error, but it can equally be applied for assessing a systematic component as distinct from its error of measurement. If σ is the standard deviation of the observed values of the variate and σ_s is the standard deviation of the error component the coefficient of individuality is given by

$$i = \sqrt{\frac{\sigma^2 - \sigma_s{}^2}{\sigma^2}}.$$

[See **Reliability Coefficient, Attenuation.**]

Inductive Behaviour A term introduced by Neyman (1937) to indicate the adjustment of a course of action based upon a limited amount of information in relation to established ideas or 'permanencies'. In particular, when the relative merits of a number of courses of action depend upon the nature of the frequency function of some observed variates, the rule of inductive behaviour is equivalent to a test of a **Statistical Hypothesis.** [See also **Decision Function.**]

Inefficient Statistic A statistic with less than the maximum possible efficiency in the sense of having the minimum sampling variance. It is customary, though perhaps not entirely satisfactory, to define relative efficiency by reference to the magnitude of sampling variance only, and not, for instance, with reference to bias or to the form of the sampling distribution; or, what is often of equal importance, the ease and speed of calculation. In fact 'efficiency' relates to precision, irrespective of cost and time; and 'inefficiency' has the corresponding interpretation. [See also **Efficiency**.]

Inequality Coefficient A measure due to Theil (1961) of the absolute difference between the actual and the forecast change in a system. If P_i and A_i are predicted and actual values, the coefficient was originally defined as

$$U = \left\{ \frac{\Sigma (P_i - A_i)^2}{n} \Big/ \frac{\Sigma A_i^2}{n} \right\}^{\frac{1}{2}}$$

and more recently modified to

$$U = \left[\frac{\Sigma (P_i - A_i)^2}{n} \right] \Big/ \left[\left\{ \frac{1}{n} \Sigma P_i^2 \right\} + \left\{ \frac{1}{n} \Sigma A_i^2 \right\}^{\frac{1}{2}} \right].$$

The summation is over n values for which forecast and actual values can be compared, usually n consecutive years.

Infinite Population An infinite population is one which either possesses the infinite property through some limiting process or, in a sampling context, can be made to possess that property by some strategy of sampling. For example, a population of real numbers from 0 to 1 or positive integers is infinite by definition. But sampling from a finite population with replacement after each drawing makes such a finite population take on the characteristics of an infinite population.

Inflation Factor A less preferable term for **Raising Factor**.

Information The word 'information' occurs frequently in statistics with its ordinary meaning. In a specialised sense in the theory of estimation, the amount of information about a parameter θ from a sample of n independent observations drawn at random from a population with frequency function $f(x, \theta)$ is defined as

$$n\mathscr{E} \left(\frac{\partial \log f}{\partial \theta} \right)^2 \equiv n \int_{-\infty}^{\infty} \left(\frac{\partial \log f(x, \theta)}{\partial \theta} \right)^2 f(x, \theta) dx.$$

Under some general regularity conditions the reciprocal of the information gives a lower bound for the variance of unbiassed estimators of θ, so that the greater the variance the less the 'information'.

If the extremes of the distribution do not depend on θ an equivalent expression is

$$-n\mathscr{E} \left(\frac{\partial^2 \log f}{\partial \theta^2} \right).$$

The concept generalises easily to the case of several variates.

Information Matrix In generalisation of the definition of 'information' for one parameter under estimate, the information matrix of a sample of n drawn independently from a population with frequency function $f(x, \theta_1, \theta_2, ..., \theta_p)$ is the matrix for which the element in the ith row and jth column is

$$n\mathscr{E} \left(\frac{\partial \log f}{\partial \theta_i} \frac{\partial \log f}{\partial \theta_j} \right) \equiv n \int_{-\infty}^{\infty} \left(\frac{\partial \log f}{\partial \theta_i} \right) \left(\frac{\partial \log f}{\partial \theta_j} \right) f \, dx.$$

As in the case of one parameter, certain regularity conditions on f are required; and if the range is independent of the θ's an equivalent expression is

$$-n\mathscr{E} \left(\frac{\partial^2 \log f}{\partial \theta_i \partial \theta_j} \right).$$

Inherent Bias A rather loosely defined expression which in general means, or ought to mean, a bias which is due to the nature of the situation and cannot, for example, be removed by increasing the sample size or choosing a different type of estimator. An example of inherent bias is the systematic error of an observer or an instrument; a further example, in the interrogation of human population, is the distortion of truth by the respondent for reasons of prestige, vanity or sympathy with the investigator.

It is possible also to speak of the inherent bias of a method of estimation, although in this context the word 'inherent' appears redundant. For example, in the theory of index numbers it may be shown that the standard formulae of **Laspeyres** and **Paasche** possess an inherent bias due to the methods of weighting and averaging the items.

Input/Output Process A broad class of stochastic processes, the state $\xi(t)$ at time t being the resultant of two random variables x, y;

$$\xi(t) = x(t) - y(t),$$

x and y being called respectively the input and the output. Examples are queueing processes, provisioning and epidemic processes.

Inquiry (Rilevazione) The nearest English equivalent to the Italian 'rilevazione' seems to be 'inquiry'. An inquiry is complete (*completa*) if it covers all the objects under study, incomplete if it covers only part and over complete (*più che completa*) if it covers all objects once and some more than once. It is direct (*diretta*) if it measures the phenomenon under study directly (e.g. population census), indirect (*indiretta*) in the contrary case. It is continuous (*continua*) if it proceeds continuously through time, periodic (*periodica*) if repeated at fixed intervals of time and intermittent (*occasionale*) or ad hoc if it is conducted only to suit some particular purpose on certain occasions. A **Pilot Inquiry** is said to be *preliminare*. An inquiry is representative (*rappresentativa*) if every member of the population under investigation has the same chance of being chosen, non-representative (*non rappresentativa*) in the opposite case.

Inspection Diagram This term is capable of several interpretations. It may describe a diagram of a manufacturing process showing the inspection points as part of the process. It may be applied to the diagrammatic layout of a double or multiple sampling scheme which describes the stages from the inspection of the first sample to the acceptance or rejection of the **Inspection Lot**. It may refer to the graph upon which are summarised the elements of a sequential sampling plan together with the course of the actual sampling process. Finally, it may be applied to the graph of a sampling inspection plan which shows the **Operating Characteristic** curve.

Inspection Lot   A **Lot** presented for inspection, which may be carried out on each member of the lot or on a sample of members only.

Instantaneous Death Rate   A version of the **Age Specific Death Rate** where the 'next unit time period' is reckoned to be short.

Instrumental Variable   In econometrics, and generally in the analysis of the structure of a stochastic situation, an instrumental variable is a predetermined **Variable** which is used to derive consistent estimators of the parameters of the system. Its use is inefficient in relation to a complete set of equations in the sense that only a limited amount of information is employed. On the other hand it may be applied to incomplete systems. [See also **Limited Information Methods, Reduced Form Method**.]

Integer Programming   A form of **Linear Programming** where the solutions are derived in an integer form, e.g. in the scheduling of ships or men.

Integrated Data   A class of statistical data in which the values for short unit intervals can be added together to give a series of values relating to longer intervals; for example, daily rainfall can be integrated into a new series of weekly, monthly or annual rainfall figures each of which will possess a longer time base than the previous series. On the other hand, a series of, say, temperature readings cannot be integrated in this sense and series for longer time intervals must be derived by averaging or the selection of typical values.

Integrated Spectrum   A concept analogous to the cumulative distribution function in the same fashion as the spectrum density function is analogous to the probability density function.   [See **Spectrum**.]

Intensity   In the harmonic analysis of time series, a measure which provides an estimate of the amplitudes of the constituent harmonics. If the series is $u_1, u_2, ..., u_n$ the intensity for a period μ is defined as $A^2 + B^2$ where

$$A = \frac{2}{n} \sum_{j=1}^{n} u_j \cos \frac{2\pi j}{\mu}, \qquad B = \frac{2}{n} \sum_{j=1}^{n} u_j \sin \frac{2\pi j}{\mu}.$$

An entirely different use of this term occurs in the theory of extreme values, life testing, etc. Here it is equivalent to the **Force of Mortality, Hazard** or **Age Specific Death (Failure) Rate**.

Intensity Function   In an actuarial context, the 'intensity' in this second sense is an alternative term to **Hazard, Age Specific Death Rate** or **Force of Mortality**. If a number of values are calculated, i.e. for a sequence of ages, then these values can be regarded as the intensity function.

*Intensity of Transvariation   See ***Transvariation**.

*Intensive Magnitudes   See ***Extensive Magnitudes**.

Intensive Sampling   Like **Extensive Sampling** this expression may mean two different things: either (*a*) sampling in a particular area with a dense scatter of sampling points or (*b*) sampling wherein information on a restricted range of topics is sought by probing on them very deeply with an intricate schedule of questions.

Interaction   In general, when a number of individuals or items are grouped according to several factors of classification and these factors are not **Independent** there is said to be interaction between them.

The most common use of the term is in experimental design. In the **Factorial Experiment** a number of factors can be studied simultaneously, each at several levels. The interaction is a measure of the extent to which the effect upon the dependent variable of changing the level of one factor depends on the level(s) of another or others. They are often of as much interest as main effects and it is one of the advantages of factorial design that they can be estimated and tested.

Thus with 2 treatments, say N and P, each at two levels, the effects of the four treatment combinations can be written $n_o p_o$, $n_1 p_o$, $n_2 p_1$ and $n_1 p_1$. If the treatments are independent, the effect of varying N from n_o to n_1 would be the same with p_o as with p_1. The extent to which this is not so is a measure of interaction:

$$n_1 p_1 - n_o p_1 - n_1 p_o + n_o p_o,$$

which may be written symbolically as $(n_1 - n_o)(p_1 - p_o)$.

A main effect, i.e. the effect of one factor only, is regarded as an interaction of order zero. Interactions of two factors are regarded as of order one (first order interactions); those concerning three factors as of the second order; and so on.

If the total sum of squares in a variance analysis is split into sums allocated to interaction terms the respective sums are known as components due to interaction; and they may be used to estimate components of interaction in the model which is regarded as generating the data. [See also **Variance Components**.]

Interblock   See **Block**.

Intercalate Latin Square   A term proposed by Norton (1939) for any Latin square or rectangle which may be embedded in a larger **Latin square**.

Interclass Correlation This expression denotes correlation in the ordinary sense; the qualifying adjective 'interclass' is only employed to distinguish ordinary correlation from **'Intraclass Correlation'**.

Interclass Variance In the analysis of variance of data subject to multiple classification the sum of squares of all observations about their mean is expressed as the sum of squares of observations about class means plus the sum of squares of class means about the mean of the whole. The former, divided by an appropriate number of degrees of freedom, is sometimes called **Intraclass Variance**; those of the latter type, again divided by degrees of freedom, are called interclass variances. The expressions are convenient and the quantities in question estimate components of variance in the model generating the observations, but strictly speaking they are not always variances.

Sundry synonymous expressions occur such as between class variance, within class variance, external and internal variance, etc.

Intercorrelation A term used to denote the correlation of a number of variates among themselves, as distinct from the correlations between them and an 'outside' or dependent variate.

Interdecile Range Although a somewhat loose usage, this term is usually interpreted as the variate range between the 1st and 9th **Deciles**. Like the **Interquartile Range** it provides an indication of the spread of the frequency but does not appear to have come into use as a measure of dispersion. [See also **Semi-interquartile Range**.]

***Intergraduated Values** Values which are subject either to *Cograduation or to *Contragraduation.

Internal Least Squares A name given by Hartley (1948) to a method of dealing with non-linear regression. He was led to consider a regression of a variate y on the accumulated sums of y as well as on independent variables x. This was called internal regression and the estimation of parameters of the regression equation by least squares was called internal least squares.

Internal Regression See **Internal Least Squares**.

Internal Variance See **Interclass Variance**.

Interpenetrating Samples (sub-samples) When two or more samples are taken from the same population by the same process of selection the samples are called interpenetrating samples. The samples may or may not be drawn independently, linked interpenetrating samples being an example of the latter. There may be different levels of interpenetration corresponding to different stages in a multi stage sampling scheme. Thus, in a two stage sampling scheme with village as the primary and household as the second stage unit, when the sample villages are distributed into two interpenetrating sub-samples we have interpenetration at the first stage only; but when the sample of households within every sample village is broken up into two interpenetrating sub-samples we have interpenetration at the second stage; and we can have interpenetration of a mixed type, e.g. the four sub-samples obtained by combining the two earlier types. Generally, the sub-samples are distinguished not merely by the act of separation into sub-samples but by definite differences in survey or processing features, e.g. when different parties are assigned to different sub-samples, or one sub-sample is taken up earlier in time than the others.

Interquartile Range The variate distance between the upper and lower **Quartiles**. This range contains one half of the total frequency and provides a simple measure of dispersion which is useful in descriptive statistics. [See also **Semi-interquartile Range, Quartile Deviation.**]

Interval Distribution A probability distribution formed by the intervals, or gaps, in time or distance between events or occurrences.

Interval Estimation The estimation of a population parameter by specifying a range of values bounded by an upper and a lower limit, within which the true value is asserted to lie, as distinct from **Point Estimation** which assigns a single value to the true value of the parameter. The unknown value of the population parameter is presumed to lie within the specified interval either on a stated proportion of occasions, under conditions of repeated sampling, or in some other probabilistic sense. The first of these two approaches is that of **Confidence Intervals** due to Neyman (1937), which regards the value of the population parameter as fixed and the limits to the intervals as random variables. A second approach is that of **Fiducial Limits** due to R. A. Fisher (1930) where the population parameter is regarded as having a 'fiducial probability' distribution which determines the limits.

Interviewer Bias In surveys of human populations by interview, bias in the responses or recorded information which is the direct result of the action of the interviewer. This bias may be due, among other things, to failure to contact the right persons; to the failure of the interviewer to establish proper relations with the informant, with the result that imperfect or inaccurate information is offered; or to systematic errors in recording the answers received from the respondent.

Intrablock See **Block**.

Intrablock Sub-group In the orthodox design of symmetrical factorial experiments in incomplete blocks, the treatments in the same block as the control may be considered to form a group in the mathematical sense. This approach to the matter provides a simple way of specifying the treatments in the different blocks.

For example, if there are n factors each at two levels, the 2^n treatments form a group in which the square of every element is the identity element. If there are 2^c blocks available $(c < n)$ there will be $2^c - 1$ interactions confounded forming, with the identity, a group of 2^c members. The 2^{n-c} treatments having an even number of letters (identifying treatments) in common with them form the intrablock sub-group.

Intraclass Correlation A measure of correlation within the members of certain natural groups or 'families'. For example, if a variate x is measured on a number of members, say k, of a family and it is desired to ascertain the correlation between the k members, a correlation table is constructed in which each of the $\frac{1}{2}k(k-1)$ pairs of members is represented twice, according to which member of the pair is taken as providing the first, and which the second, variate. For several families the correlation tables are superimposed, and a product moment correlation computed for the resulting bivariate table. This is the intraclass correlation. In practice it is not, in fact, necessary to construct the actual table in order to compute the coefficient.

The concept is closely allied to a variance ratio in variance analysis, which has largely superseded it.

Intraclass Variance In variance analysis, where the data are classified into groups, the total variation (sum of squares about the grand mean) may be expressed as the sum of two components, expressing variation among the means of groups and variation within groups. The latter is the sum of squares about group means pooled for the various groups, and an estimate of variance within groups based on it is known as within group or intraclass variance.

Intrinsic Accuracy The intrinsic accuracy of a distribution with frequency function $f(x, \theta)$ is defined as

$$ I = \int_{-\infty}^{\infty} \left(\frac{\partial \log f}{\partial \theta} \right)^2 f(x) \, dx. $$

Strictly speaking it should be related to the particular parameter θ entering into the differentiation.

Under very general conditions the variance of any unbiassed estimator t of θ based on an independent sample of n observations cannot be less than $1/nI$. The intrinsic accuracy thus gives a limit to the variance independent of particular estimators. The term is due to R. A. Fisher (1924). [See **Cramér-Rao Inequality**.]

Invariance This term is mostly used in statistics in its mathematical sense, namely to denote a property that is not changed by a particular transformation. For example, an orthogonal transformation of a set of independent Normal variates leaves the properties of independence and normality unaffected; they are invariant under the transformation.

A suggestion was formerly made to denote the recipro-

cal of the variance by the term 'invariance' but this, fortunately, seems to have passed out of usage.

Invariance Method A principle of estimation or hypothesis testing which requires an estimator or hypothesis under test to remain invariant if the data of the problem undergo some transformation.

Inverse Correlation See **Correlation Coefficient**.

Inverse Distribution This expression is used in several senses, for example:
- (*i*) to denote the distribution of the reciprocal of a variable, such as Tweedie's (1947, 1957) 'inverse Normal' which is the distribution of the reciprocal of the square root of a Normal (Gaussian) variable;
- (*ii*) in a sense to be avoided, to denote the distribution of sample numbers where sampling is continued until a predetermined number of successes has been attained, e.g. the 'inverse hypergeometric' is the distribution of sample sizes n required to obtain k successes in sampling without replacement from a finite population;
- (*iii*) an equally regrettable sense, to denote distributions (mainly discrete) where the frequencies are reciprocal quantities, see for example, Waring Distribution;
- (*iv*) by extension of (*ii*) to denote similar expressions in more general situations, see **Inverse Pólya Distribution**;
- (*v*) instead of expressing the distribution function $F(x)$ in terms of the variable x, to denote expansions of x in terms of $F(x)$.

It is probably safer to avoid the expression altogether.

Inverse Factorial Series Distribution See **Factorial Distribution**.

Inverse Gaussian Distribution See **Inverse Distribution**.

Inverse Hypergeometric Distribution See **Inverse Distribution**.

Inverse Pólya Distribution If, in a finite population of size $n_1 + n_2$ consisting of two components of size n_1 and n_2, units are drawn at random replacing each type with c additional units of the same kind until k units of the first kind are achieved, then the distribution of the number of units of the second kind has the Inverse Pólya Distribution.

Inverse Polynomial A form of polynomial proposed by Nelder (1966) for use in multi-factor response designs. The general form is $x/y = P_n(x)$ where $P_n(x)$ is a polynomial of order n with non-negative coefficients, y is the response and x the stimulus. This form of response function is bounded and its second order form is not necessarily symmetric. In this way it overcomes two disadvantages of the ordinary polynomial as a response function.

Inverse Probability The probability approach which endeavours to reason from observed events to the probabilities of the hypotheses which may explain them, as distinct from direct probability, which reasons deductively from given probabilities to the probabilities of contingent events. A principal theorem in this connection is **Bayes' Theorem**. The word 'inverse' usually means inverse in some logical relationship but is sometimes used in the sense of 'prior in time'. For example, in connection with the analysis of stochastic processes, it is sometimes desirable to consider the past history of a system rather than its future development. The past probabilities of transition which have given rise to the present state are thus sometimes called 'inverse'. This usage is comparatively rare and not to be recommended. [See **Projection**.]

Inverse Sampling A method of sampling which requires that drawings at random shall be continued until certain specified conditions dependent on the results of those drawings have been fulfilled, e.g. until a given number of individuals of specified type have emerged. In this sense it is allied to **Sequential Sampling**. The term is not a good one.

Inverse Serial Correlation A name sometimes, and rather regrettably, used to denote the process of attempting to determine from a set of serial correlations the series which generated them. The problem is in fact indeterminate as different series may give the same autocorrelations.

Inverse Sine Transformation A transformation of a variate x to a variate y by some such formula as $y = \sin^{-1}(ax+b)$ or, more generally, any formula involving the arc sin function. It is used particularly when x is a binomial proportion p, in order to stabilise the variance. [See **Stabilisation of Variance**.]

A similar transformation is also used involving the inverse sinh, principally for variates which have great skewness.

Inverse Tanh Transformation See **Fisher's Transformation**.

Inversion An inversion of two elements occurs when they are in the inverse order as compared with some standard given order. Hence if the given order of n elements is $a_1, a_2, a_3, ..., a_n$, an element a_n produces r inversions if it precedes r elements among $a_1, a_2, a_3, ..., a_{n-1}$ in an observed order. The number of inversions in a ranking forms the basis of several tests of independence in series and of certain rank correlation coefficients.

There is an entirely different usage of the word 'inversion' in the so called 'Inversion' Theorem which proves that a frequency distribution is uniquely determined by its **Characteristic Function**. If the latter is $\phi(t)$ and is thus defined in terms of the frequency function

$$\phi(t) = \int_{-\infty}^{\infty} e^{itx} f(x)\, dx$$

the inversion theorem states that

$$f(x) = \frac{1}{2\pi} \int_{-\infty}^{\infty} e^{-itx} \phi(t)\, dt.$$

There are similar but more complicated expressions connecting distribution functions.

Inverted Beta Distribution See **Beta Distribution**.

Inverted Dirichlet Distribution Just as the **Dirichlet Distribution** is the multivariate analogue of the **Beta Distribution**, the inverted beta distribution has its multivariate analogue in the inverted Dirichlet distribution (Tiao & Guttman, 1965). It may also be obtained from $k+1$ independent randomly distributed chi-squared variates $y_1, ..., y_{k+1}$ with $2\nu_1, ..., 2\nu_{k+1}$ degrees of freedom, where the ratios $x_i = y_i/y_{k+1}$, $(i = 1, ..., k)$ have the k-dimensional inverted Dirichlet distribution.

Irreducible Markov Chain A **Markov Chain** is said to be irreducible if all pairs of states of the chain communicate, so that the chain consists of exactly one **Communicating Class**.

Irregular Kollectiv An infinite series of a finite number of characteristics obeying the following laws:

(a) The proportion of a given characteristic in the first n terms tends to a limit as n increases.

(b) Any infinite subsequence of the Kollectiv designated by some independent rule possesses the same limiting property.

The Irregular Kollectiv was taken by von Mises as the basis of his frequency theory of probability.

Irwin Distribution See **Factorial Distribution**.

Ising-Stevens Distribution If there are n_1 and n_2 objects of two kinds C_1 and C_2 arranged at random in $n = n_1+n_2$ positions along a circle then the distribution (Stevens, 1939; Ising, 1925) of the number of runs of either object is

$$\Pr(x) = \binom{n_1}{x}\binom{n_2-1}{x-1} \Big/ \binom{n_1+n_2-1}{n_2}$$
$$= \binom{n_1-1}{x-1}\binom{n_2}{x} \Big/ \binom{n_1+n_2-1}{n_1}.$$

Isokurtosis See **Kurtosis**.

Isometric Chart A chart which attempts to depict three-dimensional material on a plane. It is a form of **Axonometric Chart** where the distances on the three axes are measured on an equal scale. There are various conventional combinations for the angles at which two out of three axes are drawn *vis-à-vis* the horizontal.

Isomorphism The logical equivalence of two theories, in the sense that one theory can be obtained from the other by a translation or reinterpretation of basic notions and symbols. For example the mathematical part of the theory of probability is isomorphic to a branch of the theory of additive set functions.

Isotropy A contingency table is isotropic when the associations in all tetrads of any four frequencies for two rows and two columns are of the same sign. An isotropic contingency table remains isotropic in whatever way the table may be condensed by grouping adjacent rows or columns, even if it be condensed to a **Fourfold Table**. The case of complete independence is a special case of isotropy since the association is zero for every tetrad of the **Independence Frequencies**. The expression was introduced by Yule in a discussion of contingency tables based on an underlying Normal distribution.

Isotype Method See **Pictogram**.

Iterated Logarithm, Law of If S_n is the number of successes in a sequence of n Bernoulli trials with a probability p of success at each trial, Khintchine's (1924) law of the iterated logarithm states that

$$\limsup_{n \to \infty} \frac{S_n - np}{(2npq \log \log n)^{\frac{1}{2}}} = 1$$

where $q = 1 - p$. It is possible to formulate stronger theorems in terms of a more general sequence of variates.

J-shaped Distribution An extremely asymmetrical frequency distribution with the maximum frequency at the initial (or final) frequency group and a declining or increasing frequency elsewhere. The shape of this distribution roughly resembles the letter J or its reverse. Among theoretical curves referred to as J-shaped are the **Pareto** and certain of the **Pearson** system of frequency curves.

'Jacknife' A method proposed by Tukey (1958), extending an idea of Quenouille (1949), which reduces bias in estimation and provides approximate confidence intervals in cases where ordinary distribution theory proves difficult. It is based upon the idea that if t_n is a statistic biassed to order $1/n$, t_{n-1} is the mean of the n statistics derivable by omitting one member of the sample, then $nt_n - (n-1)t_{n-1}$ is biassed only to order $1/n^2$.

Jensen's Inequality If X is a random variable defined on an interval I, and g is any convex function on I, then the inequality

$$g(\mathscr{E}[x]) \leqslant \mathscr{E}[g(x)]$$

is known as Jensen's Inequality.

Jírina Sequential Procedure A procedure for the sequential estimation of tolerance limits due to Jírina (1952). Under certain conditions it yields a larger probability for a stated coverage than the standard procedure.

Jittered Sampling A term sometimes used in the analysis of time series (Parzen, 1967) to denote the sampling of a continuous series where the intervals between points of observation are the values of a random variable.

John's Cyclic Incomplete Block Designs A form of incomplete block experiment proposed by John (1966) where the t treatments in blocks of k treatments per block can be divided into k cyclic sets of t blocks each. This experiment design was shown (Clatworthy, 1967) to be equivalent to the balanced **Incomplete Block Design** or the partially balanced incomplete block with m **Associate Classes**.

Johnson's System A system of frequency curves based upon transformations of the variable. The original designation by Edgeworth (1898) was Method of Translation. The idea was extended by Johnson (1949) and developed in later papers. [See also S_B, and S_U, **Distributions**.]

Joint Distribution The distribution of two or more variates. The term is equivalent to multivariate distribution and is especially used of two variates.

Joint moment See **Product moment**.

Joint Prediction Intervals The joint interval (Lieberman, 1961) which is appropriate for some $k > 1$ new predictions of the dependent variate at $k > 1$ different values of the independent variates, based upon the original sample of observations.

Joint Regression The classical regression model assumes that the dependent variate is a function, often linear, of a set of independent variables. If the linear model is inadequate, attempts are sometimes made to allow for cross-product terms in the independent variables, e.g. if there are two independent variables x_1 and x_2 the joint regression equation would be:

$$y = a_1 x_1 + a_2 x_2 + a_3 (x_1 x_2)$$

where the third term allows for any interaction between the original variables. Such regression is sometimes said to be 'joint' but the usage does not seem very desirable.

Joint Sufficiency Estimators $t_1, t_2, ..., t_k$ are said to be jointly sufficient for parameters $\theta_1, \theta_2, ..., \theta_l$ if the likelihood function can be expressed as:

$L(x_1 ... x_n; \theta_1 ... \theta_l)$
$$= L_1(t_1, ..., t_k; \theta_1, ..., \theta_l) \, L_2(x_1, ..., x_n)$$

where L_2 does not depend on the $\theta_1 ... \theta_l$, although it may depend on other parameters of the system.

Judgment Sample In the terminology of Deming (1947) a judgment sample is, in general, any sample which is not a **Probability Sample**. The dichotomy is not, perhaps, perfect; for example, the visible stars of the sky are not a 'probability sample' of the matter in the universe, nor does it appear proper to describe them as a 'judgment sample'. It seems best to confine this term to the case where some element of human judgment enters directly into the selection of the sample.

Jump Statistic A statistic used to assess the reality of an apparent jump in the mean of a stationary time series. Jowett & Wright (1959) proposed using the mean semi-squared difference (jump) between the means of the last $\alpha(\leqslant \frac{1}{2}n)$ terms of one section and the first α terms of the following section.

Just Identified Model A term which is sometimes used to indicate a model which is completely identifiable, but not overidentified. [See **Identification**.]

'k'-Class Estimator In econometric regression analysis, a general class of estimators proposed by Theil (1958) for the parameters in $By + \Gamma z = u$ where y are the endogenous variables, z are exogenous and u are random errors. The method proceeds by least squares in stages. If $k = 0$ we have the ordinary least squares form applied to individual equations, which gives the consistent estimators. The value $k = 1$ gives the two stage least squares. The **Limited Information Estimator** is also a member of the k-class, the value of k being the smallest latent root of a determinant dependent on the coefficients in the **Reduced Form**.

k-samples Problem The problem of determining whether, given k samples, one from each of k populations, the parent populations are different. The usual tests developed in this connection are homogeneity tests for means or variances, but an infinite number of tests are possible: a summary is given in Bradley (1968). [See also **Mosteller's k-sample Slippage Test**.]

k-statistics A set of symmetric functions of sample values proposed by R. A. Fisher (1928). The univariate k-statistic of order r is defined as the statistic whose mean value is the rth cumulant, κ_r, of the parent population. The statistics have semi-invariant properties and their sampling cumulants can be obtained directly by combinatorial methods.

Similarly the multivariate k-statistic, say $k(r, s, ..., v)$ is defined as the symmetric function of observations whose mean value is the corresponding cumulant $\kappa_{rs} \ldots _v$. It is more usually written $K_{rs} \ldots _v$.

Another generalisation is due to Tukey, who defines a statistic, say $k\{r, s, ..., v\}$ as the symmetric function whose mean value is $\kappa_r \kappa_s \ldots \kappa_v$. This statistic is also often written $k_{rs} \ldots _v$ and should not be confused with the multivariate k-statistic. [See also **Polykays. Bipolykays** and **Generalised Polykays**.]

K-test A distribution-free test for a trend in a series proposed by Mann (1945). If the equally spaced series is $x_1, ..., x_n$ and a decreasing trend exists each term will tend to be greater than succeeding terms. The smallest interval K for which $x_i < x_{i+K}$, $i = 1, 2, ..., n-K$ is taken as the test-statistic, the null hypothesis that no trend exists being rejected if K is small. A similar test, of course, exists for increasing trend.

Kapteyn's Distribution A generalisation of the Normal distribution using the method of **Kapteyn's Transformation** of the form

$$d\Phi_x = \frac{1}{\sigma\sqrt{(2\pi)}} \exp\left\{-\frac{1}{2\sigma^2}[G(x)-\mu]^2\right\} \left|\frac{dG(x)}{dx}\right| dx.$$

If $G(x) = x$ this reduces to the Normal distribution.

Kapteyn's Transformation A method proposed by Kapteyn (1903) and Kapteyn and van Uven (1916) for transforming the variate x of a skew frequency function into a variate z which is Normally distributed.

Kärber's Method A method for estimating the median effective dose of a stimulus from data on **Quantal Responses**, advanced by Kärber in 1931. Essentially the same method was proposed by Spearman in 1908 so that the name **Spearman-Kärber Method** is to be preferred.

Kendall's 'S' Score In rank correlation analysis, Kendall (1955) defined a score, for a pair of rankings of n items, as $+1$ if any two are ranked in the same order by the two rankings, -1 if in opposite order and zero if tied in either or both rankings. The total score S is the algebraic sum of the $\frac{1}{2}n(n-1)$ contributions from pairs of items.

Kendall's Tau (τ) A coefficient of rank correlation based on the number of **Inversions** in one ranking as compared with another. It was proposed by Kendall in 1938 as a rank correlation coefficient independent of the nature of underlying variate distributions, but had earlier been considered by Greiner (1909) and by Esscher (1924) as a statistic for the estimation of the correlation parameter in a bivariate Normal distribution.

Kendall's Terminology A system of describing particular cases in queueing theory developed by D. G. Kendall (1953). It is based upon initial letters of key words associated with the input, queue discipline and service mechanism of a congestion situation. For example, a random, Poissonian or Markov input is denoted by M and the number of service channels as 1 or $s (= 1, 2, ...)$.

Kesten's Process A variation of the **Robbins-Munro Stochastic Approximation Process** intended to reduce the bias related to the starting point. The step constant is reduced only when there is evidence that doses are straddling the **Medium Lethal Dose**.

Khintchine's Theorem Let $x_1, x_2, ...$ be a sequence of independent and identically distributed variates each with a finite mean μ. Then the variable $\bar{x} = \frac{1}{n}\sum_{i=1}^{n} x_i$ converges in probability to μ as n tends to infinity. This theorem was first proved rigorously by Khintchine in 1929. [See **Large Numbers, Law of.**]

Kiefer-Wolfowitz Process A **Stochastic Approximation Procedure** proposed by Kiefer & Wolfowitz (1952) for estimating the maximum of a regression function. This concept is related to the **Robbins-Munro Process** and has also been developed by other writers.

Knut-Vik Square A form of experimental design attributed to the Norwegian, Knut Vik. It is also known as a 'Knight's-move' square from the association of ideas between the construction of the square design and the move of the chess piece. The design may be illustrated by a 5×5 square in which it is desired to have each of five treatments once and once only in each row and column. The rows are formed by cyclic permutations of A, B, C, D, E moving forward two places instead of one:

A	B	C	D	E
D	E	A	B	C
B	C	D	E	A
E	A	B	C	D
C	D	E	A	B

The design is thus a Latin square of a particular type.

Kollectiv This word occurs in English with a more specialised meaning than 'aggregate', which is its literal translation from the German. It denotes a population of objects in which each unit bears one of a finite number of identifiable characteristics. In particular, von Mises (1919) used the infinite sequence of such characteristics which bears no systematic properties—the so-called **Irregular Kollectiv**—as the central concept of his theory of probability. [See also **Frequency Theory of Probability.**]

Kolmogorov Axioms A set of axioms given by Kolmogorov (1933) for the foundation of probability theory in terms of set- and measure theory. They form the starting point of most modern expositions of a mathematical theory of probability.

Kolmogorov's Equations Two systems of differential equations first derived by Kolmogorov (1931) each of which often uniquely determines a system of transition probabilities for a Markov process. They are known as the forward and backward equations.

Kolmogorov's Inequality A generalisation of the **Bienaymé-Tchebychev Inequality.** Let $x_i(i = 1, 2, ..., n)$ be n mutually independent variates with means a_i and variances v_i. Let $A_k = \sum_{i=1}^{k} a_i$ and $V_k = \sum_{i=1}^{k} v_i$, and let $S_k = \sum_{i=1}^{k} x_i$. Then for $t > 0$ the probability that the inequalities

$$|S_k - A_k| < t V_k$$

are simultaneously realised is at least equal to $1 - 1/t^2$.

Kolmogorov-Smirnov Test A **Significance Test** proposed by Kolmogorov (1933) and developed by Smirnov (1939) and later writers. If $F(x)$ is the population distribution function and $S_n(x)$ the observed distribution (step) function of the sample, then the test makes use of the statistic $d = \max\{F(x) - S_n(x)\}$. This has a distribution which is independent of $F(x)$ provided that the latter is a continuous distribution function. The method can be extended to discontinuous distributions by modifying the associated probability statement from an exact to a minimum level.

The test may be used as one of 'goodness of fit' and the d-statistic may also be used to set confidence limits to an unknown probability distribution. It has been extended by Smirnov to test the homogeneity of two distribution functions on the basis of a sample from each. The Smirnov test of homogeneity depends on the greatest difference of the two observed distribution (step-) functions and is distribution-free. [See also **Cramér-von Mises Test.**]

Kolmogorov's Theorem Two theorems stated by Kolmogorov (1928-30) which give the conditions under which the **Strong Law of Large Numbers** holds:

(*i*) when the random variables are independent;

(*ii*) when the random variables are independent and identically distributed.

Konyus Conditions In a paper published in Russian in 1924 (English translation, 1939) Konyus advanced the idea that a true index of the cost of living is the ratio of money expenditures which will leave the standard of living unchanged between two situations which differ only with respect to prices. Konyus showed that under some conditions **Laspeyres' Index** provided an upper limit and under others **Paasche's Index** provided a lower limit, but that the two did not provide simultaneously upper and lower limits. He then discussed the problems of finding conditions which would ensure the approximate equality of the two standards of living; and of setting limits to changes in the 'true' cost of living index. Certain conditions arising in this investigation are known as Konyus conditions.

Konyus Index Number The name given to a class of index number rather than to one specific formula. It is an index of prices based on quantities, the 'budget' of the consumer, which are optimal in some field of consumer preference. If the indifference level for consumers is specified in terms of an optimal budget at the base prices, the index is called Laspeyres-Konyus; if by an optimal budget at prices in the period under comparison with the base period, a Paasche-Konyus index. [See **Konyus Conditions, Laspeyres' Index, Paasche Index.**]

Kronecker Product of Design It can be shown that an experiment design D uniquely determines its **Incidence**

Matrix and vice versa. Thus the Kronecker product (in the sense of matrix theory) of two incidence matrices produces a new design matrix and can be used to determine the existence of certain classes of design.

Kronecker Product of Matrices The product $A \times B$ of an $m \times m$ matrix A and an $n \times n$ matrix B is the $mn \times mn$ matrix whose elements are the products of terms, one from A and one from B.

Kruskal Statistic A rank order statistic for the k-sample problem

$$H = \{12/N(N+1)\} \Sigma n_i \{\bar{R}_i - (N+1)/2\}^2$$

where $i = 1, 2, ..., k$, n_i is the number in the ith sample, N is the total Σn_i and $\bar{R}_i$ is the average rank sum in the ith ranking. This statistic, proposed by Kruskal (1952), has been generalised by Basu (1967) for a right censored sample of r observations.

Kuder-Richardson Formula A formula for estimating the **Reliability Coefficient** of a test which endeavours to overcome the disadvantages of formulae associated with **Split Half Methods**. There are numerous versions of the formula, which may be written

$$r_{tt} = \left(\frac{n}{n-1}\right) \left(\frac{\sigma_t^2 - \Sigma pq}{\sigma_t^2}\right)$$

where n is the number of items in the test, p is the proportion passing or satisfactorily responding to an item, $q = 1 - p$ and σ_t^2 is the overall variance of scores. [See also **Spearman-Brown Formula**.]

Kuiper Statistic A goodness-of-fit test statistic similar to the **Kolmogorov-Smirnov** statistic, for the single-sample problem it is

$$V_n = n^{\frac{1}{2}}[\text{Sup}_x\{F_n(x) - F(x)\} - \text{Inf}_x\{F_n(x) - F(x)\}]$$

and for the two-sample problem

$$V_{m,n} = [mn/(m+n)]^{\frac{1}{2}}[\text{Sup}_x\{F_n(x) - G_m(x)\} - \text{Inf}_x\{F_n(x) - G_m(x)\}].$$

Kullbach-Liebler Information Number A formulation proposed by Kullbach & Liebler (1951) based upon the Shannon/Wiener concept of **Information** which is different from Fisher's **Intrinsic Accuracy**. The information number $I_x(1:2)$ denotes the mean information per observation for discriminating between hypotheses H_1 and H_2 when H_1 is true:

$$I_x(1:2) = \int_{-\infty}^{\infty} f_1(x) \log \frac{f_1(x)}{f_2(x)} d\psi(x)$$

where $\psi(x)$ is some common measure.

Kurtic Curve See Regression.

Kurtosis A term used to describe the extent to which a unimodal frequency curve is 'peaked'; that is to say, the extent of the relative steepness of ascent in the neighbourhood of the mode. The term was introduced by Karl Pearson in 1906; and he proposed as a measure of

kurtosis the moment-ratio $\beta_2(= \mu_4/\mu_2^2)$. It is doubtful, however, whether any single ratio can adequately measure the quality of 'peakedness'.

In a bivariate frequency array, if the different arrays corresponding to one variate have different degrees of kurtosis they are said to be heterokurtic or allokurtic, and if they have the same degree, homokurtic or isokurtic.

If the moment ratio is adopted as a measure of kurtosis, the value it assumes for a Normal distribution, namely 3, is taken as a standard. Curves for which the ratio is less than, equal to or greater than 3 are known respectively as platykurtic, mesokurtic and leptokurtic.

A curve showing the variation in the kurtosis of one variate against values of the other in bivariate variation is called a kurtic curve. [See also **Clisy, Scedasticity**.]

L-Statistics A class of rank order statistics proposed by Dwass (1956) in connection with rank order tests in the two sample problem; this must not be confused with the Neyman & Pearson L-tests for homogeneity of sample variances. The L-statistic is defined as $\Sigma a_{N_i} b_{N R_i}$ where the a_{N_i} are $(n/mN)^{\frac{1}{2}} - i = 1, ..., m$, and $i = m+1, ..., N$ being the two samples under test—the b_N another set of constants and $R_1 ... R_N$ the ranks of the N random variables.

L-Tests Tests proposed by Neyman & E. S. Pearson (1933) for testing the homogeneity of a set of sample variances. The tests, which are based on likelihood ratios in normal variation, vary according to the precise type of hypothesis under test; for example, if the hypothesis is that the parent variances of k samples are all equal but means may differ the test statistic is

$$L_1 = \prod_{t=1}^{k} (s_t^2)^{n_t/N} \Big/ \frac{1}{N} \sum_{t=1}^{k} (n_t s_t^2)$$

where N is the total number of observations, n_t the number of observations of the tth sample and s_t^2 the variance of the tth sample. Small values of the test function lead to a rejection of the hypothesis.

Λ-criterion (λ-criterion) An alternative name for a criterion, in hypothesis testing, based on the likelihood ratio; especially the ratio of dispersion determinants given by **Wilk's Criterion**.

l-Statistics An alternative designation, proposed by Kendall & Stuart (1958), for the **Polykays**.

L_2 Association Scheme A partially balanced **Incomplete Block Design** with two **Associate Classes** is said to have an L_2 association schemes if the number of treatments (s^2) can be arranged in a square such that any two treatments in the same row or column are first associates: pairs of treatments not in the same row and column are second associates.

Ladder Indices A ladder index is the time point (epoch) in a random walk at which a **Ladder Variable** begins to occur.

Ladder Variable If, in a random walk, the sum of the mutually independent variables at time point (epoch) n is S_n then, if $S_n > S_j$ ($j = 0, 1, 2, \ldots, n-1$), the variable $\bar{X}_n$ is an ascending ladder variable. Descending ladder variables are similarly defined where $S_n < S_j$.

Lag An event occurring at time $t + k (k > 0)$ is said to lag behind event occurring at time t, the extent of the lag being k. An event occurring k time units before another may be regarded as having a negative lag.

By extension, two time series u_t and v_t are said to be lagged in relation to each other or one is said to lag behind the other if the values of one are associated with lagged values of the other; for example, if the production of a commodity at time t, q_t is regarded as dependent on the price of that commodity at a previous time $t - k$, p_{t-k}, the series q_t is said to be lagged with respect to p_{t-k}. An equation connecting them, such as $q_t = ap_{t-k} + \beta$ is said to contain a lag.

Lag Correlation The correlation between two series where one of the series has a **Lag** with reference to the other. [See **Lag Covariance**.]

Lag Covariance The first **Product Moment** between two series, one of which is lagged in relation to the other. For example, if u_t and v_t are two series defined at $t = 1, 2, \ldots, n$, the lag covariance of u_t and v_t of order $k > 0$ is

$$\frac{1}{n-k} \sum_1^{n-k} (u_t - \bar{u}_1)(v_{t+k} - \bar{v}_2)$$

where

$$\bar{u}_1 = \frac{1}{n-k} \sum_1^{n-k} u_t \quad \text{and} \quad \bar{v}_2 = \frac{1}{n-k} \sum_1^{n-k} v_{t+k}.$$

In this convention the lag k refers to the extent to which the second series lags behind the first. The lag covariance of order $-k (k > 0)$ is

$$\sum_{t=1}^{n-k} (u_{t+k} - \bar{u}_2)(v_t - \bar{v}_1),$$

where

$$\bar{u}_2 = \frac{1}{n-k} \sum_1^{n-k} u_{t+k} \quad \text{and} \quad \bar{v}_1 = \frac{1}{n-k} \sum_1^{n-k} v_t,$$

and is not in general the same as the lag covariance of order k. Where there is likely to be confusion it is desirable to specify which series leads the other, e.g. by some such phase as 'the covariance of v_t lagging behind u_t by k'.

Lag Hysteresis The word 'hysteresis' was taken from the theory of electromagnetism and introduced into econometrics by C. F. Roos in 1925. Later Jones (1937) distinguished between lag hysteresis and skew hysteresis. Lag hysteresis in econometrics is confined to cases of sinusoidal variation in the two variables: skew hysteresis refers to the more realistic case in which the oscillatory movements are asymmetrical. The value of the concept where oscillatory movements are not cyclical is unknown.

Lag Regression A regression in which the values of the dependent variate and one at least of the independent variables are lagged in relation to each other.

Laguerre Polynomials Polynomials, due to Laguerre, defined by

$$L_n(x) = e^x D^n(x^n e^{-n})/n!$$

where $D = d/dx$. They have important orthogonal properties. The polynomials up to the third order are: $L_0(x) = 1$; $L_1(x) = 1 - x$; $L_2(x) = (x^2 - 4x + 2)/2!$; $L_3(x) = (-x^3 + 9x^2 - 18x + 6)/3!$

Lambdagram A graphic device proposed by Yule (1945) in connection with the analysis of time series. For a series of values $x_1, x_2, \ldots, x_n$ the lambdagram consists of a coefficient λ_n plotted as ordinate against the sample size n. The coefficient is given by

$$\lambda_n = \frac{n-1}{n} \sum_{j=1}^{n} r_j = (n-1)\bar{r}_n$$

where r_j is the jth serial correlation and $\bar{r}_n$ is the mean of the first n serial correlations. λ_n is related to the variance of the mean in sampling from a series whose items are internally correlated:

$$\text{var } \bar{x} = \text{var } x [1 + (n-1)\bar{r}_n].$$

The coefficient λ_n may be interpreted as an index of the divergence of the samples from samples of n random observations, i.e. of the way in which the n values of the samples are linked together.

Lancaster's Partition of Chi-Squares A total chi-square value is computed to provide a gross measure of the extent to which the cell frequencies depart from expectation. This value is partitioned into additive components to show how much is attributable to individual classifications and how much is due to both first order and second order interaction. This procedure can be extended to cover contingency tables involving more than three ways of classification where higher order interactions occur. A special merit of this procedure is that it can be used with theoretical parameters or with parameters estimated from the data.

Laplace Distribution A frequency distribution of the double exponential type, expressible in the form

$$dF = \frac{1}{2\sigma} \exp\left\{-\frac{|x-m|}{\sigma}\right\} dx, \quad -\infty \leq x \leq \infty; \sigma > 0.$$

It is sometimes known as the First Law of Laplace, in contradistinction to the Second Law which is the same as the **Normal Distribution**.

Laplace Law of Succession A rule given by Laplace (1812) concerning the probability of events in further trials when certain trials have been made. If in n previous trials m have yielded an event E, the probability, according to the succession rule, that E happens on the next trial

G
81

is $(m+1)/(n+2)$. The rule is based on **Bayes' Postulate** for unknown probabilities and has been subject to much dispute and undiscriminating application.

Laplace-Lévy Theorem A name sometimes given to the **Central Limit Theorem**, which was known to Laplace in its essentials but was not proved rigorously under necessary and sufficient conditions until the beginning of the 20th century.

Laplace's Theorem This limit theorem, of which the **Bernoulli Theorem** is a corollary, states that if there are n independent trials, in each of which the probability of an event is p, and if this event occurs k times, then

$$\Pr\left\{z_1 \leqslant \frac{k-np}{\sqrt{(npq)}} \leqslant z_2\right\} \to \frac{1}{\sqrt{(2\pi)}}\int_{z_1}^{z_2} e^{-\frac{1}{2}z^2}\,dz$$

as $n \to \infty$ whatever the numbers z_1 and z_2.

Generally speaking, the theorem states that the number of successes k in n trials is Normally distributed for large n.

Laplace Transform If a function $g(t)$ is related to a second function $f(x)$ by the equation

$$g(t) = \int_0^\infty e^{-tx} f(x)\,dx$$

then $g(t)$ is the Laplace transform of $f(x)$.

In statistical theory it is more customary to use the **Fourier Transform**, which has certain advantages over the Laplace form; e.g. if $f(x)$ is a frequency function the Fourier transform always exists whereas the Laplace transform may not do so for real t.

Large Numbers, Law of A general form of this fundamental law relating to random variables may be stated as follows:

If x_k is a sequence of mutually independent variates with a common distribution and if the expectation $\mu = \mathscr{E}(x_k)$ exists, then for every $\epsilon > 0$ as $n \to \infty$ the probability

$$\Pr\left\{\left|\frac{X_1 + \ldots + X_n}{n} - \mu\right| > \epsilon\right\} \to 0.$$

In this form, the law was first proved by Khintchine (1929), but less general forms were known from the time of James Bernoulli onwards. The above form is the so-called 'weak' law. [See also **Strong Law of Large Numbers**.]

Laspeyres' Index A form of index number due to Laspeyres (1871). If the prices of a set of commodities in a base period are $p_o, p_o', p_o'', \ldots$ and those in a given period $p_n, p_n', p_n'', \ldots$; and if q_o, q_o', q_o'' are the quantities sold in the base period, the Laspeyres' price index number is written

$$I_{on} = \frac{\Sigma(p_n q_o)}{\Sigma(p_o q_o)}$$

where the summation takes place over commodities. In short, the prices are weighted by quantities in the base period.

Generally, an index number of the above form is said to be of the Laspeyres' type even when p and q do not relate to prices or quantities; the characteristic feature being that the weights relate to the base period, as contrasted with the **Paasche Index** in which they relate to the given period. [See **Lowe Index, Palgrave Index, Crossed Weights**.]

Laspeyres-Konyus Index See **Konyus Conditions**.

Latent Root (Vector) See **Characteristic Root**.

Latent Structure In general this phrase refers to a structure expressed in terms of variates or variables which are 'latent' in the sense of not being directly observable. Certain econometric relations e.g., in terms of 'utility' are of this type, and the models used in **Factor Analysis** as used in psychology also may be regarded as a kind of latent structure.

More recently the term has been applied to studies of attitudes by questionnaire (Lazarsfeld, 1950). The observed replies to questionnaires are expressed in terms of 'latent' distributions of attitude.

Latent Variable A variable which is unobservable but is supposed to enter into the structure of a system under study, such as demand in economics or the 'general' factor in psychology. Unobservable quantities such as errors are not usually described as latent.

Latin Cube An extension of the principle of the **Latin Square** so that the layers, and sections in two directions, are effectively Latin Squares.

Latin Rectangle An experimental design derived from the **Latin Square**. It consists of a Latin square with one or more adjacent rows or columns added or omitted. This particular design is one form of the **Incomplete Latin Square** or **Youden Square**.

Latin Square One of the basic statistical designs for experiments which aim at removing from the experimental error the variation from two sources, which may be identified with the rows and columns of the square. In such a design the allocation of k experimental treatments in the cells of a k by k (Latin) square is such that each treatment occurs exactly once in each row or column. A specimen design for a 5×5 square with five treatments, A, B, C, D and E is as follows:

A	B	C	D	E
B	A	E	C	D
C	D	A	E	B
D	E	B	A	C
E	C	D	B	A

The earliest recorded discussion of the Latin square was given by Euler (1782) but it occurs in puzzles at a much earlier date. Its introduction into experimental design is due to R. A. Fisher.

Lattice Design   See **Quasi-factorial Design, Square Lattice.**

Lattice Sampling   A method of sampling in which sub-strata are selected, for the sampling of individuals, according to some pattern analogous to the allocation of treatments on a lattice experimental design. For example, if there are two criteria of stratification, each p-fold, so that there are p^2 sub-strata, it is possible to choose p sub-strata so that none occurs in more than one 'row' or 'column' of the array representing the p^2 possible sub-strata; in short, in the manner of a **Latin Square.** Similar schemes are possible for three- or more-way classification. Various schemes of the lattice type are known under the name of 'deep stratification'.

Laurent Process   A stochastic process for which the covariance generating function is a Laurent series; the important case is that in which the process is generated by a moving average of a random process.

Least Favourable Distribution   Given the composite hypothesis $H_0:f_\theta$, $\theta\epsilon\omega$, to be tested against the simple alternative H, the general procedure is to reduce the composite hypothesis H_0 to a simple one $H_{0\lambda}$, where λ is a probability distribution function over ω. The maximum power β_2 that can be obtained against H_1 is that of the most powerful test of $H_{0\lambda}$ against H_1. The probability distribution λ is said to be least favourable (for a given level α) if, for all λ' the inequality $\beta_\lambda \leqslant \beta_{\lambda'}$ holds.

Least Significant Difference Test   A test for comparing mean values arising in analysis of variance. It is an extension of the standard t-test for the difference between two specified mean values. Because the tests between pairs are not independent the error rate is difficult to assess exactly. [See also **Multiple Comparisons.**]

Least Squares Estimator   An estimator obtained by the **Method of Least Squares.**

Least Squares Method   A technique of estimation by which the quantities under estimate are determined by minimising a certain quadratic form in the observations and those quantities. In general, the method may be regarded as possessing an empirical justification, in that the process of minimisation gives an optimum fit of observation to theoretical models; but for certain more restricted cases it has demonstrable optimum properties. The two cases of statistical importance are (a) where linear unbiased estimators with minimal variance are sought (see **Gauss-Markov Theorem**); (b) where the model involves errors which are normally distributed and least-squares estimation becomes equivalent to **Maximum Likelihood Estimation**; (c) where residuals are auto-correlated and the method provides estimates that are linear, unbiased, consistent and distribution free.

Least Variance Difference Method   This is a method of generalised linear estimation of structural parameters. The technique is formulated exactly as Aitken's (1935) generalisation of Gauss's method of treating independent observations of unequal precision to the case of inter-dependent observations.

Legendre Polynomials   A set of polynomials due to Legendre (1785). They are the coefficients of $P_n(x)$ in the expansion
$$(1-2xh-h^2)^{-\frac{1}{2}} = \sum_{n=0}^{\infty} P_n(x)h^n.$$
In particular
$$P_0(x) = 1, P_1(x) = x,$$
$$P_2(x) = \tfrac{1}{2}(3x^2-1), P_3(x) = \tfrac{1}{2}(5x^3-3x).$$
They have important orthogonal properties and can be transformed linearly so that they are **Orthonormal** on the interval $(0, 1)$.

Legit   A transform of quantal data used in genetics. A proportion p, regarded as a gene ratio, in certain circumstances is connected with a variate x by the differential equation
$$\frac{\partial^2 p}{\partial x^2} = 4pqx$$
where $q = 1-p$, and with the boundary conditions that $p = \frac{1}{2}$, $x = 0$ and $p\rightarrow0$, $x\rightarrow\infty$. x as a function of p is called a legit.

Lehmann Alternatives   A class of nonparametric alternative hypotheses used by Lehmann (1953) in developing power functions of **Rank Order Tests.**

Lehmann's Test   A two sample nonparametric test for variances of the **Wilcoxon-Mann-Whitney** type proposed by Lehmann (1951). It is based upon all possible differences between the observations in the two samples.

Leptokurtosis   See **Kurtosis.**

Level Map   A graph showing curves in an (x, y) plane corresponding to various values of the constant k in a defining equation $f(x, y) = k$. It is similar to the contours of equal height on a geographical map.

Level of a Factor   See **Factorial Experiment.**

Level of Interpenetration   See **Interpenetrating Samples.**

Level of Significance   Many statistical tests of hypotheses depend on the use of the probability distributions of a statistic t chosen for the purpose of the particular test. When the hypothesis is true this distribution has a known form, at least approximately, and the probability $\Pr(t\geqslant t_1)$ or $\Pr(t\leqslant t_0)$ can be determined for assigned t_0 or t_1. The acceptability of the hypothesis is usually discussed, *inter alia*, in terms of the values of t observed; if they have a small probability, in the sense of falling outside the range t_0 to t_1, $\Pr(t\geqslant t_1)$ and $\Pr(t\leqslant t_0)$ small, the hypothesis is rejected. The probabilities $\Pr(t\geqslant t_1)$ and

$\Pr(t \leqslant t_0)$ are called levels of significance and are usually expressed as percentages, e.g. 5 per cent. The actual values are, of course, arbitrary, but popular values are 5, 1 and 0·1 per cent. Thus, for example, the expression 't falls above the 5 per cent level of significance' means that the observed value of t is greater than t_1 where the probability of all values greater than t_1 is 0·05; t_1 is called the upper 5 per cent significance point, and similarly for the lower significance point t_0.

Lévy-Cramér Theorem This is the converse of the **First Limit Theorem**, proved simultaneously by Lévy and Cramér about 1925. Let $\{\phi_n(t)\}$ be a sequence of characteristic functions corresponding to a sequence of distribution functions $\{F_n(x)\}$. Then if $\phi_n(t)$ tends to $\phi(t)$ uniformly in some finite t-interval, $\{F_n(x)\}$ tends to a distribution function $F(x)$ and $\phi(t)$ is the characteristic function of $F(x)$.

Lévy-Pareto Distribution The work of Lévy in connection with stable distributions showed that the **Pareto Distribution** can be derived from a version of the Central Limit Theorem in which the individual random variables do not have finite variance.

Lévy's Theorem A synonym for the **First Limit Theorem** which was first proved rigorously by P. Lévy and H. Cramér independently about 1925.

Lexis Ratio This ratio provides a measure for distinguishing the three kinds of variation in sampling for attributes: Bernoullian, Lexian and Poissonian.

If k samples of $n_1, n_2, \ldots, n_k$ members bear observed proportions of the attribute $p_1, p_2, \ldots, p_k$ the Lexis Ratio Q is defined by

$$Q^2 = \frac{\sum\limits_{j=1}^{k} n_j (p_j - p)^2}{(k-1)pq}$$

where p is the proportion of the attribute in all samples together and $q = 1-p$. If the Lexis ratio is equal to unity within sampling limits the sampling is regarded as Bernoullian; if greater than unity, as of Lexian type; if less than unity, as Poissonian. They are also said to possess normal, hypernormal (or supernormal) and subnormal dispersion. In Italian usage the terms hyperbinomial, binomial and hypobinomial are sometimes employed.

Lexis Theory A general term to describe the theory of sampling for attributes (see **Lexis Ratio**) developed by Lexis (1879). In modern terminology it is part of the analysis of variance applied to dichotomised material.

Lexis Variation A type of sampling variation considered by Lexis (1877). On each of k occasions let n members be drawn at random, and let the probability of success be the same for any member of a set, but vary from one occasion to another, the probabilities being $p_1, p_2, \ldots, p_k$. The mean proportional frequency of occurrence of successes over all occasions is $p = \sum\limits_{i=1}^{k} p_i/k$ and the variance of the number of successes is $npq + n(n-1)$ var p_i where $q = 1-p$ and var p_i is the variance of p_i in the k sets. If all the p_i are equal this reduces to **Bernoulli Variation**. In other cases the Lexian is larger than the Bernoullian variance. This effect is encountered in sampling from non-homogeneous strata. The dispersion is said to be supernormal or hypernormal. [See also **Lexis Ratio, Poisson Variation**.]

Liapounov's Inequality An inequality due to Liapounov (1901) concerning the relations between the **Absolute Moments** of a frequency distribution. If $a \geqslant b \geqslant c \geqslant 0$ are three real numbers and ν_a, ν_b, ν_c the absolute moments of orders a, b and c for some arbitrary distribution, then:

$$\nu_b^{a-c} \leqslant \nu_c^{a-b} \nu_a^{b-c}.$$

Liapounov's Theorem A form of the **Central Limit Theorem** which assumes the existence of absolute third moments. If x_j ($j = 1, 2, \ldots, n$) is a sequence of independent variates with means m_j, variances σ^2 and absolute third mean-moments ρ_j^3 the sum $\sum\limits_{j=1}^{n} x_j$ is asymptotically Normal provided that $\lim\limits_{n \to \infty} \rho/\sigma = 0$ where

$$\rho^3 = \sum\limits_{j=1}^{n} \rho_j{}^3 \text{ and } \sigma^2 = \sum\limits_{j=1}^{n} \sigma_j{}^2.$$

Life Table A table showing the number of persons who, of a given number born or living at a specified age, live to attain successive higher ages, together with the numbers who die in the intervals.

Likelihood If the distribution function of continuous variates $x_1, \ldots, x_n$, dependent on parameters $\theta_1, \ldots, \theta_k$, is expressed as

$dF = f(x_1, x_2, \ldots, x_n; \theta_1, \theta_2, \ldots, \theta_k)dx_1, \ldots dx_n$

the function $f(x_1, x_2, \ldots, x_n; \theta_1, \theta_2, \ldots, \theta_k)$, considered as a function of the θ's for fixed x's, is called the likelihood function. Likewise, for discontinuous variation, the likelihood function emanating from the population specified by $f(x_1, \ldots, x_n; \theta_1, \theta_2, \ldots, \theta_k)$ is that frequency function itself, considered as a function of the θ's.

The likelihood function is usually denoted by L, but for certain purposes a more useful function is the logarithm of the likelihood, which is also sometimes denoted by L.

If a sample of n independent values $x_1, x_2, \ldots, x_n$ is drawn from a univariate population with frequency function $f(x, \theta_1, \ldots, \theta_k)$; the likelihood of the sample is $\prod\limits_{i=1}^{k} f(x_i, \theta_1, \ldots, \theta_k)$; with obvious extensions to the multivariate case.

Likelihood Ratio If $x_1, x_2, ..., x_n$ be a random sample from a population $f(x; \theta_1, \theta_2 ... \theta_k)$ the likelihood of this particular sample is:

$$L = \prod_{i=1}^{n} f(x_i, \theta_1, \theta_2, ..., \theta_k).$$

This will have a maximum, with respect to the θ's somewhere in the parameter space Ω which can be written $L(\hat{\Omega})$. For a sub-space ω of the parameter space, i.e. the set of populations corresponding to some restrictions on the parameters, there will also be a corresponding maximum value $L(\hat{\omega})$. The **Null Hypothesis** H_0 that the particular population under test belongs to the sub-space ω of Ω may be tested by using the Likelihood Ratio

$$\lambda = \frac{L(\hat{\omega})}{L(\hat{\Omega})}, \quad 0 \leqslant \lambda \leqslant 1$$

or some simple function of it. The method is due to Neyman & E. S. Pearson (1928), and can be generalised to the multivariate case.

Likelihood Ratio Dependence If, in the concept of **Regression Dependence**, the requirement that the conditioned variable $(y|x)$ to be stochastically increasing is replaced by one of monotone likelihood ratio in x, then Lehmann (1966) termed this to be likelihood ratio dependence. For example a bivariate Normal distribution $(\rho \geqslant 0)$ has positive likelihood ratio dependence.

Likelihood Ratio Test A test of a hypothesis H_0 against an alternative H_1, based on the ratio of two likelihood functions, one derived from each of H_0 and H_1.

Limited Information Methods In econometrics, methods of deriving estimates of parameters in a stochastic system which do not use all the information available. The term is usually confined to those methods which give consistent estimates, i.e. are unbiassed for large samples. One such method involves the employment of **Instrumental Variables**. A second is the method of **Reduced Forms** which applies to systems which are exactly **Identifiable**. Another is the limited information maximum likelihood method which is applied to over identified systems and incorporates also a reduced form technique. Loosely speaking, this method ignores certain restrictions on the parameters imposed by the structural equations but still produces consistent estimates. Its great advantage is that it can be used without a complete specification of all the equations of the system.

Lincoln Index One method of estimating the size of populations of mobile units, e.g. animal populations, relies on the capture, marking, release and recapture of samples from the population under investigation. The technique appears to have been used first by Lincoln (1930) and the statistic formed by dividing the total number of marked units released by the proportion of marked units recaptured—the Lincoln Index—can be used to estimate the total population size. The Lincoln Index should properly be restricted to populations not subject to birth or immigration. It is slightly biassed but the bias can be removed by modification in the estimator.

Lindeberg-Feller Theorem A form of the **Central Limit Theorem** which gives a necessary and sufficient condition for the distribution of a sum of independent random variables to be asymptotically Normal.

Lindeberg-Lévy Theorem A particular case of the **Central Limit Theorem** when all the variates concerned have the same distribution.

Line of Equal Distribution The straight line on the graph of a **Lorenz Curve** which passes through the origin and the upper extreme of the curve; when, as is customary, the two variables vary from 0 to 1, this is the line making an angle of 45° with the coordinate axes. It provides a reference line of equal distribution, production, concentration, etc. The gradual approach of a succession of Lorenz curves to this diagonal line indicates that the unequal concentration of distribution is being reduced. This interpretation is similar to that afforded by the change in slope of the **Pareto Curve** as indicated by the change in the **Pareto Index**.

Line Sampling A method of sampling in a geographical area. Lines are drawn across the area and all members of the population falling on the line, or intersected by it, are included in the sample. If the lines are straight parallels equally spaced across the area concerned, then the sampling becomes one form of **Systematic Sampling**. If, instead of all intercepts on the lines, a series of evenly spaced points are chosen on each line, the sampling is equivalent to choosing the points on a lattice and may also be regarded as two stage line sampling.

Line Spectrum A term in spectrum analysis of time series denoting the form of the spectrum diagram when the variance or power is concentrated at distinct frequencies.

Line-up The American equivalent of the English meaning of the French word 'queue'. [See **Queueing Problem**.]

Linear Constraint A condition imposed on certain variate values or frequencies which is linear in form. For example, if samples of variates $x_1, x_2, ..., x_n$ are drawn their mean, in general, will vary; but if only those samples are considered which have a mean equal to zero, all other samples being ignored, the variates are subject to the linear constraint $\sum_{i=1}^{n} x_i = 0$. The distribution under constraint may be regarded as **Conditional**.

Linear Correlation An obsolete expression once used to denote either (a) the product moment correlation in cases where the corresponding regressions were linear or (b) a coefficient of correlation constructed from linear functions of the observations. The expression is best avoided altogether.

Linear Discriminant Function A **Discriminant Function** which is a linear function of observed variate values or frequencies.

Linear Estimator An estimator which is a linear function of the observations.

Linear Hypothesis Logically, this expression ought to relate to any statistical hypothesis concerning the parameters of a distribution which can be expressed linearly in terms of them. For example, with the distribution

$$dF = e^{-(x-m)/\sigma} \, dx/\sigma, \quad m \leqslant x \leqslant \infty$$

the hypotheses $m = m_0$ or $m - \sigma = 0$ are linear, whereas $m^2 - \sigma = 0$ is not linear, although it could be placed into a linear form by transforming the parameter m to m^2.

More particularly, the expression '*the* linear hypothesis' relates to Normal variation. If there are p independent Normally distributed variates with a common variance whose means $\mu_i (i = 1, 2, ..., p)$ are connected with parameters $\theta_i (i = 1, ..., p)$ by linear equations

$$\mu_k = \sum_{j=1}^{p} c_{jk}\theta_j$$

a hypothesis which specifies r of the n parameters is a linear hypothesis. Many of the commonly occurring hypotheses of statistical analysis are of this form, e.g. those concerning the difference between two means where variances are equal, certain of the hypotheses underlying the analysis of variance tests and those concerning regression coefficients.

Linear Maximum Likelihood Method A method of parameter estimation in multi-equation models in which the Taylor series linear approximations of the partial derivatives of the likelihood function are equated to zero rather than the actual partial derivatives.

Linear Model A **Model** in which the equations connecting the variates or variables are in a linear form.

Linear Process A stochastic process defined by the formal expression

$$x_t = \sum_{u=-\infty}^{t} g_{t-u}w_u$$

for the discontinuous case or the analogous integral

$$x(t) = \int_{-\infty}^{t} g(t-u)dw(u)$$

for the continuous case, where w_u or $dw(u)$ represent independent and stationary disturbances.

Linear programming The procedure used in maximising, or minimising, a linear function of several variables when these variables, or some of them, are subject to constraints expressed in linear terms: these may be equations or inequalities. The term 'programming' in this context indicates a schedule of actions.

Linear Regression See **Regression**.

Linear Structural Relation A relationship in linear form among the observed **Endogenous Variables**.

Linear Sufficiency A term proposed by Barnard (1963) for linear parametric functions in the Gauss-Markov estimation model, and extended by Godambe (1966) to the case of sampling from finite populations, where some problems of estimation do not admit a sufficient statistic. It is a weaker form of sufficiency and restricted to linear estimators.

Linear Systematic Statistic A systematic statistic which is a linear function of the observations. [See **Systematic Statistic**.] Most 'systematic' statistics in current use are, in fact, linear and the linearity is sometimes taken as understood in describing the linear statistic simply as 'systematic'.

Linear Trend A trend for which the value is a linear function of the time variable, e.g. $u(t) = a + bt$ where a and b are constants.

Link Relative In index number theory, the value of a magnitude in a given period divided by the value in the previous period. [See **Chain Index**.]

Linked Blocks A class of incomplete designs proposed by Youden (1951) for the purpose of reducing the number of replications normally required in such designs and of restoring the symmetry and simplicity of analysis which results from using lattice designs. For example, a design for 10 treatments involving 5 blocks with 4 treatments in a block is as follows:

1	1	2	3	4
2	5	5	6	7
3	6	8	8	9
4	7	9	10	10

where any two blocks have one linked treatment in common.

Linked Paired Comparison Designs An experiment design, proposed by Bose (1956) for comparing n objects with a number of judges (m). Each judge compares r pairs of objects and among these pairs each object appears equally often (α-times). Each pair is compared by k (> 1) judges and, given any two judges, there are exactly λ pairs compared by both judges. If a linked paired comparison design exists then there must be a corresponding balanced incomplete block design with m treatments and $\frac{1}{2}n(n-1)$ blocks.

Linked Samples Two samples of same size in which there is a one-one correspondence between their respective sample units. The link between a pair of corresponding units may be rigid in the sense that one of them uniquely determines the other, or it may be semi-rigid in that one of them restricts the choice of the other, e.g. a pair of linked grids may be separated by a fixed distance. Linking among three or more samples is also possible. [See also **Method of Overlapping Maps**.]

Lipschitz Condition If, say, a probability density function has a continuous derivative at each point in a closed interval then it satisfies a Lipschitz condition. This point occurs in connection with the distribution of sums of independent random variables.

List Sample A sample selected by taking entries from a list of the items constituting the population under review. The usual method of selecting entries is to take them at equal intervals, the starting point being selected at random.

Loading See **Factor Loading**.

Local Asymptotic Efficiency A development by Konijn (1956) of the concept of **Asymptotic Relative Efficiency** concerned with two-sided tests in which the tails are not necessarily equal.

Local Statistic A statistic, estimator or test statistic, which is derived from short term comparisons within a time series. The concept is due to Jowett (1955) and includes the **Jump Statistic**.

Locally Asymptotically Most Powerful Test A test of a composite statistical hypothesis defined by Neyman (1959); it is available only for a one dimensional parameter. [See also **Locally Asymptotically Most Stringent Tests, Uniformly Most Powerful Test**.]

Locally Asymptotically Most Stringent Test A development of the **Locally Asymptotically Most Powerful Test** to the case of a k-dimensional parameter (Bhat & Nagnur, 1965). [See also **Most Stringent Test**.]

Locally Most Powerful Rank Order Test This test for shift in location parameter with symmetry, as defined by Fraser (1957), is of the form $\sum_{i=1}^{N} z_i \, \mathscr{E}(X_{N_i})$ where X_{N_i} is the ith order statistic from the Normal distribution; the z_i, $i = 0, 1$ fall according to y_i being a negative or positive member derived from the y_N ranking of the x_N observations.

Location See **Measure of Location**.

Location Parameter A parameter which 'locates' a frequency distribution in the sense of defining a central or typical value such as a mean or mode.

Location Shift Alternative Hypothesis If we have a random sample of observations from K ($\geqslant 2$) populations, in testing the hypothesis that these populations are identical, the alternative hypothesis might be that the only difference between the parent distribution is one of location. This alternative hypothesis is known as the location shift alternative hypothesis.

If the alternative hypothesis further specifies that the k population means form an ordered sequence, it is called an ordered alternative hypothesis.

Lods A term introduced by Barnard (1949) in connection with certain developments in statistical inference. It is a contraction of the term 'logarithmic-odds', the basic probabilities being expressed on a logarithmic scale in terms of odds in favour of or against an event.

Log-Chi Squared Distribution A transformed distribution used in the analyses of heterogeneous variances (Bartlett & Kendall, 1946) and the analyses of Poisson processes (Cox & Lewis, 1966).

Logarithmic Chart A graph whereon one or both axes are scaled in terms of logarithms of the variables. The chart may be called a **Semi-** or **Double Logarithmic Chart** according to whether only the ordinate or both the ordinate and abscissa are on a logarithmic scale. In general, the logarithmic method of plotting is used when relative changes are important, since equal linear displacement on a logarithmic scale indicates equal proportional changes in the variable itself.

Logarithmic Normal (Lognormal) Distribution If the logarithms of a set of variate values are distributed according to the **Normal Distribution** the variate is said to have a logarithmic Normal distribution, or be distributed 'lognormally'.

Logarithmic Series Distribution A frequency distribution developed by R. A. Fisher (1941) in connection with the frequency distribution of species. It is a limiting form of the **Negative Binomial Distribution** with the zero class missing, the frequency of the values 1, 2, 3, ..., being

$$\alpha x, \tfrac{1}{2}\alpha x^2, \tfrac{1}{3}\alpha x^3, \ldots,$$

where $1/\alpha = -\log(1-x)$ and x is some parameter; that is to say, the frequency of the value r is the coefficient of x^r in the expansion of $-\alpha \log(1-x)$.

Logarithmic Transformation In general, a transformation of a variable x to a new variable y by some such relation as $y = a + b \log(x-c)$. There are a number of contexts in which such transformations are useful in statistics, e.g. to normalise a frequency function, to stabilise a variance, and to reduce a curvilinear to a linear relationship in regression or probit analysis.

Log Convex Tolerance Limits A class of tolerance limits proposed by Hanson & Koopmans (1964) to deal with a position intermediate between those for, say, the Normal distribution and non-parametric conditions. They use the **Pólya Frequency Function of Order Two** which includes a broad group of distributions.

Logistic Curve   See **Growth Curve**.

Logistic Distribution   A frequency distribution of the form $\beta\{e^r/(1+e^r)^2\}$ with $\beta > 0$. It closely approximates the Normal distribution and may also be shown to be the asymptotic distribution of the mid-range of an exponential-type parent distribution. In its cumulative form it has also been used as a **Growth Curve**.

Logistic Process   A stochastic process associated with the logistic law of growth (see **Growth Curve**). It is a particular case of a birth and death process in which the rates for these two phenomena are linearly dependent upon population size. If, for a finite constant population, the condition is also made that individuals not possessing some characteristic will eventually do so we have the epidemic model proposed by Bartlett (1946). It may be noted in this connection that 'logistic' in statistical usage has nothing to do with the military use of the word as meaning the provision of material.

Logit   In some problems relating to the proportion of subjects responding to different doses of a stimulus the model based on a **Tolerance Distribution** is not appropriate. A logistic relation
$$P = \{1 + e^{-(\alpha + \beta x)}\}^{-1}$$
may approximately represent the dependence of probability of response P on dose x. Berkson (1944) defined the logit of P as
$$Y = \log_e\{P/(1-P)\},$$
and analysis can then be based upon a linear regression of logit on dose similar to that used with **Probits**. Essentially the same transformation was proposed earlier by Fisher & Yates and by Wilson & Worcester.

Loglog Transformation   The transformation of a probability P to a **Response Metameter** Y according to the formula
$$Y = \log_e(-\log_e P).$$
This was first suggested by Mather (1949) and adapted by Finney (1951) to the estimation of bacterial densities from **Dilution Series**.

Loop Plan   A name given by Deming (1950) to a method of estimating the variance of an estimator derived from a systematic sample. If a population arranged in a line is sampled by taking units at a fixed interval k apart along the line, starting at random within the initial interval of width k, there is no theoretically valid method of deriving an estimate of the sampling variance from the sample itself. The units are therefore paired (or 'looped together') and each pair is regarded as a sample of two chosen at random within an artificial stratum of length $2k$. On this assumption an estimate of sampling error can be made; although the method gives biassed estimators unless the population is arranged at random along the line.

Lorenz Curve   A graphical method of showing the concentration of ownership of economic quantities, such as income and wealth. If the cumulative distribution of the amount of the variable concerned is plotted as ordinate against the cumulative frequency distribution of the individuals possessing the amount, the resultant curve is a Lorenz Curve. The cumulation is usually expressed as a percentage of the total quantity or total number of individuals, as the case may be, and from the curve it is possible to make statements of the kind: 'x per cent of the people receive y per cent of the income'. It is also possible to study the variations of these figures through time, or between different areas, by plotting successive curves on the same graph. The same technique can be applied to other variables, such as production against numbers of producing units.

Loss Function   In the making of decisions on the basis of observations on a variate x, disadvantage may be suffered through ignorance of the true distribution of x. The extent of the disadvantage is often a function of the true distribution and of the decision which is actually made. This function is called the loss function.

Loss Matrix   In the theory of decision functions, a matrix specifying the economic loss or gain incurred according to the various decisions which can be taken and the various situations which can in reality exist.

Loss of Information   This term is used in two entirely different senses: (*a*) to denote the actual loss of information in the ordinary sense, e.g. by the destruction of records; (*b*) to denote failure to extract all the information which exists in the available data about a particular matter. In the second case the failure may be due to avoidable causes, such as the use of inefficient statistics; or it may, in the technical sense of the word 'information' be due to the fact that no single estimator exists embodying all the 'information' which exists in the sample under scrutiny. [See **Ancillary Estimators, Information, Sufficiency**.]

Lot   A term used in quality control in the sense of aggregate, collection or batch, but usually with a somewhat more specialised meaning. A lot is a group of units of a product produced under similar conditions and therefore, in a sense, of homogeneous origin; e.g. a set of screws produced by a lathe or a set of electric light bulbs produced by a number of similar machines. It is sometimes implicit that the lot is for inspection.

Lot Quality Protection   See **Average Quality Protection**.

Lot Tolerance Per Cent Defective   The proportion of defective product allowable, or acceptable, in each lot submitted for inspection under a scheme designed for **Lot Quality Protection**. This is sometimes called 'Lot Tolerance Fraction Defective'.

Lottery Sampling   A method of drawing random samples from a population by constructing a miniature of the

88

population, e.g. by inscribing the particulars of each member on to a card and drawing members at random from it, e.g. by shuffling the cards and dealing a set haphazardly. It is the method usually employed at a lottery, hence its name, but suffers from the disadvantage that the preparation of the cards entails considerable labour and strict precautions must be taken in the shuffling process to guard against bias.

Lowe Index An index number proposed by Lowe (1823) in which average weights are used. If the prices of a set of commodities in either a base or given period are p_o, p_o', p_o'', ... $(p_n, p_n', p_n'', ...)$ and q are the weights the Lowe price index number is written

$$I_{on} = \frac{\Sigma(p_n q)}{\Sigma(p_o q)}$$

where the summation takes place over commodities. The set of periods over which quantities are averaged to obtain the weight q is to some extent at choice. If q relates only to the base period the index number is that of **Laspeyres**; if it relates only to the given period it is that of **Paasche**; if it is the arithmetic mean of the quantity in the base and given period the index number is that of Marshall, Edgeworth and Bowley. [See also **Crossed Weights, Marshall-Edgeworth-Bowley Index.**]

Lower Control Limit See **Control Chart.**

Lower Quartile See **Quartile.**

Lumped Variance Test A one sided test of a rectangular hypothesis proposed by Broadbent (1955) in connection with the analysis of multi-modal data involving a **Quantum Hypothesis.**

Lyttkens' Correction A correction to the standard error of the coefficients of lagged endogenous variables which are autocorrelated, proposed by Lyttkens (1962) based upon Wold (1951) who covered the cases of autocorrelated exogenous variables and residuals.

McNemar's Test A non-parametric test (1947) for the difference between proportions, or percentages, derived from correlated samples.

m-rankings, Problem of Given m rankings of, say, n objects the problem arises of finding some measure of the general agreement between the rankings and of testing its significance. One such measure is termed the **Coefficient of Concordance.**

m-Statistic See **Wood's W-test.**

mth Values The mth values of a set of n observations are the mth largest or mth smallest when the values are arranged in order of magnitude. They are particular cases of **Order Statistics.** Thus, if $m = 1$ the mth values are the two **Extreme Values.**

Macaulay's Formula A moving average formula proposed by Macaulay (1931) to carry out long period smoothing. It has 43 terms and reduces the variance of a random series to a degree equivalent to a nine-term moving average with equal weights.

Madow-Leipnik Distribution An approximation by Madow (1945) and Leipnik (1947) to the distribution of the first serial correlation in samples of n from a Markov Normal process with known mean.

Magic Square Design A square array $(n \times n)$ with n^2 integers placed in the cells in such a way that the row, column and two principal diagonal sums are the same; namely $\frac{1}{2}n(n^2+1)$. These squares may be used for balancing out linear trend from main effects and lower order interactions in some factorial designs and in some Latin and Graeco Latin square designs.

Mahalanobis Distance A particular development in the general topic of discrimination in this case concerned with the relationship or 'distance' between two populations. The work of Mahalanobis (1930) resulted in the D^2-**statistic** for which the population parameter ($\varDelta^2$) is referred to as the generalised squared distance. It bears a simple relationship to the T^2-**statistic** developed by Hotelling (1931) and the **Discriminant Function** developed by Fisher (1936) in connection with problems of classification.

Mahalanobis' Generalised Distance See D^2-**statistic.**

Main Effect An estimate of the effect of an experimental variable or treatment measured independently of other treatments which may form part of the experiment. Thus, in a balanced experiment involving three factors, A, B, C, each at two levels applied and not applied, the main effect due to A would be the average of the four effects where A was applied, and B and C were applied at each of the two levels, less the average of the four effects where A was not applied. One of the reasons for introducing orthogonality into an experimental design is to enable main effects to be separately estimated.

Manifold Classification If a population is divided into a number of mutually exclusive classes according to some given characteristic and then each class is divided by reference to some second, third, etc. characteristic, the final grouping is called a manifold classification. While some aspects of experimental design in the factorial form are akin to manifold classification, the term most often occurs with reference to two characteristics, where the manifold classification gives rise to a **Contingency Table.**

Mann-Whitney Test See **Wilcoxon's Test.**

Marginal Category One of the frequency classes of a **Marginal Classification.**

Marginal Classification In a bivariate frequency table it is customary to show, as row and column totals, the univariate frequencies of the two variates separately. This is sometimes called marginal classification. Similarly, for a multivariate frequency array the arrays of one lower dimension formed by summing one of the variates are occasionally said to be marginal in relation to the original array. The frequencies are said to be marginal.

'Marker' Variable A name sometimes given (e.g. D. G. Kendall, 1950) to a two valued variable introduced in order to assist the analysis of a situation involving two states, e.g. birth or death. [See also **Dummy Variable**.]

Markov Chain This expression is used in two different senses, both relating to a **Markov Process**. In one sense a process $[x_t]$ is called a chain if the time parameter is discontinuous. In the other it is called a chain if the values of x are discontinuous. The former appears preferable.

Markov Estimate An estimate of a parameter derived from an estimator given by the so-called Markov or **Gauss-Markov Theorem**.

Markov Inequality If a variate x is non-negative and has mean equal to a then for $t > 0$ the Markov inequality states that

$$\Pr\{x \geqslant t\} < \frac{a}{t}.$$

Some writers credit this inequality to Tchebychev. [See **Bienaymé-Tchebychev Inequality**.]

Markov Process A stochastic process such that the conditional probability distribution for the state at any future instant, given the present state, is unaffected by any additional knowledge of the past history of the system.

Markov Renewal Process A type of stochastic process closely related to the **Semi-Markov Process** and first proposed by Pyke (1961). This process $\{N(t); t \geqslant 0\}$ is determined by (m, A, Q) where A is a vector of initial probabilities, Q a matrix of transition distributions and $N(t)$ a series of counting functions. Where $m = 1$ we have the ordinary renewal process.

Marshall-Edgeworth-Bowley Index An index number formula, proposed by Marshall, Edgeworth and Bowley as an alternative to the standard formulae of **Laspeyres** and **Paasche**. It may be written, in terms of prices and quantities, as:

$$I_{on} = \frac{\Sigma\{p_n(q_o + q_n)\}}{\Sigma\{p_o(q_o + q_n)\}}$$

and where the suffix o refers to the base period and n to the period to which the index relates. The index represents a compromise with no general bias. However, it suffers from the disadvantage of lack of comparability, between different years owing to the shifting pattern of weights.

Martingale Originally, a process known to gamblers under which the loser at a fair game doubled his stakes for the next, and so on at each loss, the paradox being that in the long run he appeared certain to win sooner or later and at that point would have a net gain.

More recently the term has been given a precise significance in the theory of stochastic processes.

A stochastic process $\{x_t\}$ is called a martingale if $\mathscr{E}\{\,|\,x_t\,|\,\}$ is finite for all t, and

$$\mathscr{E}\{x_{t_{n+1}} \mid x_{t_1}, \ldots, x_{t_n}\} = x_{t_n}$$

with probability unity for all $n \geqslant 1$ and $t_1 < \ldots < t_{n+1}$.

If the equality sign is replaced by $\geqslant$ the process becomes sub-martingale and semi-martingale if replaced by an inequality.

Master Sample A sample drawn from a population for use on a number of future occasions, so as to avoid *ad hoc* sampling on each occasion. Sometimes the Master Sample is large and subsequent inquiries are based on a sub-sample from it.

Matched Samples A pair, or set of, matched samples are those in which each member of a sample is matched with a corresponding member in every other sample by reference to qualities other than those immediately under investigation. The object of matching is to obtain better estimates of differences by 'removing' the possible effects of other variables. For example, if it is desired to investigate acuity of vision for a sample of smokers as compared with a sample of non-smokers, better comparisons can usually be made if, to every member of one sample, there can be associated a member of the other sample of the same sex and about the same age.

Difficulties arise in the assessment of significance, however, if the members of a second sample have to be chosen purposively in order to match the first, instead of being chosen at random.

Matching If two sequences of a finite number of characteristics $A_1, A_2, \ldots, A_k$ are compared, the jth of one sequence against the jth of the other, a comparison in which both members exhibit the same characteristic is called a match. The number of matches in an observed pair of sequences provides a test of various hypotheses concerning the system which generated them.

More generally, p sequences instead of two may be considered and the occurrence of the same characteristic in the jth member of each sequence is also called a match or a multiple match.

A distinct interpretation is used in communication theory, where there occurs the problem of matching the message source to the communication channel, in order to secure the maximum efficiency in transmitting messages.

Matching Coefficient A variant of the **Similarity Index** proposed by Sokal & Michener (1958) and developed by Goodall (1967).

Matching Distribution Suppose two sets having n objects each are numbered 1, 2, ..., n and each is arranged in a random order so as to form n pairs. Then the number of pairs on which the two numbers are the same has the matching distribution. More general distributions are studied, in which there are several qualitative characteristics, e.g. colour or suit instead of numbering in the comparison of two packs of cards or more than two sets are available.

Matrix Sampling The process of taking a **Bi-sample** from an $R \times C$ population matrix whose elements are $\| x_{IJ} \|$ consists of two samples, r of rows and c of columns, and forming a matrix whose elements are those at the intersections of the selected rows and columns.

Maverick A term encountered in the literature of industrial statistics to denote an observation lying so far outside the usual range that it is suspected of not belonging to the population under inquiry.

Maximum F-ratio In testing the homogeneity of a set of variances, the ratio of the largest to the smallest, proposed by Hartley (1950) as a simple test alternative to **Bartlett's**.

Maximum Likelihood Method A method of estimating a parameter or parameters of a population by that value (or values) which maximises (or maximise) the **Likelihood** of a sample. For instance, if the likelihood is $L(x_1, ..., x_n, \theta)$ the parameter θ is estimated as the function of the x's, $\hat{\theta}$, for which, under certain regularity conditions,

$$\left(\frac{\partial L}{\partial \theta}\right)_{\theta = \hat{\theta}} = 0, \left(\frac{\partial^2 L}{\partial \theta^2}\right)_{\theta = \hat{\theta}} < 0.$$

Maximum Probability Estimator A method of estimation proposed by Weiss & Wolfowitz (1967), for which the generalised **Maximum Likelihood** estimator is a special case, designed to answer problems posed by ordinary maximum likelihood estimation and not covered by the generalised case.

Maxwell Distribution A fortunately rare expression for a chi-squared distribution with three degrees of freedom, based presumably on Clark Maxwell's discussion of the energy of particles moving at random in three dimensions.

Maxwell-Boltzmann Statistic See **Bose-Einstein Statistics**.

*Mean (Media) Italian usage of 'media' in the senses of 'mean' corresponds very closely to English usage, but there are a number of Italian terms which have no current English counterparts. A mean is said to be basal (basale) or exponential (esponenziale) according as the terms figure as bases or exponents in the definition. If they do both, it is a media baso-esponenziale (basic-exponential mean). A mean is said to be stable (ferma) if it depends on all the terms x. If it does not depend on all the values it is said

to be relaxed (lasca). If it does not coincide in value with any of the x's it is called a media di conto (a computing mean) as contrasted with a media effettiva (effective mean) which takes one of the values of the x's. A mean M_k of the form

$$M_k = \left\{\frac{1}{n} \sum_{i=1}^{n} x_i{}^k\right\}^{\frac{1}{k}}$$

is called a power mean (media potenziata). If of the form
$$T_k = \Sigma(x_i{}^k)/\Sigma(x_i{}^{k-1})$$
it is called a mean of power-sums (media di somma di potenza). If given by

$$C^E = \frac{1}{n} \sum_{i=1}^{n} c^{x_i}$$

when E is the mean and C is some positive constant it is called exponential (esponenziale). A mean in the wider sense (in senso lato) is one which does not coincide with the value of a member of the series; in the contrary case it is a mean in the strict sense (in senso stretto). A mean is objective (oggettiva) if there exists a concrete object of which the observed values represent divergent measurements; subjective (soggettiva) when it is an abstraction such as the mean number of children per family. [See also ***Combinatorial Power Mean.**]

Mean Absolute Error An alternative but much less preferable name for the **Mean Deviation**.

*Mean Density, Curve of (Curva di Densità Media) In Italian usage, a **Curve of the Concentration** type, especially for incomes. It shows the mean density of income for earners with income less than or equal to x against the relative frequency of earners with income less than or equal to x. [See **Lorenz Curve**.]

Mean Deviation A measure of dispersion derived from the average deviation of observations from some central value, such deviations being taken absolutely, i.e. without reference to algebraic sign. The central value may be the arithmetic mean or the median. Expressed formally the mean deviation is the **First Absolute Moment**.

Mean Difference A measure of dispersion due to Gini (1912) and based upon the average of the absolute differences of all possible pairs of variate values. For a continuous variate x with distribution functions $F(x)$ it may be written:

$$\Delta_R = \int_{-\infty}^{\infty} \int_{-\infty}^{\infty} |x - y| \, dF(x) \, dF(y).$$

For a discontinuous variate x with frequency function $f(x)$

$$\Delta_R = \frac{1}{N^2} \sum_{j=-\infty}^{\infty} \sum_{k=-\infty}^{\infty} |x_j - x_k| f(x_j) f(x_k),$$

with the factor $\frac{1}{N^2}$ reduced to $\frac{1}{N(N-1)}$ if repetition is not required, i.e. if a variate value is not regarded as occurring with itself.

Mean Likelihood Estimator An estimator proposed by Barnard (1959) which minimises the mean square error when the prior distribution is uniform.

Mean Linear Successive Difference The same as **Mean Successive Difference**.

Mean Probit Difference A measure of the difference between two series of observations giving parallel **Probit Regression Lines** proposed by Finney (1943). It is defined as the constant vertical difference between the lines. A disadvantage of the mean probit difference is that it measures the difference in effects of equal doses rather than comparing the sizes of equally effective doses.

Mean Range The arithmetic mean of the ranges of a set of samples of the same size. The mean range in repeated sampling may be used as an estimator for the population standard deviation.

Mean Semi-Squared Difference The statistic used in the study of **Serial Variation**.

Mean Square In general, the mean square of a set of values is the arithmetic mean of the squares of their differences from some given value, namely their second moment about that value.

When the mean square is regarded as an estimator of certain parental variance components the sum of squares about the observed mean is usually divided by the number of degrees of freedom, not the number of observations. It is still known as a mean square, an expression which is convenient if somewhat inaccurate.

Mean Square Consecutive Fluctuation Estimator A generalised mean square successive difference proposed by Ruben (1963) as an estimator for the interaction or migration parameter in an emigration/immigration process.

Mean Square Contingency See **Contingency**.

Mean Square Deviation The second moment of a set of observations about some arbitrary origin. If that origin is the mean of the observations the mean square deviation is equivalent to the **Variance**.

An equivalent expression, especially when the observations are variate values, is mean square error. This latter term also occurs in older writings in the sense of variance but should not be employed in that sense.

Mean Square Error See **Mean Square Deviation**.

Mean Square Successive Difference An estimate of the population variance may be based upon the first difference of a series of independent observations, $x_1, x_2, ..., x_n$ by the formula

$$\delta^2 = \frac{1}{n-1} \sum_{t=1}^{n-1} \{x_{t+1} - x_t\}^2$$

δ^2 is called the mean square successive difference. Its mean value in random samples from a Normally distributed population is $2\sigma^2$, whence $\frac{1}{2}\delta^2$ affords an unbiassed estimator of σ^2 in the absence of serial correlation in the x's. In this connection it is related to **von Neumann's Ratio**: alternative forms using second order differences developed by Kamat & Sathe (1962). [See also **Modified Mean Square Successive Difference**.]

Mean Successive Difference In a time series, the arithmetic mean of the difference of successive values, i.e. for a series $x_1, x_2, ..., x_n$ is

$$d = \frac{1}{n-1} \sum_{i=1}^{n-1} |x_i - x_{i+1}|.$$

The intractability of absolute quantities leads to a preferred use of the mean square of differences, e.g. Variate difference Method.

*Mean Trigonometric Deviation (Scarto Trigonometrico Medio) A measure of variability appropriate to **Cyclical Series** developed by Salvemini in Italy. If $x_1, x_2, ..., x_s$ are the quantitative values of a cyclical series and the corresponding frequencies are $y_1, y_2, ..., y_s$ the coefficient in question is given by

$$S_r = 1 - \frac{1}{n} |A \sin \bar{x} + B \cos \bar{x}|$$

where $A = \sum_{i=1}^{s} y_i \sin x_i$, $B = \sum_{i=1}^{s} y_i \cos x_i$, $\bar{x} = \arctan A/B$ and n is the sum $\sum_{i=1}^{s} y_i$.

Mean Values Mean values are a general class of functions of distributions of which **Moments** constitute a special case. If a variate has a distribution function $F(x)$ and $t(x)$ be some function defined within the range of the distribution, the mean value, or mathematical expectation, is defined as

$$\mathscr{E}[t(x)] = \int_{-\infty}^{\infty} t(x) \, dF(x),$$

subject, of course, to existence. The definition may readily be generalised to n-dimensional variation. The various moments of a distribution are derived by substituting the appropriate function for $t(x)$. [See **Expectation**.]

Measure of Location A quality which purports to locate a distribution, or a set of sample values derived therefrom, by means of a value which is, in some sense, central or typical, e.g. the arithmetic mean, the median or the mode.

Medial Test A graphical test of association between two variates. The **Scatter Diagram** for the pairs of observations is divided into four quadrants by lines, parallel to abscissa and ordinate, passing through the medians of the variates. Association is judged by the number of points falling into the positive quadrant as compared with the number expected, namely one quarter of the observations, on the hypothesis of no association.

If the total number of points is n and the total number in the positive quadrant and its opposite quadrant is d, the coefficient $2d/n-1$ has been termed (Quenouille, 1952) the medial correlation coefficient.

A test by Olmsted and Tukey (1947), based on the outlying members in each quadrant, is known as the Corner Test.

Median The median is that value of the variate which divides the total frequency into two halves. As a partition value it may be defined for a continuous frequency distribution by the equation

$$\int_{-\infty}^{M} f(x)dx = \int_{M}^{\infty} f(x)dx = \tfrac{1}{2},$$

M being the median value. For a discontinuous variate ambiguity may arise which can only be removed by some convention. For a total frequency of $2N+1$ items the median is the variate value of the $(N+1)$th item: for $2N$ items it is customary to take the average of the Nth and $(N+1)$th item.

*Median Centre (Centro Mediano) In Italian usage, a point such that the sum of distances from a given set of points is a minimum; as distinct from the *centro medio* or centre of gravity, for which the sum of squares of distances is a minimum.

Median Effective Dose A term proposed by Trevan (1927) to characterise the potency of a stimulus by reference to the amount which produces a response in 50 per cent of the cases where it is applied. The median effective dose is sometimes written ED_{50}; so, by a natural extension, many other effective doses for different quantile values may be styled, e.g. ED_{75} or ED_{90}.

Median F-Statistic If a sample of $2m+1$ items is ranked and transformed to a **Uniform Distribution** the normalised random variable

$$M_{2m+1} = 2\sqrt{(2m+3)}\{F(x'_{m+1})-\tfrac{1}{2}\}$$

converges in distribution to the standardised Normal variable. The statistic $F(x'_{m+1})$ proposed by Birnbaum & Tang (1964) is the sample median and M_{2m+1} is distribution free with respect to the distribution of the original observations.

Median Lethal Dose A particular name for the **Median Effective Dose** when the response is death. It is often written LD_{50}.

*Median Line (Linea Media) In Italian usage, any line in a plane which divides a given set of points into two equally numerous sets.

Median Regression Curve The type of regression line or curve derived from the **Mood-Brown Procedure**.

Median Test A rank-order test proposed by Mood (1950) which rejects the hypothesis of identity of two populations, in the one sided case, when there are too few observations from one sample larger or smaller than the median of the combined sample.

Median Unbiassed Confidence Interval A $(1-\alpha)$ confidence interval bounded by two confidence limits θ' and θ'' at $(1-\alpha)/2$ (see Birnbaum, 1961).

Median Unbiassedness A concept, one of a group proposed by Brown (1947) and subsequently developed, for example, by Van der Vaart (1961), whereby the expected value of an estimator being equal to the parameter is replaced by the median value as a demonstration of unbiassedness.

Mellin Transform A transform of a function, in statistics usually a frequency function, traceable to Riemann but first rigorously discussed by Mellin (1896). The transform of a function $f(x)$ may be written

$$F(s) = \int_{0}^{\infty} f(x)\, x^{s-1}\, dx,\ x \geqslant 0$$

and there exists a reciprocal relation

$$f(x) = \frac{1}{2\pi i} \lim_{c \to \infty} \int_{c-i\infty}^{c+i\infty} F(s)x^{-s}ds.$$

Mellin's formula may also be derived from the **Fourier Transform**.

Merrington-Pearson Approximation An approximation to the non-central t-distribution proposed by Merrington & Pearson (1958) based upon the **Pearson Type IV Distribution**.

Mesokurtosis See **Kurtosis**.

Metameter A transformed value of a dose or a response, e.g. logarithm or probit, obtained by using a transformation equation that is independent of all parameters. It is adopted mainly to simplify the analysis or the expression of the dose response relationship. The word was apparently coined by Hogben.

Method of Overlapping Maps A device for the selection of **Linked Samples**. Thus to reduce travel costs in a multipurpose survey where it is desired to select a sample of villages with probability proportional to (1) population, for population inquiry; and (2) to area, for direct physical observation of fields for land utilisation inquiry, the villages may be ordered in a serpentine manner, and represented twice on a straight line of fixed length which is completely covered twice, once by segments proportional to population and a second time to area. A point thrown at random on this overlapping map will select a linked pair of villages, each with the desired probability and the two are likely to be in the same neighbourhood, if not identical.

Mid-range For a set of values, $x_1, x_2, ..., x_n$ arranged in order of magnitude the mid-range is defined as $\tfrac{1}{2}(x_n+x_1)$.

It is synonymous with 'Centre', but 'mid-range' usually refers to a sample and 'centre' to the parent distribution.

Mid-rank Method See **Tied Ranks.**

Mills' Ratio The ratio of the area of the 'tail' of a distribution to the bounding ordinate. For a Normal deviate x, the function

$$e^{\frac{1}{2}x^2} \int_x^\infty e^{-\frac{1}{2}u^2}\, du.$$

The ratio occurs naturally in the computation of values of the Normal integral and was considered by Laplace, who gave a continued fraction for it. It was tabulated by J. P. Mills in 1926.

Minimal Sufficient Statistics A vector of statistics is minimal sufficient if it has a minimal number of components, in which case it is a function of all other sufficient vectors for the parameters in question.

Minimax Estimation The estimation of parameters by the application of the **Minimax Principle** to a **Risk Function.** In particular it may be shown that a **Bayes' Estimator** which has a constant risk function is also a minimax estimator.

Minimax Principle A principle introduced into decision function theory by Wald (1939). It supposes that decisions are taken subject to the condition that the maximum risk in taking a wrong decision is minimised. The principle has been criticised on the grounds that decisions in real life are scarcely ever made by such a rule, which enjoins 'that one should never walk under a tree for fear of being killed by its falling'. In the theory of games it is not open to the same objection, a prudent player being entitled to assume that his adversary will do his worst.

Minimax Regret Principle An alternative to the **Minimax Principle** in which the function minimised is the 'regret' defined as the excess of the actual value of the risk function over the least possible value of this function.

Minimax Strategy If a strategy is selected from a group of **Admissible Strategies** as being the one which, on a basis of the expected loss, has the smallest maximum loss this strategy will be a minimax strategy.

Minimum Chi-squared A method of estimation based upon the χ^2 **Goodness of fit Statistic.** The method determines values of the parameters so as to minimise χ^2 calculated from observed frequencies and 'expected' frequencies expressed in terms of the parameters. The method is troublesome to apply in general because of the difficulty of expressing the observed frequencies explicitly in terms of the parameters under estimate. A modified minimum chi-squared method (Jeffreys, 1938) simplifies the method to some extent by minimising the statistic $\chi'^2 = \Sigma(\lambda_j - l_j)^2/l_j$ where λ is the theoretical and l the observed frequency in the jth group. For large samples the estimators from the two methods are asymptotically equivalent and they also tend to the values of **Maximum Likelihood** estimators.

Minimum Discrimination Information Statistic This statistic is based upon the principle of **Information** in a sample and can be considered as the 'divergence' of the alternative hypothesis from the null hypothesis. The distribution of this statistic involves central and non-central χ^2 and it has additive properties. For a wide class of problems concerning contingency tables, this statistic takes the form of $-2 \ln \lambda$, where λ is a likelihood ratio.

Minimum Logit Chi-squared A method of estimation in bio-assay proposed by Berkson (1944) and subsequently developed by him and other authors. The **Logistic Function** is used as the **Tolerance Distribution** and the principle of minimum chi-squared rather than maximum likelihood. It has also been shown by Taylor (1953) that such estimators are **Regular Best Asymptotically Normal.**

Minimum Normit Chi-square Estimator A method of estimating the cumulative Normal distribution proposed by Berkson (1955) which involves minimising the following quantity, distributed asymptotically as χ^2

$$\chi^2 \text{ (normit)} = \Sigma n_i \frac{z_i^2}{p_i q_i} (v_i - \hat{v}_i)^2$$

where n_i is the number exposed to stimulus x_i; $p_i = 1 - q_i$ is proportion affected, v_i and $\hat{v}_i$ the observed and estimated **Normit** with z_i the Normal curve ordinate at point where the area divides p_i and q_i.

Minimum Variance As applied to estimators, this term denotes the property of possessing the least variance among the members of a defined class. A minimum variance estimator exists only where there exists a **Sufficient Estimator.** [See also **Cramér-Rao Inequality.**]

Minimum Variance Linear Unbiassed Estimator In the case of the **Linear Hypothesis** the **Method of Least Squares** provides estimators that are unbiassed, linear in the observations, and with minimum variance. [See also **Gauss-Markov Theorem.**]

Missing Plot Technique The name given by Allen & Wishart (1930) and Yates (1933) to the process of analysing material which was designed to conform to an experimental pattern but from which certain values are missing through circumstances beyond the control of the experimenter. The use of the word 'plot' arose from the agricultural background of the original investigations noted above but methods are applicable to missing values generally.

Mixed Autoregressive-Regressive Systems A system of equations which could represent a particular econometric model might be an autoregressive set of y's regressed upon fixed x's:

$$\sum_{j=0}^{k} \alpha_j y_{t-j} = \sum_{l=1}^{q} \beta_{lt} x_t + \epsilon_t.$$

This may be rewritten in the form:

$$y_t = \sum_{l=1}^{q} \beta_{lt} x_t - \sum_{j=1}^{k} \alpha_j y_{t-j} + \epsilon_t$$

giving the first term on the R.H.S. as an ordinary regressive component and the second term a component in autoregressive form.

Mixed Distribution A term frequently used in modern statistical literature for the distribution which results from the parameter of a given distribution being itself a random variable. [See also **Compound Distribution, Generalised Distribution.**]

Mixed Exponential Response Law There are a number of situations where the distribution of the response exhibited by members of a population is negative exponential but the presence of definite strata creates a mixture of such distributions. The form of a 'mixed' law depends critically on the weight function used to combine the separate distributions.

Mixed Factorial Experiments An experiment in factorial form where the number of levels for the factors varies from one factor to another. For example, an experiment involving one factor at two levels, one at three levels and one at four levels would be a three factor experiment of the 'mixed' type.

Mixed Model This term is used in at least four slightly different senses; a model is termed 'mixed' if

(*i*) its equations contain both determinate and stochastic elements;

or (*ii*) its equations contain both difference and differential terms;

or (*iii*) it contains **Endogenous** as well as **Exogenous** elements;

or (*iv*) in an analysis of variance context for a two way layout, the rows correspond to a **Model I** factor, i.e. fixed effects, and the columns correspond to a **Model II** factor, i.e. random effects. This version is sometimes known as Model III but this is best avoided (see Plackett, 1960).

Mixed Sampling Where a sampling plan envisages the use of two or more basic methods of sampling it is termed mixed sampling. For example, in a multi-stage sample, if the sampling units at one stage are drawn at random and those at another by a systematic method, the whole process is 'mixed'.

Usage is not uniform, but where samples at one stage were drawn at random with replacement and at another stage were drawn at random without replacement, it would seem better not to describe the whole process as 'mixed', the essential basic method of random selection being employed throughout.

Mixed Spectrum A **Spectral Density Function** of a (general stationary) stochastic process which contains both discrete and continuous components.

Mixed Strategy See **Strategy**.

Mixed-up Observations A phrase which has been used to describe a situation where the identity of some observations may be lost but their total value is known. It is too imprecise for general use in view of the many ways in which observations can be mixed up.

Mixture of Distributions A process of obtaining a weighted average of a group of distribution functions which is a new distribution function. It is to be distinguished from the **Compound Distributions.**

*Modality (Modalità) In Italian usage, *modalità* refers to the particular value assumed by a characteristic. The English equivalent modality, though used in logic, is not employed in statistics. A characteristic which can assume different quantitative values (*diverse modalità quantitative*) is called *variabile* and its susceptibility to do so is called *variabilità*. On the contrary a characteristic which can assume different qualitative values (*diverse modalità qualitative*) is called *mutabile* and its susceptibility to do so is called *mutabilità*. The English words 'variability' and 'variable' may refer to qualitative as well as to quantitative variation.

Mode The mode was originally conceived of as that value of the variate which is possessed by the greatest number of members of the population. Although the idea of the most frequently encountered or fashionable value of the variate is probably very old, it was not generally used in statistics until popularised by K. Pearson (1894). The concept is essentially of use only for continuous distributions, although it can be extended to the discontinuous case. More formally, if $f(x)$ is a frequency function, a mode is a value of x for which

$$\frac{df(x)}{dx} = 0, \quad \frac{d^2f(x)}{dx^2} < 0.$$

There may thus be more than one mode of a distribution, though the practical occurrence of multimodality is comparatively rare.

Model A model is a formalised expression of a theory or the causal situation which is regarded as having generated observed data. In statistical analysis the model is generally expressed in symbols, that is to say in a mathematical form, but diagrammatic models are also found. The word has recently become very popular and possibly somewhat overworked.

Model I (or First Kind) A term introduced by Eisenhart (1947) to denote analysis of variance based upon the

95

least squares analysis of the general linear model. Essentially it is an analysis of mean values and is frequently termed the fixed effects or constants model. [See also **Variance Component**.]

Model II (or Second Kind)   A term in the general scheme proposed by Eisenhart (1947) to denote analysis of variance based upon a vector of random variables instead of a vector of parameters (means). It is also termed the components of variance model. See also **Variance Component**.]

Modified Binomial Distribution   Consider a sequence of n trials with constant probability p of success in an individual trial with the restriction that as soon as a success is experienced, the subsequent $m-1$ trials result in failures. The number of successes in the above sequence has a modified form of the binomial distribution.

A further modification assumes that the probability of a success in the first trial was p and regards the first observation in the experiment as a random start in an infinite sequence of similar trials.

Modified Control Limits   In the case of statistical quality control using the **Shewhart Chart**, the use of **Reject Limits** as the control limits has been termed 'modified control limits'.

Modified Exponential Curve   See **Growth Curve**.

Modified Latin Square   A generalisation of the **Semi-Latin Square** described by Rojar & White (1957).

Modified Mean   This term is found in two different senses. In the first, which is very bad practice, it refers to the mean of the highest and lowest values of a set of values, or what is more generally known as the **Mid-range**. In the second, it refers to the mean of a set of observations from which certain values have been rejected as atypical.

Modified Mean Square Successive Difference   For a series of $2m$ independent observations this quantity is

$$\delta_0^2 = \{1/4(m-1)\} \sum_{i=1}^{2m-1} (x_{i+1}-x_i)^2 \quad i \neq m.$$

[See **Variate Difference**.]

Modified von Neumann Ratio   A minor modification of the **von Neumann Ratio** due to Geisser (1957) which is twice the ratio of the **Modified Mean Square Successive Difference** to the pooled variance of observations 1 to m and $m+1$ to $2m$.

Moment   In general, a moment is the **Mean Value** of a power of a variate; for a univariate value x with distribution $dF(x)$ the rth moment of the variate $g(x)$ is

$$\int_{-\infty}^{\infty} \{g(x)\}^r dF(x).$$

More generally, for a multivariate distribution $dF(x_1, x_2, ..., x_p)$, the moment of order $(r_1, r_2, ..., r_k)$ of the functions $g_1, g_2, ..., g_k$ is the expectation

$$\int_{-\infty}^{\infty} \int_{-\infty}^{\infty} g_1^{r_1} ... g_k^{r_k} dF(x_1, ..., x_p).$$

In particular the moment of a variate x is given by

$$\mu'_r = \int_{-\infty}^{\infty} x^r dF(x)$$

and the moment about a particular fixed value a by

$$\int_{-\infty}^{\infty} (x-a)^r dF(x),$$

again with obvious generalisations to the multivariate case.

Moment Coefficient   The older and obsolete term for what are now called **Moments**.

Moment Estimator   Estimator(s) of the unknown parameter(s) of a distribution by fitting **Moments**.

Moment Generating Function   A function of a variable t which, when expanded as a power series in t yields the moments of a frequency distribution as coefficients of the powers. For example, the characteristic function is a moment generating function in virtue of the formal expansion

$$\phi(t) = \int_{-\infty}^{\infty} e^{itx} dF(x) = \sum_{r=0}^{\infty} \frac{(it)^r}{r!} \mu_r'.$$

By an easy extension from the univariate case there may be derived moment generating functions for multivariate distributions, the characteristic function being one such.

Moment Matrix   If p variates $x_1, x_2, ..., x_p$ have second order moments typified by μ_{ij} as the product moment of x_i and x_j, the matrix whose jth row and jth column is μ_{ij} is called the moment matrix. It is, in fact, not a display of all the joint moments of the variates, but only those of the second order. If the moments are taken about the respective variate means the matrix becomes the covariance or dispersion matrix.

Moment Ratio   A ratio in which the numerator and the denominator are moments or simple functions of moments. In certain cases the moment ratios may be interpreted as characteristics of the frequency distribution. For example, the most common of the moment ratios are those referring to the shape of the distribution:

a measure of skewness: $\beta_1 = \mu_3^2/\mu_2^3$

a measure of kurtosis: $\beta_2 = \mu_4/\mu_2^2$.

More general ratios of this type, due to K. Pearson, are

$$\beta_{2n+1} = \frac{\mu_3 \mu_{2n+3}}{\mu_2^{n+3}}$$

$$\beta_{2n} = \frac{\mu_{2n+2}}{\mu_2^{n+2}}.$$

[See also **Beta Coefficients**, **g-statistics**, **Gamma Coefficients**.]

Moments, Method of A method of curve fitting which proceeds by identifying the lower moments of the observed data with those of the particular curve form being fitted. This method has been generally associated with the fitting of frequency distributions of the Pearson type. Where sampling questions are involved the method is generally not the most efficient.

Monotone Likelihood Ratio If there is a family of density functions $p_\theta(x)$ and if, for $\theta < \theta'$, the distributions P_θ and $P_{\theta'}$ are distinct then if $p_{\theta'}(x)/p_\theta(x)$ is monotone increasing we have a family which is monotone likelihood ratio.

Monotonic Structure An alternative term proposed by Barlow & Proschan (1965) for **Coherent Structure**.

Monte Carlo Method A term which has been used with several different meanings:
(a) to denote the approximate solution of distributional problems by sampling experiments; this usage is not to be recommended.
(b) to denote the solution of mathematical problems arising in a stochastic context by sampling experiments. For example, the **Fokker-Planck Equation** arises in several physical problems, but it also arises in a probability problem, and hence sampling can be used to obtain approximate solutions applicable to the physical case.
(c) by extension of (b), the solution of any mathematical problem by sampling methods; the procedure is to construct an artificial stochastic model of the mathematical process and then to perform sampling experiments upon it.

Monthly Average By analogy with annual averages and **Moving Averages** generally this term ought to refer to the average of values of a time series occurring within a month, the resulting figure being representative of that particular month. In practice the phrase is sometimes used to denote the averaging of monthly values occurring in the same month, e.g. January from year to year, the object being to provide a pattern of seasonal fluctuation. This is objectionable and a better expression would be 'seasonal average by months'.

Mood's W-test A distribution free procedure for dispersion proposed by Mood (1954) for the two sample problem. If there are two samples m, n from distributions $F(x)$ and $G(y)$, the observations in the combined sample $m+n$ can be ranked and the statistic W formed as follows

$$W = \sum_{i=1}^{n} \left(r_i - \frac{m+n+1}{2}\right)^2$$

where r_i is the rank of the ith observation from the sample of n from $G(y)$. It should be noted that subsequent writers tend to refer to this as the M-test or statistic.

Mood-Brown Estimation (of a line) A method proposed by Mood & Brown (1951) for ascertaining the para-

meters in the line $\hat\alpha_n + \hat\beta_n x$ by determining them so that
$$\underset{(i:\, x_i \leq M)}{\text{median}} (y_i - \hat\alpha_n - \hat\beta_n x_i) = \underset{(i:\, x_i > M)}{\text{median}} (y_i - \hat\alpha_n - \hat\beta_n x_i) = 0.$$
This method may be generalised to fitting curves of higher degree.

Mood-Brown Median Test A distribution free test of the difference between k-populations proposed by Brown & Mood (1951) based upon the overall median of the k samples. If A and B are the numbers of observations above and below the grand median ($\tilde M$) then the
Mood-Brown statistic is $1/AB\{ \sum_{i=1}^{k} (n_i^\dagger N' - n_i' A)^2/n_i'\}$
where $n_i^\dagger$ are the sample observations above $\tilde M$ and $N' = A + B = \Sigma n_i'$.

Moran's Test Statistic A test (1951) of the null hypothesis that data from an observed series of events arise from a **Poisson Process** against the alternative hypothesis that they arise from a **Renewal Process**. [See also **Sherman's Test Statistic**.]

Mortara Formula A method proposed by Mortara (1949) to enable age specific fertility rates to be calculated from census results
$$f(y) = \frac{K_{y+1}}{L_{y+1}} - \frac{K_y}{L_y} = F(y+1) - F(y)$$
where L_y denotes the number of women ($y \pm 0.5$) years old at the census date and K_y the total number of live children born to these women at that date.

Moses Test A two sample distribution free test based upon the hypergeometric distribution proposed by Moses (1952) for similarity of proportions in two populations.

Most Efficient Estimator An unbiased estimator whose sampling variance is not greater than that of any other unbiased estimator is called a most efficient estimator, the qualification 'unbiassed' being understood. For biassed estimators the same expression is sometimes used to denote an estimator for which the mean square error is minimal. If an estimator is consistent, that is to say asymptotically unbiassed, the mean square error is asymptotically equal to the variance and the two usages coincide. Such an estimator, if its variance is minimal, is called asymptotically most efficient.

Most Powerful Critical Region The **Critical Region** which has the highest **Power** in testing a hypothesis.

Most Powerful Rank Test A two sample rank order test which is most powerful under some alternative hypothesis; for example, Normal distributions differing only in the mean.

Most Powerful Test A test of a hypothesis which is most powerful against an alternative hypothesis. [See **Power, Uniformly Most Powerful Test**.]

Most Selective Confidence Intervals An alternative name proposed by Kendall (1946) for what Neyman (1937) designated as 'shortest' confidence intervals, the objection to Neyman's term being that such intervals were not necessarily shortest in terms of length. Whereas 'shortest' confidence intervals should be concerned only with the narrowness of the intervals the concept of 'most selective' confidence intervals gives due weight to the frequency with which alternative values of the parameter are covered. The most selective set of confidence intervals covers false values of the parameters with minimum frequency.

Most Stringent Test A test of a statistical hypothesis (H_0) is said to be most stringent if it minimises the maximum difference by which the test falls short, with respect to a particular class of alternative hypotheses (H_1), of the power that could be attained with respect to these alternatives. A **Uniformly Most Powerful Test** is necessarily most stringent.

Mosteller's k-sample Slippage Test A statistic proposed by Mosteller (1948) to detect the existence of an extreme population. It is based upon the number of observations in one sample greater than all observations in the remaining $k-1$ samples. It was extended to the case of unequal sample sizes by Mosteller & Tukey (1950).

Mover-Stayer Model This generalisation of the Markov chain model assumes two types of individuals in the population under consideration. First, the 'stayer' who, with probability one, remains in the same category during the entire period of study and secondly, the 'mover' whose changes in category over time can be described by a Markov chain with constant transition probability matrix.

Moving Annual Total A series derived from an observed time series in which each term consists of the current observation and those immediately preceding it for the period of a year, e.g. the moving annual total of a monthly series would consist of the sums of twelve consecutive monthly values. It may be regarded as the first stage in the computation of a simple **Moving Average** for which the span or extent is one year. This derived series, however, has an existence in its own right since the upper curve of the three curves on a **Z-chart** is a moving annual total.

Moving Average If a time series is $x_1, x_2, ..., x_n$ and there are chosen a set of weights $w_0, w_1, ..., w_k$ ($\sum_{i=0}^{k} w_i = 1$) the series of values

$$u_t = \sum_{j=0}^{k} w_j x_{t+j}, \ t = 1, 2, ..., n-k,$$

are the moving averages of the series. In practice it is usual to choose k to be odd, say $2p+1$, and to locate the

corresponding u_t at the middle of the span of $2p+1$ values which contribute to it. By a suitable choice of weights the series can be represented locally by the values of a polynomial and hence 'smoothed'.

If all the weights are equal to $1/k$ the moving average is said to be *simple* and can be constructed by dividing the **Moving Total** by k. [See also **Trend, Smoothing.**]

Moving Average Disturbance In an equation expressing a relationship between variates it is sometimes convenient to include a final term which serves to summarise the effect of factors not separately specified. If such a term z_t takes the form of a **Moving Average Process**—say,

$$z_t = \alpha_0 \epsilon_t + \alpha_1 \epsilon_{t-1},$$

then the equation is said to possess a moving average disturbance.

Moving Average Method A method for estimating the **Median Effective Dose** of a stimulus from data on quantal responses suggested by Thompson (1947). A moving average of span k formed from the proportions of subjects responding to the various doses of the stimulus is associated with the corresponding average doses. The method then proceeds by linear interpolation between the successive values of the first moving average to estimate the value for which the smoothed response would be 0·50.

This method is valid only when the tolerance distribution is symmetrical.

Moving Average Model The representation of a stationary stochastic process in terms of a moving average of infinite length. The observation x_t is a linear combination of residuals z_t of the form

$$\{x_t\} = \sum_{j=0}^{\infty} b_j z_{t-j}.$$

Moving Average Process A special case of the **Moving Summation Process**. If $\{\epsilon_t\}$ be a random process the process $\{\xi_t\}$ defined by

$$\xi_t = \sum_{t=0}^{k} \alpha_j \epsilon_{t-j}$$

will exist and be stationary if $\{\epsilon_t\}$ is so. It is called a moving average process.

Moving Observer Technique A method of enumerating a moving population in which the observer himself moves among the population. If, for example, it is required to estimate the number of people in a street the observer walks in one direction making a net count of people he passes in whatever direction they are moving, deducting those who overtake him. This process is repeated in the reverse direction and the average of these two counts gives an estimate of the average number of people in the street during the time of the count.

Moving Range A concept similar to that of moving average. If the total number of measurements on a variable x is N, by taking absolute differences of successive pairs $|x_i - x_{i-1}|$, we have $N-1$ values which can be recorded as (moving) ranges of successive samples, each of size two. This can be generalised to $|x_{i+k} - x_i|$, $i = 1, 2, ..., n-k$ in order to afford a greater degree of smoothing.

Moving Seasonal Variation A pattern of seasonal variation which changes with time. It is usually obtained by determining a seasonal pattern for a certain number k of consecutive years and 'moving' the set of k years along the series as for a **Moving Average**, so obtaining a seasonal pattern for each year based on the previous k years. The method has the advantage that it avoids the under and over correction that might be induced by any fixed seasonal pattern. However, it may well become too flexible and remove more than the true seasonal movement. It is also difficult to project into the future except under assumptions which render it little different from the fixed seasonal pattern.

Moving Summation Process If a random stochastic process be written as $\{\epsilon_t\}$ then the sums formed by

$$\xi_t = \alpha_0\epsilon_t + \alpha_1\epsilon_{t-1} + \alpha_2\epsilon_{t-2} + ... - \infty < t < \infty$$

form a stationary stochastic process $\{\xi_t\}$ subject to certain convergence conditions on the coefficients α. Such a process is defined by Kolmogorov (1941) as a moving summation process. Two special cases are the **Moving Average Process** and the **Autoregressive Process**.

Moving Total For a series of ordered terms $x_1, x_2, ..., x_n$ the sums

$$\sum_{i=1}^{k} x_i, \sum_{i=2}^{k+1} x_i, \sum_{i=3}^{k+2} x_i, ...$$

are called moving totals. When divided by k they provide a **Moving Average** with equal weights.

Moving Weights In most cases where a moving average is taken of a time series the weights composing the average are constants independent of time. For certain purposes, however, it is desirable to have weights which themselves reflect changing circumstances, as for instance in index numbers of prices where the quantities purchased may alter as time goes on. In such cases the weights themselves may be moving averages of time series and are said to be moving weights. More generally the term can be used to describe any set of weights which change with time.

Multi-binomial Test A term introduced by Bradley (1953) to describe a test of the hypothesis that, by the method of **Paired Comparisons**, a set of t objects A_i, $i = 1, 2, ..., t$ are of equal merit. The test consists of pooling the $\frac{1}{2}t(t-1)$ independent binomial tests which can be made into a **Preference Table** by means of the statistic $4n \sum_{i<j} (p_{ij} - \frac{1}{2})^2$ which, under Bradley's alternative

hypothesis, has asymptotically a non-central χ^2 distribution with $\frac{1}{2}t(t-1)$ degrees of freedom. The p_{ij} are the binomial probabilities in the right hand upper triangle in the preference table.

Multicollinearity In regression analysis, a situation in which there exists a linear relation connecting the predicated ('independent') variables. The coefficients of the regression on these variables are then indeterminate and their standard errors become infinite.

The same word is used when the predicated variables are subject to error. In a sample there may occur a collinearity due to sampling accident; or there may be a linear relation among the variables which, in general, does not become a linear relation in the observed values owing to the observational errors. The regression line in such a case is determinate but unreliable in the sense that a different sample might give entirely different results. The detection of underlying linearities of this kind is one of the objects of **Confluence Analysis**.

Strictly speaking, perhaps, the existence of one linear relation should be described as collinearity and the existence of several as multicollinearity.

Multi-equational Model A model of a system where the variables are interconnected by more than one equation. An alternative name is '**Simultaneous Equations Model**'.

Multi-factorial Design A vague phrase which ought to refer to any experimental design involving more than one factor. Since, however, most factorial experiments are of this type the term, in such a sense, is practically equivalent to 'factorial design'. Some writers appear to use the expression in cases where several (say at least three) factors are involved.

Multi-level Continuous Sampling Plans A type of sampling for control of a continuous process which permits sampling at more than one partial level and complete inspection. In fact, the plan proposed by Lieberman & Solomon (1955), was a random walk with reflecting barriers.

Multilinear Process A type of stochastic process, introduced by Parzen (1957), which may be represented as: $X(t) = \Sigma a(v_1.........v_k) W_1(t-v_1)... W_k(t-v_k)$ where k is a positive integer, $a(v_1.........v_k)$ constants subject to certain conditions.

Multi-modal Distribution A frequency distribution with more than one modal value.

Such distributions are comparatively rare when they are derived from homogeneous material and, in fact, multimodality is often accepted as presumptive evidence that the underlying variation is a mixture of different distributions.

Multi-phase Sampling It is sometimes convenient and economical to collect certain items of information from

99

the whole of the units of a sample and other items of usually more detailed information from a sub-sample of the units constituting the original sample. This may be termed two phase sampling, e.g. if the collection of information concerning variate, y, is relatively expensive, and there exists some other variate, x, correlated with it, which is relatively cheap to investigate, it may be profitable to carry out sampling in two phases. At the first phase, x is investigated, and the information thus obtained is used either (*a*) to stratify the population at the second phase, when y is investigated, or (*b*) as **Supplementary Information** at the second phase, a **Ratio** or **Regression** estimate being used. Two phase sampling is sometimes called 'double sampling'. Further phases may be added if desired. It may be noted, however, that multi-phase sampling does not necessarily imply the use of any relationships between the variates x and y. The expression is not to be confused with **Multi-stage Sampling**.

Multinomial Distribution   The discrete distribution associated with events which can have more than two outcomes: it is a generalisation of the **Binomial Distribution**. If there are k possible incompatible and exhaustive results of some chance event for which the separate probabilities are $p_i(i = 1, 2, ..., k)$ then in n trials the distribution of x_1 events of the first kind, x_2 of the second kind ... x_k of the kth kind is

$$f(x_1, x_2, x_3, ..., x_k) = n! \prod_{i=1}^{k} \left\{ \frac{p_i{}^{x_i}}{x_i!} \right\}, \ 0 \leqslant x_i \leqslant n.$$

that is to say, is the term involving $\Pi p_i{}^{x_i}$ in the multi-nomial expansion of $(p_1 + p_2 + ... + p_k)^n$.

Multiple Bar Chart   A chart depicting two or more characteristics in the form of bars of length proportional to the magnitude of the characteristics. For example, a chart comparing the age and sex distribution of two populations may be drawn with sets of pairs of bars, one bar of each pair for each population, and one pair for each age group. [See also **Component Bar Chart**.]

Multiple Classification   An alternative name for **Manifold Classification**. Sometimes the expression is restricted to the case of quantitative variates.

Multiple Comparisons   Apart from the obvious colloquial meaning, this term occurs mainly in variance analysis. If the data are categorised into a number k of groups it is usual to test the difference of a pair of group means against the residual variance. However, the tests of the $\frac{1}{2}k(k-1)$ possible pairs are not independent, and problems arise as to how to test, say, the difference between the largest and the smallest difference. [See also **Duncan's Test, Gabriel's Test, Newman-Keuls Test, Scheffé's Test, Tukey's Test**.]

Multiple Correlation, Coefficient of   The product moment correlation between the actual values of the 'dependent'

variate in multiple regression and the values as given by the regression equation. It measures the closeness of representation by the regression line and may also be regarded as the maximum of the correlation coefficient between the 'dependent' variate and all linear functions of a set of two or more of the 'independent' variates. The coefficient is usually denoted by R but is regarded as essentially non-negative, the quantity R^2 being the one which occurs in practice.

Multiple Curvilinear Correlation   A regrettably inexact term used sometimes to describe the relationship between variates which do not have linear regression one on another but are not independent. There seems to be nothing to recommend the use of this term.

Multiple Decision Methods   A general term to describe statistical techniques for dealing with problems where the possible outcomes or decisions are more than two.

Multiple Decision Problem   The problem of choosing one hypothesis or decision from a set of k mutually exclusive and exhaustive hypotheses or decisions on the basis of some observations on a random variable. The simple problem of testing a statistical hypothesis against a single alternative is a special case of this, when $k = 2$.

Multiple Factor Analysis   In current usage this is equivalent to **Factor Analysis**. Historically, the analysis of psychological material into factors grew from one factor to two factor and then to m-factor complexes and the latter was called 'multiple' to distinguish it from the simpler forms. This no longer seems necessary.

Multiple Markov Process   A stochastic process in which the transition probabilities depend on previous values at more than one point. In most contexts the expression is equivalent to '**Autoregressive Process**'. The expression is to be distinguished from the Multivariate Markov Process, which is a **Markov Process** in more than one variate.

Multiple Partial Correlation, Coefficient of   An extension of the classical correlation coefficient by Cowden (1952), based on a suggestion of Hotelling (1926), to the case of multiple correlation between the 'dependent' variate and two or more 'independent variates' when all these have been adjusted for the effect of one or more other variates.

Multiple Phase Process   A stochastic birth process (D. G. Kendall, 1948) in which an individual, after being "born', passes through k successive phases and can only subdivide or give birth itself after the kth phase. The lifetimes in each phase are usually taken to be independently distributed. If $k = 1$ the process becomes the **Simple Birth Process**.

Multiple Poisson Distribution   This distribution is the joint distribution of independent Poisson variates. It may also be regarded as an approximation to the **Multinomial Distribution** when m is large and $np_i = \lambda_i$ in the form:

$$e^{-(\lambda_1 + \ldots + \lambda_{r-1})} \frac{\lambda_1^{k_1} \lambda_2^{k_2} \ldots \lambda_{r-1}^{k_{r-1}}}{k_1! k_2! \ldots k_{r-1}!}.$$

Multiple Poisson Process A term sometimes used for a generalisation of a **Poisson Process** where there are simultaneous occurrences at the event points.

Multiple Random Starts An idea for selection procedure under systematic sampling originally due to Tukey and developed by others, e.g. Gautschi (1957). The method is to choose a random sample of s without replacement from the first k elements and subsequently every kth multiple element of those selected.

Multiple Range Test A method of comparing mean values arising in analyses of variance by using the range of subsets of means, or of ranges of sets of observations contributing to mean values. [See also **Duncan's, Gabriel's, Newman-Keuls', Scheffé's** and **Tukey's Tests.**]

Multiple Recapture Census A sequential method of **Capture/Release Sampling** where the method of capture does not kill or affect the behaviour of the sample unit and where the experimenter has full control over the sampling and marking. [See **Capture-Recapture.**]

Multiple Regression The regression of a dependent variate on more than one 'independent' or 'predicated' variable.

Multiple Sampling See **Double Sampling, Sequential Analysis.**

Multiple Smoothing Method A generalisation by Brown & Meyer (1961) of the simple model for **Exponential Smoothing** to take account of polynomial trends. This approach should not be confused with the iterated application of a moving average to a time series.

Multiple Stratification If a sample is stratified according to two or more factors it is said to be multiply stratified. In practice multiple stratification is difficult to carry out because the information with which to divide the population into sub-strata is often not available. [See **Control of Sub-strata.**]

Multiplicative Process A synonym of **Branching Process.**

Multi-stage Estimation A generalisation to n stages (Blum & Rosenblatt, 1963) of the principle of estimation first proposed for two stages by Stein (1945).

Multi-stage Sampling A sample which is selected by stages, the sampling units at each stage being subsampled from the (larger) units chosen at the previous stage. The sampling units pertaining to the first stage are called primary or first stage units; and similarly for second stage units, etc. Where the sampling frame has to be constructed in the course of the sampling operation, multi-stage sampling has the additional advantage that only the parts of the population selected at any stage need to be listed for sampling at the next stage.

Multi-temporal Model See **Dynamic Model.**

Multi-valued Decision The ordinary test of a statistical hypothesis involves according to some, a two valued decision: accept or reject, though other writers contend that failure to reject is not the same thing as acceptance. Where there are more than two possible decisions, for example, to sell/to stock/to destroy, the decision is said to be multi-valued. The simple plan for acceptance sampling by sequential methods is a case of a multi-valued decision problem. At each successive stage in the sampling a decision has to be taken: accept/reject/continue sampling.

Multivariate Analysis This expression is used rather loosely to denote the analysis of data which are multivariate in the sense that each member bears the values of p variates.

The principal techniques of multivariate analysis, beyond those admitting of a straightforward generalisation, e.g. regression, correlation and variance analysis, are **Factor** and **Component Analysis, Classification, Discriminatory Analysis, Canonical Analysis** and various generalisations of homogeneity tests, as illustrated by the D^2 statistic. [See also **Hotelling's T, Wishart's Distribution** and **Wilks' Criterion.**]

Multivariate Beta Distribution If there are two symmetric matrices $\mathbf{A}$ and $\mathbf{B}$ of order p with independent Wishart distribution (f_1, f_2 degrees of freedom), there exists a lower triangular matrix $\mathbf{C}$ so that $\mathbf{A} + \mathbf{B} = \mathbf{CC}'$. Let $\mathbf{A} = \mathbf{CLC}'$ then the distribution of $\mathbf{L}$ is the Multivariate Beta Distribution (Kshirsagar, 1961). It has the form

$$f \propto |\mathbf{L}|^{\frac{1}{2}(f_1 - p - 1)} |\mathbf{I} - \mathbf{L}|^{\frac{1}{2}(f_2 - p - 1)}.$$

Multivariate Binomial Distribution An s-variate generalisation of the binomial distribution. For example, if a member can bear the value of two dichotomised variables with probabilities

		Variable 1	
		Present	Absent
Variable 2	Present	p_{11}	p_{10}
	Absent	p_{01}	p_{00}

the bivariate binomial in samples of n is given by $(p_{11} + p_{10} + p_{01} + p_{00})^n$.

Multivariate Burr's Distribution A multivariate form of the **Burr Distribution** developed by Takahasi (1965).

Multivariate Distribution The simultaneous distribution of a number p of variates ($p > 1$): or equivalently, the probability distribution of p variates.

101

Multivariate Exponential Distribution A distribution of considerable importance in life assessment of systems. The complement of the distribution function is given by $\Pr\{X_1 > x_1, X_2 > x_2, ..., X_n > x_n\}$ which is equal to

$$\exp\left[-\sum_{s \in S} \lambda_s \max(x_i s_i)\right],$$

where S is a set of vectors; marginal distributions are ordinary negative exponential (Marshall & Olkin, 1967).

Multivariate 'F' Distribution If there are two non-symmetric matrices $m \times p$ and $m \times n$ ($p \leqslant m \leqslant n$) denoting matrix variates X and Y and all columns are Normally and independently distributed with covariance Σ, the multivariate F-distribution is that of $X' (YY')^{-1}X$ provided $\mathcal{E}(X) = \mathcal{E}(Y) = 0$.

Multivariate Hypergeometric Distribution Suppose n balls are drawn at random without replacement from an urn which contains N balls of $(s+1)$ different colours, M_i being of the ith colour, $i = 0, 1, 2, ..., s$. If X_i is the number of balls of the ith colour $i = 1, 2, ..., s$ drawn in the n draws then $(X_1, X_2, ..., X_s)$ has the s-dimensional hypergeometric distribution.

This distribution is also called the factorial multinomial distribution.

Multivariate Inverse Hypergeometric Distribution This is the limiting form of the **Negative Multinomial Distribution** when $N \to \infty$ and the $M_i \to \infty$ so that $M_i/N \to p_i$ of the negative multinomial. [See also **Inversion**.]

Multivariate Moment See **Product Moment**.

Multivariate Multinomial Distribution A multivariate distribution of the multinomial type. It is difficult to write down in generality but an example will be found under **Bivariate Binomial**.

Multivariate Negative Binomial Distribution An alternative name for the **Negative Multinomial Distribution**.

Multivariate Negative Hypergeometric Distribution An alternative name for the **Multivariate Inverse Hypergeometric Distribution**.

Multivariate Normal Distribution A generalisation of the univariate Normal distribution to the case of p-variates ($p \geqslant 2$). If the ith variate x_i has mean m_i and the covariance (dispersion) matrix of the variates is (v_{ij}) $i, j = 1, 2, ..., p$ with an inverse V_{ij}, the multivariate Normal distribution has the frequency function

$$\frac{|V_{ij}|^{\frac{1}{2}}}{(2\pi)^{\frac{1}{2}p}} \exp\left\{-\tfrac{1}{2} \sum_{i,j=1}^{p} V_{ij}(x_i - m_i)(x_j - m_j)\right\}.$$

Multivariate Pareto Distribution An extension by Mardia (1962) of the two forms of bivariate Pareto distribution. The conditional distributions are of Pareto form with displaced origins.

Multivariate Pascal Distribution If a population of individuals can be described by one or more of characters $A_1 ... A_s$ with probabilities of individuals as $p_1, p, ... p_s$, then random observations are taken until k individuals possessing none of the characteristics are recorded, the x_i relating to occurrences of A_i have an s-variate Pascal distribution. It is a special case of the multivariate negative binomial distribution.

Multivariate Poisson Distribution An s-variate generalisation of the Poisson distribution and the limiting distribution as $n \to \infty$ and $p_i \to 0$ of the multivariate binomial distribution.

Multivariate Pólya Distribution Suppose an urn contains N balls of $(s+1)$ different colours, a_i being the ith colour, $i = 1, 2, ..., s$ and b of the $(s+1)$th colour. Suppose n balls are drawn one after another, with replacement, such that at each replacement c new balls of the same colour are added to the urn. If X_i denotes the number of balls of the ith colour in the sample, $i = 1, 2, ..., s$ then $(X_1, X_2, ..., X_s)$ has the s-variate Pólya distribution.

Multivariate Power Series Distribution Let

$$f(\theta_1, \theta_2, ..., \theta_s) = \sum_{x_1, x_2, ..., x_s} a_{x_1 x_2 ... x_s} \theta_1^{x_1} \theta_2^{x_2} ... \theta_s^{x_s}$$

be a convergent power series in $\theta_1, \theta_2, ..., \theta_s$ such that $a_{x_1 x_2 ... x_s} \theta_1^{x_1} \theta_2^{x_2} ... \theta_s^{x_s} \geqslant 0$, $x_i = 0, 1, 2, ...$; $i = 1, 2, ..., s$ for all $(\theta_1, \theta_2, ..., \theta_s) \in \Theta$, the s-dimensional parameter space. Then the s-variate power series distribution with the series function $f(\theta_1, \theta_2, ..., \theta_s)$ is defined by the probability function

$$\Pr(x_1, x_2, ..., x_s) = \frac{a_{x_1 x_2 ... x_s} \theta_1^{x_1} \theta_2^{x_2} ... \theta_s^{x_s}}{f(\theta_1, \theta_2, ..., \theta_s)}$$

Multivariate Processes A class of **Stochastic Processes** involving more than one random variable. Also known as simultaneous or vector processes and not to be confused with multi-dimensional processes which are concerned with more than one parameter.

Multivariate Quality Control Control of quality in which each item for inspection must conform to standards in respect of more than one variable, as, for example, the length, breadth and height of a metal block.

Multivariate Signed Rank Test Generalisations of the **Wilcoxon Test** for the case of matched bivariate and trivariate properties (Bennett, 1965).

Multivariate Tchebyshev Inequalities A development by Olkin & Pratt (1958) based upon the bivariate form (Berge, 1937). For example, for p uncorrelated random variates ($\rho = 0$)

$$\Pr(|y_i| \geqslant k_i \sigma_i \text{ for some } i) \leqslant \Sigma k_i^{-2} \text{ and } \rho = -1/(p-1)$$
$$\Pr(|y_i| \geqslant k_i \sigma_i \text{ for some } i) \leqslant (p-1)/k^2.$$

Murthy's Estimator In sampling without replacement, an estimator of the population mean and variance proposed by Murthy (1957) and based upon order statistics. It is of limited use for sample size greater than three owing to the heavy computation involved.

*Mutability (Mutabilità) See **Modality**.

Nearly Best Linear Estimator A form of estimator proposed by Blum (1956) making use of approximations for the problem of minimum variance which is central to the 'best' linear estimator.

Negative Binomial Distribution A distribution in which the relative frequencies (probabilities) are arrayed by a binomial with a negative index. For example, if the variate values are 0, 1, 2, etc., the frequency at $x = j$ is the coefficient of t^j in the expansion of $(1-pt)^{-n}(1-p)^n$ in powers of t. The distribution is sometimes called the Pascal distribution.

Negative Exponential Distribution A synonym for the **Exponential Distribution**.

Negative Factorial Multinomial Distribution An alternative name for the **Multivariate Inverse Hypergeometric Distribution**.

Negative Hypergeometric Distribution The analogue of the **Negative Binomial** when sampling is from a finite population of attributes, without replacement, i.e. it is the distribution of the sample number required in such circumstances to reach a pre-assigned number of 'successes'.

Negative Moments The negative moments of a frequency distribution are the moments of reciprocal powers of the variate; for example, $\mathcal{E}(1/x^k)$ where k is the order of the moment. Their existence, of course, depends upon convergence and they are little used. In fact, owing to possible confusion with ordinary moments, which have negative sign, the preferable expression would be 'moments of negative order'.

Negative Multinomial Distribution If there is a sequence of independent trials in each of which there can be $s+1$ mutually exclusive outcomes $A_0, ..., A_s$ with probabilities $p_0, ..., p_s$ ($\Sigma p_i = 1$) and $x_1, ..., x_s$ are the occurrences of $A_1, ..., A_s$ before A_0 occurs k times, then $(x_1, ..., A_s)$ has the negative multinomial distribution.

This distribution can also arise as a compound of Poisson variates where the parameter has a gamma distribution: in this form it is also known as the multivariate negative binomial distribution.

Nested Balanced Incomplete Block Design An experiment design with two systems of blocks, the second within the first such that ignoring either system leaves a balanced incomplete design whose blocks are those of the other system. In the nesting system each block from the first part contains m blocks from the second.

Nested Design A class of experimental design in which every level of a given factor appears with only a single line of any other factor. Factors which are not nested are said to be crossed. If every level of one appears with every level of the others, the factors are said to be completely crossed: if not, they are partly crossed.

Nested Hypotheses A sequence of hypotheses in which the hypothesis at any stage is contained in all hypotheses later in the sequence. Thus, if Ω represents the set of underlying assumptions common to the whole sequence, and if $H_1, H_2, ..., H_k$ are further hypotheses, the sequence of nested hypotheses is $\Omega \prod_{i=1}^{k} \cap H_i$.

Nested Sampling A term used in two somewhat different senses: (1) as equivalent to **Multi-stage Sampling** because the higher-stage units are 'nested' in the lower-stage units; (2) where the sampling is such that certain units are imbedded in larger units which form part of the whole sample, e.g. the **Entry-Plots** of clusters are 'nested' in this sense.

Net Correlation An alternative but obsolescent term for **Partial Correlation**. The name is derived from the fact that a partial correlation is the correlation between two variates of a larger group when the remaining variates are held constant giving, in a sense, a 'net' correlation. [See also **Total Correlation**.]

Network of Samples An alternative name for a set of **Interpenetrating Samples**.

*Neutral Curve (Curva Neutra) The frequency curve of a distribution with neutral **Abnormality**.

Newman-Keuls Test A version of a multiple range comparison procedure by Newman (1939) and Keuls (1952) where the sample ranges are tested against the studentised range of the subsets rather than the ranges of mean values. It is a step-by-step procedure subsequently modified by Duncan.

Neyman Allocation A method of allocating sample numbers to different strata in order to secure unbiassed estimators of parent mean values with minimal variance. The numbers allocated, for large samples, are proportional to the standard deviations of the variable under examination in the respective strata. This method was advanced by Neyman (1934). [See also **Optimum Allocation**; **Proportional Sampling**.]

Neyman Model A term proposed by Ogawa (1963) to denote experiment designs which involve situations containing technical errors in the sense that variations in replicate observations are solely due to experimental techniques. [See also **Fisher Model**.]

Neyman-Pearson Theory A general theory of testing hypotheses, due to J. Neyman and E. S. Pearson. It is based upon the consideration of two types of errors which may be incurred in judging a statistical hypothesis. Since the 1930s the theory has been developed by other writers and led to the more general theories, e.g. the statistical decision functions of Wald.

The probabilities of errors incurred by rejecting an hypothesis H_0 when it is true and accepting it when it is false—the **Errors of First and Second Kind**—are usually written α and $1-\beta$. The function $\beta(H_1)$, namely the probability with which the hypothesis H_0 is rejected when some alternative hypothesis H_1 is true, is called the **Power Function** of the particular test. It is the concept of the power function coupled with that of the two kinds of errors that form the principal features of the Neyman-Pearson theory.

Neyman-shortest Unbiassed Confidence Intervals An optimum set of confidence intervals which, among all unbiassed α-confidence intervals, it uniformly minimises the probability of covering false values.

The word 'shortest' may be misleading in that the intervals do not necessarily have minimal length. The alternative 'most selective' has been proposed.

Neyman's ψ^2 Test The first of the 'Smooth' tests of goodness-of-fit, developed as alternatives to the χ^2 test initially by Neyman (1937). The test statistic is

$$\psi_k^2 = \sum_{r=1}^{k} u_r^2 = \frac{1}{n} \sum_{r=1}^{k} \left\{ \sum_{i=1}^{n} \pi_r(\gamma_i) \right\}^2$$

where the γ_i are transformed observations from x_i, by probability integral transforms, and $\pi_r(\gamma)$ linearly transformed Legendre polynomials.

Noise A convenient term for a series of random disturbances borrowed, through communication engineering, from the theory of sound. In communication theory noise results in the possibility of a signal sent, x, being different from the signal received, y, and the latter has a probability distribution conditional upon x. If the disturbances consist of impulses at random intervals it is sometimes known as 'shot noise'. [See also **White Noise**.]

Nomic See **Clisy**.

Nomogram A form of line chart upon which appears scales for the variables involved in a particular formula in such a way that corresponding values for each variable lie on a straight line which intersects all the scales. In non-elementary statistical work the nomogram is not as extensively used as tables.

Noncentral Beta Distribution The ratio of two quantities, each distributed independently of the other as is the variance in Normal samples, has a beta distribution of the second kind. If the numerator is a sum of squares of Normal variables about some arbitrary value θ, the corresponding distribution is called the Noncentral Beta Distribution.

Noncentral Confidence Interval A confidence interval which is not **Central**.

Noncentral χ^2 Distribution The distribution of the sum of squares of independent Normal variates with unit variance but not with zero mean. The distribution was obtained by R. A. Fisher in 1928 as a special case of the distribution of the multiple correlation coefficient. It has several applications and in particular is required in order to determine the power function of the chi-squared test. The difficult problem of tabulation was investigated by Patnaik (1949) who derived certain approximations which used existing tabled functions.

Noncentral F-distribution The distribution of the ratio of a noncentral χ^2 to a central χ^2. Fisher gave the distribution in 1928. Wishart (1932) considered in it the form of the distribution of the correlation ratio. In a particular case the form reduces to that of the **Noncentral t-Distribution**.

The use of the distribution in evaluating the power of analysis of variance tests was extended by the approximations investigated by Patnaik (1949), a development of work by Tang (1938) and Hsu (1941).

Noncentral Multivariate Beta Distribution If, in a **Multivariate Beta Distribution** the matrix **B** has a **Noncentral Wishart Distribution** then the distribution of **L** is of the Noncentral Multivariate Beta form.

Noncentral Multivariate 'F' Distribution The distribution of $\mathbf{X}'(\mathbf{YY}')^{-1}\mathbf{X}$, which depends upon $\mathbf{M}'\Sigma^{-1}\mathbf{M}$, where the matrix variates **X** and **Y** are $m \times p$ and $m \times n$ respectively ($p \leqslant m \leqslant n$) with columns all independently Normally distributed with covariance Σ and $\mathscr{E}(\mathbf{X}) = \mathbf{M}$, $\mathscr{E}(\mathbf{Y}) = 0$.

Noncentral t-distribution The distribution of the ratio $(x-\alpha)/s$ where x is a Normal variate with zero mean and variance σ^2, α is a non-zero constant and s is distributed as $x\sigma/\sqrt{\nu}$ with ν degrees of freedom independently of x. The distribution is a simple transform of a particular case of the **Noncentral F-distribution** and is used, among other things, to determine the power function of 'Student's' t-test of the significance of the mean.

Noncentral Wishart Distribution The distribution of the second order dispersions, about arbitrary origins, of a multivariate Normal distribution; reducing to a Wishart distribution when the origins are the sample means. [See **Wishart Distribution**.]

Noncircular Statistic See **Circular Formula.**

Nondetermination, Coefficient of The square of the **Coefficient of Alienation,** that is to say, if r is the correlation between two variates, the coefficient of nondetermination is $1 - r^2$. [See also **Total Determination, Coefficient of.**]

Nonlinear Correlation This term is meant to relate to the correlation between variates where the regression is not linear. It is thus a misnomer; correlation, being a pure number, cannot be nonlinear. The usage is not to be recommended. [See also **Correlation Ratio.**]

Nonlinear Regression An alternative name for **Curvilinear Regression.**

Non Normal Population A population for which the frequency distribution is not the **Normal Distribution.** The term does not mean abnormal in the sense of unusual.

Nonnull Hypothesis In general, a hypothesis alternative to the one under test; the **Null Hypothesis.** In some contexts, however, it is given the meaning of a hypothesis under test where the effect is not equal to zero.

Nonorthogonal Data This expression originates in the use of **'Orthogonal'** to denote independence. Data are said to be nonorthogonal if they lead to estimates of various effects which are not independent of one another or, perhaps, of other features such as block differences which are a nuisance in the analysis. The disadvantage of such material is that effects thus mixed up may be genuinely inextricable or may require a more complicated technique for their disentanglement.

Nonparametric See **Distribution Free Method.**

Nonparametric Tolerance Limits **Tolerance limits** which do not depend on the parameters of the parent population from which a sample is drawn.

It seems possible, but is by no means universal practice, to draw a distinction between nonparametric limits, in which the parent distribution is known in form, and distribution free limits in which the form of the parent is unknown. [See also **Distribution free Method.**]

Nonrandom Sample A sample selected by a nonrandom method. For example, a scheme whereby units are selected purposively would yield a nonrandom sample. Again, a sample obtained by taking members at fixed intervals on a list is a nonrandom sample unless the list was arranged in a random order. [See also **Quasi-random Sampling.**]

Nonregular Estimator See **Regular Estimator.**

Nonresponse In sample surveys, the failure to obtain information from a designated individual for any reason (death, absence, refusal to reply) is often called a nonresponse and the proportion of such individuals of the sample aimed at is called the nonresponse rate. It would be better, however, to call this a 'failure' rate or a 'non-achievement' rate and to confine 'nonresponse' to those cases where the individual concerned is contacted but refuses to reply or is unable to do so for reasons such as deafness or illness.

Non-availability of information in other situations, e.g. arrival of the investigator for crop cutting experiments after harvesting, may also be termed nonresponse, or better, non-achievement.

When several items of information are to be collected for the same sample unit, it may so happen that information is not available for some of the items but available for others. The term nonresponse is usually not applied in such a situation; but incomplete response or incomplete achievement may be used.

Nonsampling Error An error in sample estimates which cannot be attributed to sampling fluctuations. Such errors may arise from many different sources such as defects in the frame, faulty demarcation of sample units, defects in the selection of sample units, mistakes in the collection of data due to personal variations or misunderstandings or bias or negligence or dishonesty on the part of the investigator or of the interviewee, mistakes at the stage of the processing of the data, etc.

The term 'response error' is sometimes used for mistakes in the collection of data and would not, strictly speaking, cover errors due to nonresponse. The use of the word 'bias' in the place of error, e.g. 'response bias' is not uncommon. The term 'ascertainment error' (Mahalanobis) is preferable as it would include errors due to nonresponse and also cases of collection of data by methods other than interviewing, e.g. direct physical observation of fields for crop estimates.

Nonsense Correlation See **Illusory Correlation.**

Nonsingular Distribution A multivariate distribution in, say, p variates which cannot, by a linear transformation of the variates, be converted into a distribution with fewer than p variates. A distribution is nonsingular if and only if the dispersion matrix or the **Correlation Matrix** is of rank p.

Normal Deviate The value of a deviate of the Normal distribution. [See also **Normal Equivalent Deviate.**]

Normal Dispersion See **Lexis Ratio.**

Normal Distribution The continuous frequency distribution of infinite range represented by the equation

$$dF = \frac{1}{\sigma \sqrt{(2\pi)}} e^{-\frac{1}{2}(x-m)^2/\sigma^2} \, dx, \; -\infty \leqslant x \leqslant \infty,$$

where m is the mean and σ the standard deviation.

In continental writings the distribution is often known as the Gaussian, the Laplacean, the Gauss-Laplace or the Laplace-Gauss distribution, or the Second Law of Laplace. It was apparently first discovered by De Moivre (1753) as the limiting form of the binomial distribution.

Normal Equations The set of simultaneous equations arrived at in estimation by the **Method of Least Squares**.

Normal Equivalent Deviate (N.E.D.) If P is a proportion or a probability and Y is defined by

$$P = \int_{-\infty}^{Y} \frac{1}{\sqrt{(2\pi)}} e^{-\frac{1}{2}x^2} \, dx$$

then Y is termed the N.E.D. of P. This quantity or the **Probit** are often used in the analysis of a stimulus-response relationship.

Normal Inspection The amount of inspection required by the initial application of a sampling inspection plan. It is undertaken so long as the quality of the product is close to the acceptable quality level laid down by the plan. If the actual quality improves consistently reduced inspection may be introduced, and conversely if the quality deteriorates tightened inspection is required.

Normal Probability Paper A specially ruled graph paper with a variate x as abscissa and an ordinate y scaled in such a way that the graph of the distribution function, y, of the Normal distribution, is a straight line.

Normal Scores Test A test of equality between the location of two distributions on the basis of a sample from each using the expected values of the order statistics from a standard Normal distribution. This was proposed by Fisher & Yates (1938) and a corresponding rank test by Hoeffding (1951). A closely related, and asymptotically equivalent, test was proposed by **Van de Waerden**.

Normalisation of Frequency Function The transformation of a variate so that its frequency function becomes Normal, or approximately so.

Normalisation of Scores In the analysis of data resulting from educational or psychological tests it is often desirable to convert each set of original scores to some standard scale. The process of doing so is called a normalisation of the scores, but this may mean the reduction to a norm or common standard, not necessarily to the scale of a Normal (Gaussian) distribution. Nevertheless, one method in common use is to determine the percentiles of the scores and then express these as corresponding deviations from the mean of a Normal distribution. This particular device is assisted by the use of **Normal Probability (Graph) Paper**. [See also *T*-score, *z*-score.]

Normalising Transform Transformations of variables with the object of reducing the distribution to the Normal or very closely so. It may be noted that a transformation to achieve, say, stabilisation of variance may also yield a large measure of normalisation. Usually, this expression denotes a transformation of a random variable so that its distribution approximates, or is exactly equal to the Normal (Gaussian) form. It also occurs in the sense of a transformation which reduces the distribution to some other standard form which is regarded as the norm. This usage is to be avoided.

Normit A contraction, or diminutive term, for **Normal Deviate** proposed by Berkson (1955) for use in connection with a method of analysing quantal response data due to Urban (1910). [See also **Probit**.]

Nuisance Parameters In the theories both of estimation and of tests of significance there arises the problem of finding a sampling distribution which is independent of certain unknown parameters of the population. Although these parameters are essential to the specification of the population they are a 'nuisance' in the formulation of exact statements about certain other parameters. The classical case of a nuisance parameter arises in the setting of confidence intervals for the mean, which depend on the unknown parent variance when the distribution of the mean is itself used; the difficulty in this case is overcome by the use of 'Student's' distribution which does not depend on the parent variance.

Null Hypothesis In general, this term relates to a particular hypothesis under test, as distinct from the alternative hypotheses which are under consideration. It is therefore the hypothesis which determines the probability of the **Type I Error**. In some contexts, however, the term is restricted to a hypothesis under test of 'no difference'.

Nyquist Frequency In connection with data consisting of observations recorded at equal distances of time, the frequency of a sinusoidal term whose period is twice the time interval between successive observations.

Nyquist Interval For time series consisting of a series of harmonic components with frequencies in a limited range, the Nyquist interval is such that the highest admissible frequency has a period of two intervals.

Nyquist-Shannon Theorem A theorem with applications in communications engineering, where it is known as 'the sampling theorem' and variously ascribed to Nyquist or Shannon. The theorem states that, if the probability density f has a characteristic function vanishing outside the interval $-a, a$, then f is uniquely determined by the values

$$\tfrac{1}{2}\pi f(n\pi/\lambda)$$

for any fixed $\lambda \geqslant a$, and these values induce an **Arithmetic Probability Distribution**.

'O' Statistics Estimation of quantities introduced by Loynes (1966) and derived from randomised order statistics.

ω^2-test An alternative name for the **Cramér-von Mises Test.**

Oblimax See **Factor Rotation.**

Oblique Factor In educational or psychological testing a factor which is correlated with one, or more, other factors is said to be oblique. In one geometrical representation of the situation the vectors which represent them are no longer orthogonal one to another. The use of oblique factors is confined almost entirely to psychology. [See also **Factor Analysis, Factor Pattern.**]

Observable Variable A mathematical or stochastic variable, the values of which can be directly observed, as distinct from unobservable variables which enter into structural equations but are not directly observable.

Observational Error This term ought to mean an error of observation but sometimes occurs as meaning a response error. [See **Nonsampling Error.**]

Occupancy Problems The general class of problems in probability which deal with the random distribution of r objects over n cells with particular reference to the numbers of objects falling in particular cells. [See also **Bose-Einstein Statistics.**]

Odds Ratio An alternative term for **Relative Risk.**

Ogive A general name for the **Galton Ogive.** [See also **Distribution Curve.**]

One Sided Test A test of a hypothesis for which the region of rejection is wholly located at one end of the distribution of the test statistic; that is to say, if the statistic is t the region is based on values for which $t >$ some t_1 or for which $t <$ some t_0, but not both.

One Way Classification When a set of variate values can be classified according to the k classes of a single factor such a classification is termed a 'one way' classification and forms the basis for the simplest case of **Variance Analysis.**

Open Sequential Scheme A sequential sampling scheme which does not impose an ultimate limit to the size of the sample. Many sequential sampling schemes will terminate with a high degree of probability even when no limit to the sample size is imposed; and such probability may be so high as to be 'almost certain'. Such schemes are nevertheless called open. [See **Closed Sequential Scheme.**]

Open Ended Classes If, in a frequency distribution, the initial class interval is indeterminate at its beginning and/or the final class interval is indeterminate at its end, the distribution is said to possess 'open ended' classes. This feature is undesirable owing to its effect upon certain calculations which require the central value of the class interval, e.g. the power moments. On the other hand, it usually has no effect upon the calculation of the quantiles and in particular the median value of the variate.

Open Ended Question A question which does not admit of a limited number of definite answers, as for example, 'What do you think of the present Government?'; as opposed to a closed ended question such as 'Are you in favour of the present Government?', the replies to which can be classified as 'Yes', 'No', 'Don't know'. The distinction is important and useful in practice, although it may be criticised from certain logical and psychological standpoints.

Operating Characteristic In the theory of decisions, and especially in quality control and sequential analysis, a description of the behaviour of a decision rule which provides the probability of accepting alternative hypotheses when some null hypothesis is true. For example, if the hypothesis is specified as $H(\theta)$, dependent on a parameter θ, an operating characteristic function might show, with θ as variable, the probability of accepting $H(\theta)$ when the true value is θ_0. The graph of this probability as ordinate against θ as abscissa is called the operating characteristic curve or OC curve. The OC function may be regarded as the complement of the **Power Function** in the theory of testing hypotheses.

The expression 'Performance Characteristic' also occurs, sometimes synonymously and sometimes in a more general sense as describing the consequences of the decision rule for different null hypotheses.

Opinion Survey A sample survey which aims at ascertaining or elucidating opinions possessed by the members of a given human population with regard to certain topics.

Optimal Asymptotic Test An alternative name for a **Locally Asymptotically Most Powerful Test.**

Optimum Allocation In general, the allocation of numbers of sample units to various strata so as to maximise some desirable quantity such as precision for fixed cost. Secondarily, allocation of numbers of sample units to individual strata is an optimum allocation for a given size of sample if it affords the smallest value of the variance of the mean value of the characteristic under consideration. Optimum allocation in this sense for unbiassed estimators requires that the number of observations from every stratum should be proportional to the standard deviation in the stratum.

Optimum Linear Predictor A predictor for future observations of a stochastic process which is constrained to be a linear combination of past observations and which minimises the mean square error of prediction.

Optimum Statistic An expression which is usually synonymous with **Best Estimator**. If it refers to a statistic used for testing hypotheses it is usually known as an optimum test statistic.

Optimum Stratification In a general sense a system of stratification in sampling which optimises some given criteria. Usually the criteria relates to a set of estimated means and requires the minimisation of their **Generalised Variance.**

Optimum Test A test which can be shown to possess a certain desirable characteristic or group of characteristics to a greater degree than any other test of the same class.

Ord-Carver System A system of discrete distributions analogous to the Pearson system for continuous variates first proposed by Carver (1919) and developed by Ord (1967).

Order of Coefficients Correlation and regression coefficients are referred to as being of a given 'order' according to the number of independent variates held constant. For example, in multiple regression, a simple correlation between two variates x_1 and x_2, r_{12} is called a zero order coefficient and the partial regression coefficient $b_{12 \cdot 345}$ between x_1 and x_2 when x_3, x_4 and x_5 are held constant is a third order coefficient. The number of secondary subscripts, those to the right of the point, denotes the order of the coefficient.

Order of Interaction See **Interaction.**

Order of Stationarity A stochastic process $\{x_t\}$ is said to be stationary to the rth order if the expectation

$$\mathscr{E}\{x_{t_1}^{\alpha_1} x_{t_2}^{\alpha_2} \dots x_{t_n}^{\alpha_n}\}$$

exists for any sequence $\{t_i\}$ and all $\alpha_1 + \alpha_2 + \dots + \alpha_n \leqslant r$, and if such expectation depends only on the $n-1$ differences $t_2 - t_1, \dots, t_n - t_1$.

Ordered Alternative Hypothesis See **Location Shift Alternative Hypothesis.**

Ordered Categorisation A categorisation in which, although the variable is not expressible in terms of an underlying measurable variable, it may nevertheless be arranged in order. Where even this is not possible the categorisation is said to be unordered. Thus, the classification of persons according to social classes is ordered; that according to the type of crime they commit is not.

Ordered Series A set of variate values which possess a natural sequence in time or space. In another sense, an ordered series is a set of variate values which have been arrayed in some specific manner related to their values, e.g. from the lowest value to the highest value, or from the earliest available value to the latest available value.

Order Statistics When a sample of variate values are arrayed in ascending order of magnitude these ordered values are known as order statistics. Examples are the smallest value of a sample and the median. More generally, any statistic based on order statistics in this narrower sense is called an order statistic, e.g. the range and the interquartile distance.

Organic Correlation The correlation of measurements made upon the different parts of a living organism. There appears to be no point in making a distinction between data of biometric type and other data for the purposes of defining correlation and the term seems unnecessary.

Ornstein-Uhlenbeck Process A stochastic process x_t for which

$$x_{t+k} = \alpha_k x_t + \epsilon_{k, t},$$

where $\alpha_k = e^{-\beta k} (\beta > 0)$, $\epsilon_{k,t}$ is Normally distributed with zero mean and variance $\sigma^2 (1 - e^{2\beta k})$ and the e's are independent of each other if the intervals t_j, $t_j + k_j$ do not overlap.

Orthant Probabilities Generally, for a n-dimensional distribution the probabilities of individuals falling into the 2^n orthants into which the sample space is divided by the coordinate planes. In particular, the probability that the components of a multivariate Normal distribution with zero means are all positive or all negative.

Orthogonal This word occurs in several distinct but related senses: (*a*) in its mathematical sense as meaning perpendicular, e.g. in relation to a pair of orthogonal coordinate axes; (*b*) in relation to a set of mathematical functions [see **Orthogonal Functions**]; (*c*) in relation to two variates or two linear functions of variates, which are said to be orthogonal if they are statistically independent; (*d*) in relation to an experimental design, which is called orthogonal if certain observed variates, or linear combinations of them, can be regarded as statistically independent.

Orthogonal Arrays A concept introduced by Rao (1946) in connection with the design of factorial experiments. Orthogonal arrays are a generalisation of **Orthogonal Latin Squares.** For example, a $(s^2, k, s, 2)$ orthogonal array is equivalent to $k-2$ mutually orthogonal $s \times s$ Latin squares.

Orthogonal Design See **Orthogonal.**

Orthogonal Functions A set of real functions $f_1(x)$, $f_2(x)$, ..., are orthogonal within the range (a, b) if

$$\int_a^b f_m(x) f_n(x) dx = 0 \quad m \neq n.$$

Frequently, by convention, the functions are standardised so as to make the integral equal to unity if $m = n$.

In statistics functions are sometimes said to be orthogonal in relation to a distribution $F(x)$ if

$$\int_a^b f_m(x)f_n(x)dF(x) = 0 \quad m \neq n$$

the range of x being from a to b.

Orthogonal Polynomials If $P_i \equiv P_i(x)$ is a polynomial which the coefficient of x^i is not zero and $F(x)$ is a distribution function, then P_0, P_1, ..., P_n form a set of orthogonal polynomials if $\int P_i P_j F(x)dx = 0$ $(i \neq j)$ where the integral may include summation of discrete values. If, also, $\int P_i^2 F(x)dx = 1$ the set is said to be orthonormal.

Orthogonal Process A stochastic process $\{x_t\}$ such that $\mathscr{E}\{|x_t|^2\} < \infty$ and for which $\mathscr{E}\{x_s \bar{x}_t\} = 0$, $(s \neq t)$ $\bar{x}$ being the complex conjugate of x.

A stochastic process with orthogonal increments is one for which $F\{|x_t - x_s|^2\} < \infty$ and for which $F\{(x_{t2} - x_{s2})(\bar{x}_{t1} - \bar{x}_{s1})\} = 0$, where s_1, t_1 and s_2, t_2 do not overlap.

Orthogonal Regression Given a set of bivariate values (x, y) represented as a set of points with Cartesian coordinates (x, y), the so-called 'orthogonal regression' is the straight line such that the sum of squares of perpendiculars from the points on to the line is a minimum.

The term 'regression' is open to objection in this context. It is true that the 'orthogonal regression' minimises the sum of squares in the direction perpendicular to itself, whereas the ordinary regressions minimise those sums in the direction of the coordinate axes. But **Regression** is a property of conditional variates and no such interpretation can be given to the 'orthogonal regression'. In particular it is not invariant under a change of scale.

Orthogonal Squares If two **Latin Squares** can be superimposed so that every letter of the first and every letter of the second occupy somewhere the same position the original squares are said to be orthogonal squares. For example, the two squares:

```
ABCD            ABCD
BADC    and     CDAB
CDAB            DCBA
DCBA            BADC
```

combine to give:

```
AA    BB    CC    DD
BC    AD    DA    CB
CD    DC    AB    BA
DB    CA    BD    AC
```

If this form is written with the second letter in Greek the design is said to be a **Graeco-Latin Square**. The second letter may be identified with a further source of variation to be considered in an experiment.

More generally, if a number of squares are orthogonal in pairs they are said to be mutually orthogonal. [See **Hyper-Graeco-Latin Square**.]

Orthogonal Tests Another name for tests which are independent, and one which seems better avoided.

Orthogonal Variate Transformation A linear transformation of variates x_1, x_2, ..., x_n to variates y_1, y_2, ..., y_n the form

$$y_i = \sum_{j=1}^n d_{ij}x_j$$

such that

$$\sum_{j=1}^n d_{ij}d_{kj} = 0, \ i \neq k.$$

That is to say, a linear transformation with an orthogonal matrix. The most usual type also has

$$\sum_{j=1}^n d_{ij}^2 = 1, \text{ all } i,$$

in which case the transformation is equivalent to a rotation of axes in an n-dimensional Euclidean space.

Orthonormal System See **Orthogonal Polynomials**.

Oscillation An oscillation in a time series or, more generally, in a series ordered in time or space is a more or less regular fluctuation about the mean value of the series. In this sense it is to be sharply distinguished from a cycle, which is strictly periodic; thus, while a cyclical series is oscillatory an oscillatory series is not necessarily cyclical.

*Oscillation, Index of (Indice di Oscillazione) In Italian usage, a measure of oscillation in a time series obtained as a mean of absolute first differences or the root mean square of first differences. It is sometimes standardised by division by the maximum value which the index can have in all possible permutations of the series.

Oscillatory Process A class of stochastic processes, usually nonstationary but which includes all second order stationary processes, where the covariance function may be represented as

$$R_{s, t} = \int_{-\infty}^{\infty} \phi_s(\omega)\phi_t^*(\omega)d\mu(\omega)$$

and the family of oscillatory functions $\phi_t(\omega)$ exists. The * denotes a complex conjugate and $\mu(\omega)$ a measure on the real line involving ω the angular frequency.

Outliers In a sample of n observations it is possible for a limited number to be so far separated in value from the remainder that they give rise to the question whether they are not from a different population, or that the sampling technique is at fault. Such values are called outliers. Tests are available to ascertain whether they can be accepted as homogeneous with the rest of the sample.

Over-all Estimate A vague but convenient term used to denote an estimate for a whole population, as distinct, for example, from one for a stratum or other sub-section of it. The expression is also used for an estimate derived from the whole of a sample instead of only from a part of it.

Over-all Sampling Fraction It is sometimes necessary to qualify the term 'sampling fraction' when more than one act of sample selection is involved. Thus in a three stage sampling scheme if the (sub-) sample within a given first stage sample unit is **Self-weighting** for the estimation of the first stage unit total then the reciprocal of the corresponding raising-factor is called the over-all sampling fraction for the specified first stage unit. Over-all sampling ratio (or rate) is also used. [See also **Sampling Fraction.**]

Over Identification See **Identifiability.**

Overlap Design An alternative name proposed by Thompson & Seal (1964) for **Serial Designs** in the context of routine quality control and experiments.

Overlapping Sampling Units Usually the population of elementary units or basic cells is broken up for purposes of sampling into clusters or grids of units or cells which are mutually exclusive; that is, every elementary unit or basic cell belongs to one and only one sampling unit. It is, however, possible to have a system of sampling units in which the same elementary unit or cell may occur in more than one sampling unit, in which case we have an overlapping system. If properly used such a system provides unbiassed estimates.

Paasche Index A form of index number due to Paasche (1874). If the prices (quantities) of a set of commodities in a base period are $p_0, p_0', p_0'', \ldots (q_0, q_0', q_0'', \ldots)$ and those in the given period are $p_n, p_n', p_n'', \ldots (q_n, q_n', q_n'', \ldots)$, the Paasche price index number is written

$$I_{on} = \frac{\Sigma(p_n q_n)}{\Sigma(p_0 q_n)}$$

where the summation takes place over commodities. In short, the prices are weighted by the quantities of the given period, as distinct from the **Laspeyres' Index** where they relate to the base period.

Generally, an index number of the above form is said to be of the Paasche type even when p and q do not relate to prices and quantities; the characteristic feature being that the weights relate to the given period. [See also **Lowe Index, Palgrave Index, Crossed Weights.**]

Paasche-Konyus Index See **Konyus Index.**

Paired Comparison The comparison of a set of objects in pairs, each pair AB being placed in a preference relationship: A preferred to B or B preferred to A or, in more general conditions, neither preferred to the other. The method is used where order relations are more easily determined than measurements, e.g. in investigating taste preferences. More generally, the expression is used to

denote the comparison of two samples of equal size where members of one can be paired off against members of the other.

Palgrave's Index An index number sometimes attributed to Palgrave. If the prices of a set of commodities in the base period (or the given period) are represented by $p_0, p_0', p_0'', \ldots$ (or $p_n, p_n', p_n'', \ldots$) and the corresponding quantities by $q_0, q_0', q_0'', \ldots$ (or $q_n, q_n', q_n'', \ldots$) Palgrave's index is given by

$$I_{on} = \frac{\Sigma p_n q_n \left(\dfrac{p_n}{p_0}\right)}{\Sigma p_n q_n}$$

where the summation takes place over the commodities. It is thus an **Index of Price Relatives** weighted by the total value of commodities in the given period. [See also **Laspeyres' Index, Paasche's Index.**]

Palm Function A set of functions introduced by Palm (1943-44) in connection with certain queuing problems. The functions $\phi_k(t)$ may be interpreted as the conditional probability that k requests for service occur in a period of length t assuming a single request has occurred during the first moment, or small finite time period τ, of the period.

Parallel Line Assay An important method for the bioassay of a test preparation against a standard. If the expected response is linearly related to the logarithm of dose the regressions of response on log dose for the two preparations will often be parallel and the distance between them will estimate the logarithm of the relative potency.

Parameter This word occurs in its customary mathematical meaning of an unknown quantity which may vary over a certain set of values. In statistics it most usually occurs in expressions defining frequency distributions (population parameters) or in models describing stochastic situation (e.g. regression parameters). The domain of permissible variation of the parameters defines the class of population or model under consideration.

Parameter of Location (or Scale) A parameter of a frequency function which can be identified with some measure of location (or scale).

Parameter Point If a class of frequency functions depends on certain parameters; e.g. if the univariate function of x is $f(x, \theta_1, \theta_2, \ldots, \theta_k)$, the domain of variations of the θ's is called the parameter space and any particular set of θ's determines a point in that space.

Parametric Hypothesis A **Statistical Hypothesis** concerning the parameter(s) of a distribution.

Parametric Programming A development of **Linear Programming** in which the parameters of the objective function, formed by the constraints, are allowed to vary in a determinate fashion.

Pareto Curve/Distribution An empirical relationship describing the number of persons y whose income is x, first advanced by Pareto (1897) in the form

$$y = Ax^{-(1+\alpha)}, \ 0 \leqslant x \leqslant \infty.$$

The expression is now used to denote any frequency distribution of this form, whether related to incomes or not. The variable x may be measured from some arbitrary value, not necessarily zero.

Pareto Index The coefficient α in the expression for the Pareto curve is generally referred to as the Pareto Index. It affords evidence of the concentration of incomes, or, more generally, of the concentration of variate values in distributions of the Pareto type. [See also **Concentration**.]

Pareto-type Distribution A loosely used expression to denote any distribution shaped similarly to that of **Pareto**. It is probably better avoided unless the distribution is actually Paretian apart from origin and scale.

Part-correlation, Coefficient of A term which is sometimes regarded as synonymous with the **Multiple Partial Correlation Coefficient**. Strictly, however, the part-correlation coefficient refers to the case where the relationship is between y and x_1, with x_1 adjusted for the effects of $x_2, x_3, ..., x_n$.

Partial Association The measure of association in sub-populations of qualitative characteristics analogous to the partial correlations between quantitative variates. For example, if A, B and C are three attributes the partial association of A and B with respect to C would be the association of A and B in the sub-population of members bearing the attribute C.

Partial Confounding If, in a **Factorial Experiment** with several replicates, there are interactions which are **Confounded** in some replicates but not in others these interactions are said to be partially confounded.

Partial Contingency A term introduced by K. Pearson (1916) and analogous to partial correlation in the sense that contingency is investigated when certain variables are held constant.

Partial Correlation

(1) The correlation between two variates in a conditional distribution for which one or more other variates are held fixed. Specifically, the product moment correlation coefficient in the conditional distribution.

(2) The correlation between the deviations of the values of a variate from their least square estimates by a regression function linear in terms of an external set of variates, with the corresponding deviations of another variate from its own regression function linear in the same external set.

Ordinarily the product moment correlation coefficient is used. The second definition is applicable to samples in

the usual cases, while ordinarily the first is not. The two definitions are equivalent for multivariate Normal distributions.

Partial Rank Correlation Attempts have been made to carry over into ranking theory the notion of partial correlation of ordinary variate theory, for example, by measuring the correlation between two rankings when the effect of some other ranking on which they both depend is removed. For this purpose Kendall (1942) defined a coefficient of partial rank correlation based upon his general coefficient τ. The notion of partial rank correlation is, however, a somewhat elusive one. [See **Kendall's Tau**.]

Partial Regression In regression analysis, the coefficient of an 'independent' variable in the complete regression equation, that is to say, the regression equation involving all the variables under consideration. It is called partial, not a very fortunate expression, because it differs from the coefficient which would be obtained if certain, perhaps all, other independent variables were ignored and a simple regression of the dependent variate calculated in the one independent variable.

Partial Replacement See **Sampling with Replacement**.

Partial Serial Correlation Coefficient Correlation within a series after allowing for the influence of observations within the limits of the lag; it is analogous to the ordinary concept of **Partial Correlation**. Thus, the partial correlation of x_t and x_{t+2} would be that obtained after eliminating their common dependence on x_{t+1}.

Partially Balanced Arrays Generalisations of **Orthogonal Arrays** introduced by Chakravarti (1956). They permit a multifactorial design to deal with a given number of factors by requiring a reduced number of assemblies, i.e. columns in the matrix $\mathbf{A} = (a_{ij})$ where the rows are the factors and the elements a_{ij} the levels of those factors.

Partially Balanced Incomplete Block Design An experimental design in incomplete blocks for which the layout, though not completely balanced, is partially balanced in the sense that each treatment is tested the same number of times and certain other symmetries exist. This class of design, introduced by Bose and Nair (1939), avoids the large number of replicates which may be required by a completely balanced design. [See also **Block, Balanced Incomplete Block**.]

Partially Balanced Lattice Square In certain cases it is possible to arrange a **Lattice Square** design so that each effect or interaction is confounded in only some of the replicates constituting the whole design. The lattice square is then sometimes described as semibalanced or partially balanced. [See also **Square Lattice**.]

Partially Balanced Linked Block Design The condition for a **Partially Balanced Incomplete Block Design** also to be a **Linked Block Design** is that its dual, i.e. the design created by interchanging blocks and treatments, is also balanced.

Partially Consistent Observations A term proposed by Neyman and Scott (1948) in the problem of deriving consistent estimators. If a set of observations depend partly on parameters which are common to all and partly on parameters which are specific to the individual observation they are said to be partially consistent. More precisely, if the probability laws of the variables x_i depend (*a*) on a finite number of parameters which appear in an infinity of variates of the sequence x_i and (*b*) on an infinity of parameters each of which appears in the probability law of only a finite number of the variates, the situation is described as one of partially consistent variates.

The parameters in the first class are called structural; those in the second class are called incidental.

Partially Linked Block Design An extension by Nair (1966) of the **Linked Block Design** to yield new **Partially Balanced Incomplete Block Designs** in a dual relationship, i.e. the treatments in the one design become the blocks in the other.

Partition of Chi-squared (χ^2) In certain circumstances the sum of squares of standardised Normal variates about their mean, which is distributed as χ^2, can be divided into two or more parts each of which is also distributed as χ^2 independently of the others. This is known as a partition of χ^2.

Pascal Distribution An unnecessary alternative name for the **Negative Binomial Distribution**, presumably because some untraced individual thought that Pascal discovered it. Some writers, however, restrict the usage to the case where the parameter k from the gamma distribution in the negative binomial takes integer values. The name also occurs in combination with other compounding distributions, e.g. beta and gamma, according to which law the binomial parameter is supposed to follow.

Patch In the terminology of Mahalanobis, a compact cluster of units whose variate values all fall in a specified class interval or if quantitative in a specified category, is called a patch. A further condition is that the cluster should be complete and inextensible. The term 'contour level' is also used.

Path Coefficients, Method of A method of analysis proposed by Wright (1918) for the purpose of relating the matrix of zero order correlations between the variables in a multiple system to various functional relations which are supposed to connect the variables of that system. Each path coefficient, a function of the standardised variables, measures the fraction of the standard deviation of the dependent variable for which a designated factor is deemed responsible and the term derives from a particular diagrammatic approach used in the exposition. The method of path coefficients is related to ordinary **Multiple Regression Analysis**.

Pattern Function A function of the sample number n used in connection with the evaluations of sampling cumulants of *k*-statistics. The name derives from the fact that the function depends on the configuration of zeros in an array representing a bipartition.

Patterned Sampling An alternative name for **Systematic Sampling**.

Pay-off Matrix In the theory of games, a matrix specifying how money or its equivalent is to pass from one player to the other for all the possible outcomes of a two person game. [See also **Loss Matrix**.]

Peak An observation in an ordered series is said to be a 'peak' if its value is greater than the value of its two neighbouring observations.

Pearl-Read Curve Another name for the logistic curve, a general form of **Growth Curve**.

Pearson Coefficient of Correlation The **Product Moment** coefficient of correlation is sometimes referred to as the Pearson coefficient of correlation because of K. Pearson's part in introducing it into general use.

Pearson Criterion See **Criterion**.

Pearson Curve A distribution from the family of frequency distributions developed by K. Pearson. The basic equation of the family is:

$$\frac{df}{dx} = \frac{(x-a)f}{b_0 + b_1 x + b_2 x^2}$$

where f is the frequency function. The constants of this equation may be expressed in terms of the first four moments, if these exist. The explicit solutions are classified into types according to the nature of the roots of the equation $b_0 + b_1 x + b_2 x^2 = 0$. By appropriate transformations, many of the important distributions of statistics can be derived from this basic equation. [See also **Type I** to **Type XII Distributions**.]

Pearson Measure of Skewness A measure of skewness proposed by K. Pearson in the form:

$$\text{Skewness} = \frac{\text{Mean} - \text{Mode}}{\text{Standard Deviation}}$$

which, however, suffers from the general indeterminate nature of the mode. For distributions of the **Pearson System** it may be expressed as:

$$\text{Skewness} = \frac{\sqrt{\beta_1}\,(\beta_2+3)}{2(5\beta_2-6\beta_1-9)}$$

where β_1 and β_2 are the first two **Moment Ratios**.

Peek's Inequality An improvement (1933) of the **Camp-Meidell Inequality** making use of the mean and ratio of mean deviation to standard deviation (γ)

$$\Pr(\,|\,x-\bar{x}\,|\,\geqslant t\sigma) \leqslant \frac{4}{9}\frac{1-\gamma^2}{(t-\gamma)^2}$$

Pentad Criterion In factor analysis, an extension of the tetrad criterion developed by Kelley and Holzinger and based on sets of five correlations from the correlation matrix. [See **Hierarchy**.]

Percentage Diagram A diagram which exhibits a simple analysis of statistical data in terms of percentages. The actual form of the diagram can vary; examples are the **Bar Chart** and the **Pie Chart**.

Percentage Distribution A frequency distribution with the total frequency equated to one hundred and the individual class frequencies expressed in proportion to that figure.

Percentage Point A **Level of Significance** expressed in percentage form.

Percentage Standard Deviation A regrettable synonym for the **Coefficient of Variation**.

Percentiles The set of partition values which divide the total frequency into one hundred equal parts. This particular set of values is most used in education and psychology. Some writers prefer to use the term 'centile' rather than 'percentile'. [See **Quantiles**.]

Percolation Process A stochastic process where the physical interpretation is the dispersion of a fluid through a medium influenced by a random mechanism associated with the medium. This is opposed to a **Diffusion Process** where the random mechanism is associated with the fluid.

Performance Characteristic See **Operating Characteristic**.

Period A term used to describe regularities of recurrence in ordered series, sometimes rather vaguely. Strictly, the word should relate to a period in the mathematical sense, that is to say, a term $u(t)$ has period ω if $u(t+\omega) = u(t)$ for all t; and if a series can be analysed into a sum of such functions, the corresponding set of ω's are the periods of the series.

More loosely, the expression is used to denote the interval or average interval between identifiable points of recurrence, e.g. between peaks or troughs of the series. It is better to avoid this usage in general, since the intervals between successive peaks, etc. in most time-series are not equal and the underlying model may not generate a periodic sequence.

Period (of a Markov Chain) The period of a return state k of a Markov Chain is the greatest common divisor of the set of integers n for which $\Pr_{k,\,k}(n) > 0$.

Periodic Process If any realisation of a **Stationary Stochastic Process** yields a series which is strictly periodic then the process is a periodic process.

Periodogram A diagram used in the harmonic analysis of an oscillatory series. If the value of the series at time t is u_t the procedure is to calculate

$$A = \frac{2}{n}\sum_{t=1}^{n} u_t \cos \lambda t, \qquad B = \frac{2}{n}\sum_{t=}^{n} u_t \sin \lambda t.$$

The function $S^2 = A^2+B^2$ is known as the intensity of the frequency $\lambda/2\pi$ or of the period $2\pi/\lambda = p$, say. Graphed against p as abscissa it gives the periodogram. When multiplied by a constant involving n and graphed against λ it gives the **Power Spectrum**.

Perks' Distribution The distribution of the mean in samples from a population of the form $2\lambda\pi^{-1}(e^{\lambda x}+e^{-\lambda x})^{-1}$ is sometimes referred to as the Perks (1937) distribution.

Permissible Estimator An estimator which yields permissible estimators for all possible samples; permissible meaning that the estimates lie in the known range of the parameter e.g. an estimator which gave negative values of a variance would not be 'permissible' in this sense, although its use might be permitted on occasion.

Permutation Tests A class of distribution free tests based upon the fact that any ordering of a random sample of n items has the same probability $1/n!$.

Persistency A term applied, mainly by meteorologists, to a time series to denote regularity of recurrence. Bartels (1935) endeavoured to distinguish between 'true' persistency in the sense of periodicity and 'quasi' persistency to denote oscillatory behaviour of a less durable and regular kind.

Persistent State An alternative but less desirable name for **Recurrent State**.

Peters' Method A method of estimating the standard deviation of a distribution which is approximately Normal by multiplying the mean deviation by 1·253, this being the ratio $\sqrt{(\tfrac{1}{2}\pi)}$ appropriate to the Normal distribution.

Phase The interval between the **Turning Points** of a series which is ordered in time or space is termed a phase. The distribution of phase lengths provides one test of random order.

The expression is also used in its customary mathematical sense relating to the angle α in sine or cosine terms such as $\sin(\theta t+\alpha)$.

Phase Confounded Designs A method of reducing the block size required for a full design, e.g. in cyclic rotation experiments when **Reduced Designs** are not available: Patterson (1964) used a method of partially confounding some of the test crop comparisons.

Phase Diagram A name proposed by Frisch (1937) for a diagram showing two time series x_1 and x_2 plotted as ordinate and abscissa. If the fluctuations of these two variates keep in step then the line joining the plotted points will trace a definite pattern, for example similar to an ellipse for oscillatory series.

Phase Spectrum The sample phase spectrum indicates whether the frequency components in one of two series lead or lag the components at the same frequency in the other series; it is a description of the covariance between the two series.

Phi-coefficient An obsolescent term equivalent to the coefficient V defined under **Association, Coefficient of.** [See also **Contingency.**]

Pictogram A method of visual presentation of statistical quantities by means of drawings or pictures of the subject matter under discussion. The method is restricted to the presentation of simple relationships and in order to overcome the unsatisfactory nature of crude comparisons by the eye of objects of different size it is now customary to represent a unit value of the data by a standard symbol and present the appropriate number of repetitions of this standard symbol to depict the magnitude under discussion. This virtually changes the style of the diagram to a pictorial bar-chart. The system has become known as the Isotype Method.

Pie Diagram A more picturesque term for the **Circular Diagram or Chart.**

Pilot Survey A survey, usually on a small scale, carried out prior to the main survey, primarily to gain information to improve the efficiency of the main survey. For example, it may be used to test a questionnaire, to ascertain the time taken by field procedure or to determine the most effective size of sampling unit.

The term 'exploratory survey' is also used, but in the rather more special circumstance when little is known about the material or domain under inquiry.

Pistimetric Probability A probability measure, analogous to **Fiducial Probability**, proposed by Roy (1960) and based upon an etymology relating to 'trust, faith, belief'. Its motivation refers to decision taking in the light of scarce information rather than experimental science.

Pitman Efficiency The concept of **Asymptotic Relative Efficiency** introduced by Pitman (1949).

Pitman Estimator An estimator of the location parameter (ξ) of a distribution proposed by Pitman (1939). It

is an unbiassed estimator and optimal in the class of all estimators of θ as the general form is

$$t_n = \frac{\int \xi \prod_{i}^{n} f(x_i - \xi) d\xi}{\int \prod_{i}^{n} f(x_i - \xi) d\xi}.$$

It was one of the earliest examples of the **Principle of Invariance.**

Pitman's Tests Distribution free tests developed by Pitman (1937) for testing differences of means in two samples and homogeneity of means in several samples. [See **Concordant Sample.**]

Plaid Square The use of a uasi-Latin Square in the form of a **Split Plot Design** in such a way that different treatments are applied to whole rows and columns of the square (Yates, 1937). Thus the main effects of these treatments are confounded with rows and columns and are estimated with low precision.

Platykurtosis See Kurtosis.

Plot In experimental design this term usually refers to the basic unit of the experimental material. Although it derives from the physical unit of a plot of land in an agricultural trial, its interpretation is very much more general according to the subject matter of the particular design. [See also **Split Plot Design.**]

Point Binomial An alternative name for the **Bernoulli, or Binomial, Distribution.** The word 'point', which is unnecessary, arises from the discrete character of the variate.

Point Biserial Correlation A modification of the **Biserial Correlation** to the case where one variate, instead of being based on a dichotomy of an underlying continuous variate, is discontinuous and two valued.

Point Bivariate Distribution An alternative name for a bivariate distribution of two discrete variates.

Point Density The relative frequency, or probability mass, which may be located at the different point values of a discontinuous variate. The term is better avoided in favour of frequency or probability mass, the word 'density' being reserved for continuous probabilities. [See also **Point Binomial.**]

Point Estimation One of the two principal bases of estimation in statistical analysis. Point estimation endeavours to give the best single estimated value of a parameter, as compared with **Interval Estimation**, which proceeds by specifying a range of values. Since the point estimate is surrounded by a band of error, the distinction between the two methods is sometimes blurred and in interpretation they often amount to the same thing.

Point of Control A point on the operating **Characteristic Curve** with ordinate 0·5; used as a rough summarising quantity of the curve.

Point of First Entry See **Waiting Time.**

Point of Indifference Alternatively called the 'point of control'. It is the central point of the operating characteristic curve and is the percentage defective in the bulk that will be accepted or rejected equally often: it occurs in attribute sampling schemes.

Point Processes Statistical processes concerned with the occurrence of events at points of time determined by some chance mechanism; as distinct from processes observed at fixed intervals and either continuous in the time variable or determinate at such fixed intervals.

Point Sampling A method of sampling a geographical area by selecting points in it, especially by choosing points at random on a map or aerial photograph.

Poisson Beta Distribution A compound distribution proposed by Holla & Bhattacharya (1965) where the parameter λ of a Poisson distribution is itself distributed as (i) a beta of the first kind, (ii) beta of the second kind.

Poisson Binomial Distribution A discrete distribution of the number of successes in 'n' independent trials with probability p_j of success in the jth trial. The binomial distribution is a special case where the parameters are n, $p_1, p_2, p_3, ..., p_n$ and $p_1 = p_2 = p_3 = ... = p_n = p$.

If p_j is specified as a function of a random variable p and a constant c_j, the resulting distribution obtained by integrating over the frequency of p is sometimes called a Poisson-Lexis distribution.

Poisson Clustering Process A term proposed by Bartlett (1963) to cover a complex Poisson process where each event in the basic Poisson process is followed by a sequence of associated events, themselves forming a subsidiary process not necessarily Poisson, before the succeeding event in the main process. Where the subsidiary processes are also Poisson we have a **Doubly Stochastic Poisson Process.**

Poisson Distribution A discontinuous distribution with relative frequencies at variate values 0, 1, 2, ..., r, ... given by

$$e^{-\lambda}, \ e^{-\lambda}\frac{\lambda}{1!}, \ e^{-\lambda}\frac{\lambda^2}{2!}, ..., \ e^{-\lambda}\frac{\lambda^r}{r!}, ...$$

It is also known as Poisson's exponential limit, because it may be regarded as the limiting distribution of a **Binomial.** The observation of the operation of the law in describing the behaviour of rare events has been called the 'Law of Small Numbers' but this is better avoided as being in no way antithetical to the **Law of Large Numbers.**

Poisson Index of Dispersion An index appropriate to events obeying a **Poisson Distribution.** If k samples of the same size have frequencies of occurrence $x_1, x_2, ..., x_k$, with mean $\bar{x}$, the index is $\sum_{i=1}^{k}(x_i - \bar{x})^2/\bar{x}$. If the samples emanate from the same Poisson population this is distributed as χ^2 with $k-1$ degrees of freedom, a fact which provides a test for variation in a Poisson distribution. [See also **Binomial Index of Dispersion, Lexis Ratio.**]

Poisson's Law of Large Numbers An extension by Poisson of **Bernoulli's Theorem.** Both are special cases of results deducible from the **Bienaymé-Tchebychev Inequality.** If the probability of an event varies from one trial to another and a set of n trials is $p_1, p_2, ..., p_n$; and if there are k successes in n trials, then in repeated sampling

$$\Pr\left\{\left[\frac{k}{n} - \mathscr{E}\left(\frac{k}{n}\right)\right] > t\sqrt{\frac{\sum_{i=1}^{n} p_i q_i}{n}}\right\} \leqslant \frac{1}{t^2},$$

where $\mathscr{E}\left(\dfrac{k}{n}\right)$ is the mean proportion of successes $\sum_{i=1}^{n} p_i/n$.

Poisson-Lexis Distribution See **Poisson Binomial Distribution.**

Poisson-Markov Process A stochastic process, discussed by Patil (1957), whose probability transition matrix is that of a Markov process and at any point in time the space distribution is that of k independent Poisson variates.

Poisson Probability Paper Graph paper showing curves of the relationship between P and λ where

$$P = e^{-\lambda}\left\{\frac{\lambda^c}{c!} + \frac{\lambda^{c+1}}{(c+1)!} + ...\right\}.$$

The axes may be calibrated linearly in P and λ but other scales are in use.

Poisson Process A stochastic process which may be regarded as a particular case of a **Birth Process** in which the parameter λ_n is a constant λ.

Poisson Truncated Normal Distribution A distribution compounded of half the Normal distribution in λ $(\lambda \geqslant 0)$ and a Poisson distribution with parameter λ.

Poisson Variation A type of sampling variation considered by Poisson (1837). On each of k occasions let n members be chosen at random, and let the probabilities be the same for all occasions, but such that the probability of success at the drawing of the ith member is p_i ($i = 1, 2, ..., n$). The mean number of successes on any occasion is $\sum_{i=1}^{n} p_i$ and the variance is $npq - n \operatorname{var} p_i$ where $p = \Sigma p_i/n$, $q = 1 - p$ and $\operatorname{var} p_i$ is the variance of p_i among the possible values. If all the p_i are equal this reduces to

115

Bernoulli Variation. In other cases the Poissonian variance is smaller than the Bernoullian variance. This effect is encountered in sampling where the numbers are systematically spread over different strata. The dispersion is said to be subnormal. [See also **Lexis Ratio, Lexis Variation.**]

Pollaczek's Formula A formulation of the equilibrium situation for a single server queue with Poisson input and a general service time distribution. The Poisson input is obtained as a limiting case of the Bernoulli input.

Pollaczek-Khintchine Formula Similar to the **Pollaczek** (1930) **Formula**, but obtained directly by Khintchine (1932) and acknowledged as such.

Pólya-Aeppli Distribution The compound of a **Geometric Distribution** of a parameter λ and a **Poisson Distribution** with that value of λ.

Pólya's Distribution A discontinuous frequency distribution considered by Pólya (1923) in connection with **Contagious Distributions**. It may be generated by drawing with replacement from an urn containing b black and r red balls under the condition that as every ball drawn is returned an additional c balls of the same colour are added to the urn. It is a particular case of the **Negative Binomial Distribution**.

Pólya-Eggenburger Distribution This distribution may be derived from a **Pólya Distribution** with parameters $p = b/(b+r)$, $\gamma = c/(b+r)$ and n the sample size, as the limiting form p, $\gamma \to 0$ and $n \to \infty$ such that $np \to h$ and $n\gamma \to \theta$ which are the parameters of the limit distribution. It may also be derived from the **Negative Binomial Distribution** by putting the parameters k and p of that distribution in the form $k = h/\theta$ and $p = 1/(1+\theta)$.

Pólya Frequency Function of Order Two If there are two sets of increasing number $x_1 < x_2$ and $t_1 < t_2$ and the determinant of the matrix $\| f(x_i - t_j) \|_{1, 2} \geqslant 0$, then f will be a Pólya frequency function of order two. This group includes the Normal, exponential, gamma, beta, logistic and uniform distributions.

Pólya Process A particular case of a **Birth Process** in which the parameter λ_n is given by
$$\lambda_n(t) = (1+an)/(1+at),$$
a being a constant.

Polychoric Correlation An extension of **Tetrachoric Correlation** to the case of an m by n table where it may be assumed that the two underlying variates are jointly Normally distributed.

Polykay A generalisation by Tukey (1950, 1956), of the **Fisher k-statistics.**

Polynomial Trend A trend line of the general form:
$$y = \alpha_0 + \alpha_1 t + \alpha_2 t^2 + \alpha_3 t^2 + ... + \alpha_n t^n$$
fitted to a series which is ordered in time or space. The coefficients α_i, $i = 0, 1, 2, ..., n$ are usually estimated by the method of least squares or the method of moments.

Polyspectra The spectrum of a time series may be regarded as the Fourier transform of its autocorrelations, which depend on the product of terms $u(t)u(t+k)$. The transform of the product of more than two terms is called a polyspectrum, e.g. for three terms we have the **Bispectrum.**

Polytomic Table A contingency table with more than two categories in the row and column classifications.

Pooling of Classes The amalgamation of frequencies in a group of classes to form one frequency in a more comprehensive class. This procedure often serves to eliminate blanks or small sub-class numbers in a complex analysis.

Pooling of Error In some situations where several sets of data are regarded as generated under the same model it is possible to construct several independent 'residual' sums of squares which, under the hypothesis being examined, all provide estimators of the error variance. These sums of squares may be 'pooled' by adding them together, the resulting estimator of the error variance then being based on more degrees of freedom. This is described as 'pooling the error' or, preferably, as 'pooling the residual sums of squares'.

Population In statistical usage the term population is applied to any finite or infinite collection of individuals. It has displaced the older term 'universe', which itself derived from the 'universe of discourse' of logic. It is practically synonymous with 'aggregate' and does not necessarily refer to a collection of living organisms.

Positive Skewness See **Skewness**.

Post Cluster Sampling A term proposed by Dalenius (1957) to cover the situation where lack of information on the composition of clusters indicates selecting an initial random sample from which the clusters are then formed.

Posterior Probability The value of a probability based, *inter alia*, on observation of one or more trials; in contradistinction to prior probability, which is a probability before further trials are made.

The distinction is a relative one; a probability p_1 at the outset of an experiment might be modified to p_2 in the light of its result. It would then be posterior in relation to that experiment but would be prior in relation to further experiments.

The notion is most suitably formalised in terms of

Conditional Probabilities. The probability law of a variate conditional upon an event A may be regarded as posterior to the occurrence of A, the unconditional law being prior. [See also **Projection**.]

Power In general, the power of a statistical test of some hypothesis is the probability that it rejects the alternative hypothesis when that alternative is false. The power is greatest when the probability of an **Error of the Second Kind** is least.

Power Efficiency Alternative name for **Relative Efficiency** of a test.

Power Function When the alternatives to a null hypothesis form a class which may be specified by a parameter θ the power of a test of the null hypothesis considered as a function of θ is called the power function. Exhibited graphically with the power as ordinate against θ as abscissa it provides a clear picture of the 'performance' of the test. Comparisons among a number of different tests are made by superposing the graphs of their power functions.

*Power Mean See ***Mean, *Combinatorial Power Mean**.

Power Moment This expression is synonymous with 'moment' used without qualification. The only point of adding the word 'power' is to emphasise a distinction between the ordinary moments and such functions as factorial moments. The distinction may also be necessary where the word 'moment' is used to describe any rational symmetric function of the observations, a practice not to be recommended.

Power Spectrum An alternative name for the **Spectral Function** or spectral density function.

Power Sum The sum of a series of observations on a variate each of which has been raised to the same power. Such quantities occur most frequently in the calculation of moments or similar symmetric functions of the observations.

Precision In exact usage precision is distinguished from accuracy. The latter refers to closeness of an observation to the quantity intended to be observed. Precision is a quality associated with a class of measurements and refers to the way in which repeated observations conform to themselves; and in a somewhat narrower sense refers to the dispersion of the observations, or some measure of it, whether or not the mean value around which the dispersion is measured approximates to the 'true' value. In general the precision of an estimator varies with the square root of the number of observations upon which it is based.

Precision, Modulus of In the theory of errors of observation, the reciprocal of the standard deviation multiplied by $\sqrt{2}$. It may be interpreted as the parameter h in the general equation for the Normal distribution, or error function:

$$y = (h/\sqrt{\pi})e^{-h^2x^2}$$

where $h = 1/(\sigma\sqrt{2})$, σ being the standard deviation. As h increases, the Normal curve becomes relatively narrower, i.e. the variability is reduced and, hence, the modulus of precision measures the closeness with which the observations cluster. [See also **Probable Error**.]

Predetermined Variable In the statistical analysis of models, particularly of the economic kind, the variables may be classified as **Endogenous** or **Exogenous** according to whether they represent an integral part of the system or influences impinging on it from without. Some of these variables may also appear as 'lagged', that is to say, as values occurring at some prior point of time. A predetermined variable is one whose values at any point of time may be regarded as known, and therefore includes either an exogenous or a lagged endogenous variable. The remaining variables are sometimes known as 'jointly determined' or 'currently exogenous' variables.

Prediction In general, prediction is the process of determining the magnitude of statistical variates at some future point of time. In statistical contexts the word may also occur in slightly different meanings; e.g. in a regression equation expressing a dependent variate y in terms of dependent x's, the value given for y by specified values of x's is called the 'predicted' value even when no temporal element is involved.

Prediction Interval The interval between the upper and lower limits attached to a predicted value to show, on a probability basis, its range of error.

Predictive Decomposition See **Decomposition**.

Predictor See **Fixed Variate, Independent Variable**.

Precedence Test This test, proposed by Nelson (1963), is equivalent to the **Exceedence Test** and concerns the hypothesis whether two samples come from the same population. It consists of counting the number of observations in the sample yielding the smallest observation which precede the observation of the rth rank in the other sample.

Pre-emptive Discipline A form of **Priority Queueing** whereby the arrival of an element of higher priority can actually displace the lower priority element actually in the service channel. When this displaced item returns to service the system must distinguish between 'resumption' at point of break or 'repeat' which ignores the earlier partial service.

Prewhitening A transformation used in the measurement of spectra. Degradation of results by computational noise and distortion may often be reduced by analysing the spectrum of a transformed input, whose spectrum has been altered in a specific manner, in order to make its spectrum more nearly white. Such a transformation aids accurate computation and is termed prewhitening.

Preference-field Index Number A synonym for **Konyus Index Number.**

Preference Table The $\binom{t}{2}$ **Paired Comparisons** of t objects, where ties are not permitted, may be displayed by means of 0, 1 variables in a two way table known as a preference table.

Price Compensation Index An index number of consumers' prices, constructed as a chain index on the basis of a consumer income, which varies so as to maintain a constant standard of living. The **Laspeyres-Konyus Index** is of this type. [See **Konyus Index.**]

Price Index An index number which purports to combine several series of price data into a single series expressing an average level of prices; e.g. of retail prices or of prices of manufactured products. [See **Laspeyres, Paasche, Marshall-Edgeworth-Bowley** and **'Ideal' Index Numbers.**]

Price-Relative The ratio of the price of a commodity in the given period to the price of the same commodity in the base period; such ratios enter into price index numbers of the Laspeyres or Paasche form.

Primary Unit This term is used in at least two senses. The first concerns a statistical unit of record which is basic in the sense that it does not depend upon any derived calculations, for example: persons, miles, tons, gallons, thousands of an article. The second usage of this term arises in sample surveys. Where a population consists of a number of units which may be grouped into larger aggregates but are not subdivided the units are called primary. For example, if a town is divided into districts, each of which is divided into blocks, each block comprising a number of houses; and if a sample of houses is desired, the house would be the primary unit. [See **Multi-stage Sampling.**]

Principal Components If each member of an aggregate bears the values of p variates $1, \ldots, p$ it is, in general, possible to find a linear transformation to p new variates $\xi_1, \ldots, \xi_p$ which (a) are independent, and (b) account in turn for as much of the variation as possible in the sense that the variance of ξ_1 is a maximum among all linearly transformed variates; the variance of ξ_2 is a maximum among all linearly transformed variates orthogonal to ξ_1, and so on. Such variates are called principal components.

Principle of Equipartition The division of the range of a frequency function into a number of parts such that the frequencies corresponding to each part are equal. This method is sometimes used in tests of goodness of fit; it is also used in sample surveys for the construction of strata (Kitagawa, 1956).

Prior Probability See **Posterior Probability, Bayes' Theorem.**

Priority Queueing A queueing system where the actual order of service for the arrivals is determined by some scheme of relative priority, i.e. not a first come first served discipline.

Probability A basic concept which may be taken either as undefinable, expressing in some way a 'degree of belief', or as the limiting frequency in an infinite random series. Both approaches have their difficulties and the most convenient axiomatisation of probability theory is a matter of personal taste. Fortunately both lead to much the same calculus of probabilities.

Probability Density Function An alternative term for the **Frequency Function** when the distribution concerned is considered as one of probability.

Probability Distribution A distribution giving the probability of a value x as a function of x; or more generally, the probability of joint occurrence of a set of variates $x_1, \ldots, x_p$ as a function of those quantities.

It is customary, but not the universal practice, to use 'probability distribution' to denote the probability mass or probability density of either a discontinuous or continuous variate and some such expression as 'cumulative probability distribution' to denote the probability of values up to and including the argument x. From a frequency viewpoint the distinction is the same as between 'frequency function' and **'Distribution Function'.**

Probability Element The probability associated with a small interval of a continuous variate, written in some such form as $f(x)\, dx$, or in general:

$$f(x_1, x_2, \ldots, x_n)\, dx_1 dx_2 \ldots dx_n.$$

Probability Integral An alternative name for the **Distribution Function** or the cumulative probability function for continuous variates. For example, the probability integral of a continuous variate x is a function $F(x)$ having the property that

$$F(a) = \Pr\{x \leqslant a\} = \int_{-\infty}^{a} f(x)dx$$

where $f(x)$ is the probability (frequency) function.

Probability Integral Transformation If x is a continuous variate with frequency function $f(x)$ and distribution function $F(x)$, a transformation to a new variate y given by

$$y = \int_{-\infty}^{x} f(x)dx = F(x)$$

is called the probability integral transformation. y is

uniformly or rectangularly distributed in the range $0 \leqslant y \leqslant 1$.

Probability Limits Upper and lower limits assigned to an estimated value for the purpose of indicating the range within which the true value is supposed to lie according to some statement of a probabilistic character. For example, confidence limits, control chart limits and fiducial limits are probability limits. They may be contrasted with the numerical limitations sometimes placed upon aggregates in descriptive statistics which are indicative of possible errors of collection or compilation rather than probability statements.

Probability Mass A term which is sometimes used to describe the magnitude of a probability or the relative frequency of observations located at a particular variate value, as distinct from being spread over a continuous range.

Probability Moment A synonym of **Frequency Moment**.

Probability Paper A graph paper with the grid along one axis specially ruled so that the distribution function of a specified distribution can be plotted as a straight line against the variate as abscissa. These specially ruled grids are available for the Normal, binomial, Poisson, lognormal, extreme value and Weibull distribution.

Probability Ratio Test See Likelihood Ratio Test.

Probability Sampling Any method of selection of a sample based on the theory of probability; at any stage of the operation of selection the probability of any set of units being selected must be known. It is the only general method known which can provide a measure of precision of the estimate. Sometimes the term random sampling is used in the sense of probability sampling. [See **Non-random Sampling.**]

Probability Surface A bivariate frequency surface; that is to say, the three dimensional representation of a bivariate frequency (probability) distribution with the frequency (probability density) along one axis and the variates along the other two axes.

Probable Error An older measure of sampling variability now almost superseded in statistics by the **Standard Error**. The probable error is 0.6745 times the standard error, the reason for the choice of the numerical coefficient being that the quantiles of a Normal distribution with variance σ^2 are distant $0.6745\,\sigma$ from the mean, so that one-half of the distribution lies within the range: mean $\pm 0.6745\,\sigma$.

Probit The **Normal Equivalent Deviate** increased by 5 in order to make negative values very rare. The word was suggested by Bliss (1934) as a contraction of 'probability unit'.

Probit Analysis The analysis of quantal response data using the **Probit** transformation.

Probit Regression Line In the analysis of quantal response data the percentages or proportions of the subjects reacting to the doses of stimulus can be converted into probits and plotted as ordinates against the logarithms of the doses. A line through this scatter of points, fitted by freehand methods or by an arithmetical process, is the probit regression line. The usual arithmetic procedure for obtaining it is an iterative method of successive approximation by means of a weighted linear regression of working probits on the logarithms of the doses.

Procedural Bias A somewhat vague phrase denoting bias in a sampling inquiry attributable to the procedure followed in obtaining the information, as distinct from bias due to the use of inferior methods of estimation on the data when obtained. The expression is also used incorrectly to denote general imperfection in a sampling plan, that is to say, a failure to adopt the optimum procedure with the resources available; this may be procedural but is not bias in the statistical sense.

Process Average Fraction Defective The average of the proportion of defective items in samples from a manufacturing process; the probability that an item from a process which is statistically under control is defective.

Process with Independent Increments See **Additive Process.**

Processing Error A type of error which can occur in the processing of statistical data. In survey data, for example, processing errors may include errors of transcription, errors of coding, errors of punching on to cards and errors of arithmetic in tabulation.

Producer's Risk In acceptance inspection, the risk which a producer takes that a batch will be rejected by a sampling plan even though it conforms to requirements. It is related to the probability of an error of the first kind in the theory of testing hypotheses in that it corresponds to the probability of rejecting an hypothesis when it is, in fact, true. [See also **Consumer's Risk.**]

Product Moment If the distribution function of n variates $x_1, x_2, ..., x_n$ is given by $F(x_1, ..., x_n)$ the product moment, joint or multivariate moment of order $r, s, ..., u$ is the mean value of $x_1{}^r x_2{}^s ... x_n{}^u$:

$$\int_{-\infty}^{\infty} \int_{-\infty}^{\infty} ... \int_{-\infty}^{\infty} x_1{}^r x_2{}^s ... x_n{}^u dF(x_1 ... x_n).$$

Product Moment Correlation A product moment correlation coefficient is so termed because its numerator is the first **Product Moment** or covariance of the two variates concerned. It is defined as

$$\rho = \frac{\text{Covariance } (x, y)}{\{\text{Var } (x)\, \text{Var } (y)\}^{\frac{1}{2}}}.$$

119

Progressive Average An average, of increasing extent, taken from a fixed point, e.g. for a series x_1, x_2, ..., a simple set of progressive averages might be the sequence

$$x_1, \tfrac{1}{2}(x_1+x_2), \tfrac{1}{3}(x_1+x_2+x_3), \ldots.$$

It must not be confused with the **Moving Average**.

Progressively Censored Sampling In life and dosage response studies it is frequently desirable, or practically imposed, that some of the surviving sample units are withdrawn at an initial stage of censoring: others being withdrawn at later stages. This practice facilitates the economic use of test facilities and does provide some data on the longer life spans. This form has also been referred to as 'hypercensored samples' and 'multiple censored samples'.

Projection This term is used in two connected senses. (1) In relation to a time series it means a future value calculated according to predetermined changes in the assumptions of the environment. (2) More recently it has been used in probability theory to denote the conditional expectation of a variate. Since a regression equation gives the expectation of the dependent variate conditional upon values of the predicated ('independent') variates and such equations are used for forecasting or prediction, the usages are connected. [See also **Posterior Probability, Regression**.]

Proportional Frequency In relation to a frequency distribution, the proportional frequency in any class is the frequency of the class divided by the total frequency of the distribution.

The term sometimes occurs in a different sense in relation to bivariate or multivariate frequency arrays. For instance, if, in a table of p rows and q columns the q frequencies in each row are proportional to the q row totals and, similarly, therefore, for the columns the case is said to be one of proportional frequencies. The term is sometimes used similarly in connection with proportional sub-class numbers in analysis of variance.

Proportional Sampling A method of selecting sample numbers from different strata so that the numbers chosen from the strata are proportional to the population numbers in those strata. [See also **Uniform Sampling Fraction**.]

Proportional Sub-Class Numbers See **Proportional Frequency**.

Proximity Analysis A method used in numerical taxonomy for arranging items in a line or a plane, or in a space of higher dimensions so that like is adjacent to like and far from unlike. The method is based upon rank ordering of inter-point distances and inverse ordering of similarities. [See **Cluster Analysis**.]

Proximity Theorem An elementary theorem by H. Wold (1952) on the smallness of bias in least squares estimates of regression coefficients.

Pseudo-Factor An artificial or dummy factor used in the design of factorial experiments, generally to render the number of factors a convenient one for the application of some specified balanced design.

Pseudo-Inverse A mathematical device which attempts to overcome the fact that a square matrix which is singular does not have an inverse form. [See also **Generalised Inverse**.]

Pseudo-Spectrum A somewhat misleading term denoting the mathematical expectation of the estimates of **Spectral Density** obtained without regard to the stationarity of the process.

Psi-square Statistic A chi-squared type statistic which can be used as a test of fit with serially correlated observations. First used by Kendall & Smith (1938), and its asymptotic distribution is the convolution of the two χ^2 distributions.

p-statistics A set of statistics introduced by S. N. Roy (1939) in multivariate analysis. They are closely allied to the sample values of the characteristic roots of dispersion matrices.

Psychological Probability See **Probability**.

Pure Birth Process See **Birth Process**.

Pure Random Process The simplest example of a stationary process where, in discrete time, all the random variables z_t are mutually independent. In continuous time the process is sometimes referred to as 'white noise', relating to the energy characteristics of certain physical phenomena.

Pure Strategy See **Strategy**.

Purposive Sample A sample in which the individual units are selected by some purposive method. It is therefore subject to biasses of personal selection and for this reason is now rarely advocated in its crude form. [See **Quota Sample; Balanced Sample**.]

Q-Technique A method of analysis of similarities or relationships in which, given a matrix of n observations on m individuals, n rows and m columns, the correlations or other statistical measure are sought between the m columns down the n rows, i.e. between individuals, as distinct from the R-technique which looks for relations between variables, namely between rows along the columns.

Quad A square-shaped **Basic Cell**; also, the area of such a cell.

Quadrant Dependence If the probability of any quadrant $X \leq x$, $Y \leq y$ under the distribution $F(x, y)$ is compared with its independence probability, then the pair (x, y) or its distribution as denoted by Lehmann (1966) is positively quadrant dependent if
$$\Pr(X \leq x, \ Y \leq y) \geq \Pr(X \leq x)\Pr(Y \leq y).$$
The negative dependence holds with the inequality reversed.

Quadrat A sampling device in the form of a square lattice. It may be a framework which can be placed on the ground, e.g. for dividing a plot into sub-plots, or a square grid of some transparent material for superposition on a map.

More loosely the term is sometimes used to denote (*a*) a mesh which is rectangular, not necessarily square, and (*b*) one unit of the lattice.

Quadratic Estimator An estimator which is based upon some quadratic function of sample values. For example, the standard deviation may be estimated from the square root of the variance, a quadratic estimator, or from the mean deviation or the range, which are linear estimators.

Quadratic Form A homogeneous quadratic function of the form $Q = \sum\limits_{i=1}^{n} \sum\limits_{j=1}^{n} a_{ij}x_ix_j = \mathbf{x}'\mathbf{Ax}$ where $\mathbf{A}$ is the matrix of the quadratic form and generally assumed to be symmetric. The quadratic form is important in multivariate analysis and, in particular, analysis of variance. [See also **Cochran's Theorem**.]

Quadratic Mean An average which involves the squares of the values being averaged, or, more generally, a quadratic function of them. For example, the **Standard Deviation** is a quadratic mean. In order to reduce the mean to the same dimensions as the values being averaged a square root has to be taken after the averaging process has been carried out.

Quadratic Programming A major development of **Linear Programming** in which some or all of the constraints and the objective function are quadratic in the variables.

Quadratic Response Where the relationship between a **Quantitative Response** and the **Dose Metameter** is not linear, but is of the second degree, the response is sometimes termed a 'quadratic response'.

Quadrature Spectrum The form of spectrum which measures the covariance between the sine and cosine components, or out of phase components, of a sample time series. If the quadrature spectrum is denoted by $Q_{12}(f)$ then
$$Q_{12}(f) = -A_{12}(f)\sin F_{12}(f)$$
where $A_{12}(f)$ and $F_{12}(f)$ are the **Phase Spectrum** and **Cross-amplitude Spectrum** of the two series.

Qualitative Data See **Quantitative Data, Attribute**.

Quality Control The statistical analysis of process inspection data for the purpose of controlling the quality of a manufactured product which is produced in large numbers. It aims at tracing and eliminating systematic variations in quality, or reducing them to an acceptable level, leaving the remaining variation to chance. The process is then said to be statistically under control.

Quality Control Chart See **Control Chart**.

Quantal Response The response of a subject to a stimulus is said to be quantal when the only observable, or the only recorded, consequence of applying the stimulus is the presence or absence of a certain reaction, e.g. death. A quantal response may be expressed as a two valued variate taking values 0 and 1.

Quantiles The class of $(n-1)$ partition values of a variate which divide the total frequency of a population or a sample into a given number n of equal proportions. For example, if $n = 4$ then the $n-1$ values are the **Quartiles** although the central variate value is generally termed the **Median**. [See also **Deciles, Percentiles, Quintiles**.]

Quantitative Data Strictly, this term, as contrasted with qualitative data, should relate to data in the form of numerical quantities such as measurements or counts. It is sometimes, less exactly, used to describe material in which the variables concerned are quantities, e.g. height, weight, price as distinct from data deriving from qualitative attributes, e.g. sex, nationality or commodity. This usage is to be avoided in favour of such expressions as 'data concerning quantitative (qualitative) variables' or 'data concerning numerical variables (attributes)'.

Quantitative Response A reaction, by an experimental unit to a given stimulus, which may be measured on a variate scale. For example, the response may be measured in terms of weight, size or reaction time: in particular the survival time.

Quantity Relative The ratio of the quantity of a commodity, in the given period to the corresponding quantity in the base period; such ratios enter into quantity index numbers of the Laspeyres or Paasche form. [See **Price Relative**.]

Quantum Hypothesis A hypothesis in which the possible values of a parameter are discrete and hence increase by quantum jumps.

Quantum Index An index number which purports to show the changes in quantity, usually of goods or services produced, purchased or sold, independently of changes in prices or money values. One such index is of the **Laspeyres'**

type obtained by weighting quantities in the given and base period by prices in the base period. Quantum index numbers do not necessarily measure changes in volume or weight.

Quartile There are three variate values which separate the total frequency of a distribution into four equal parts. The central value is called the **Median** and the other two the upper and lower quartiles respectively. They are a particular set of **Quantiles.**

As thus defined the quartiles are subject to some indeterminacy for discontinuous distributions. This is usually resolved by conventionally allocating a frequency partly to the left and partly to the right of its variate value or, where a range of values between two variate values satisfies the definition, by fixing the quartile at some point in the range by an arbitrary rule, e.g. by taking the mid-point.

Quartile Deviation A measure of dispersion based upon the distance between certain representative values of the variate. In this case the representative values are the upper and lower quartiles and the quartile deviation is defined by

$$Q.D. = \tfrac{1}{2}(Q_3 - Q_1).$$

An alternative name for the quartile deviation is the semi-inter-quartile range.

Quartile Measure of Skewness If the lower quartile is Q_1, the upper quartile is Q_3 and the median is M, the quartile measure of skewness of a frequency distribution is

$$\frac{(Q_3 - M) - (M - Q_1)}{(Q_3 - Q_1)}.$$

Quartile Variation An alternative to the **Standard Deviation** as a measure of variation. If the lower quartile is Q_1, and the upper quartile is Q_3, the coefficient of quartile variation, denoted V_Q is given by

$$100 \ (Q_3 - Q_1)/(Q_3 + Q_1).$$

Quartimas See **Factor Rotation.**

Quartimin See **Factor Rotation.**

Quasi-Compact Cluster See **Cluster Sampling.**

Quasi-Factorial Design An experimental design for which a formal correspondence may be set up between the treatments and their combinations and the combinations of the treatments of a factorial set. Thus, for example, four treatments can be put in correspondence with the four combinations of two factors, each at two levels, and the design for the four treatments derived from one or more of those appropriate to the 2^2 factorial design. This class of design is useful for treatments which do not have a factorial structure and, especially with the recovery of **Information** between blocks, is in general more

efficient than designs using **Randomised Blocks** or 'control plot' techniques.

Although terminology is still somewhat fluid, 'quasi-factorial' is usually synonymous with 'lattice' in relation to a design. The so-called lattice designs owe their name to the fact that the treatments are allocated in some systematic way according to a pattern which can be represented diagrammatically on a lattice.

Quasi-Latin Square A term proposed by Yates (1937) for certain kinds of factorial designs in the form of Latin Squares. In these experimental designs certain of the interactions are confounded with the rows and columns of the squares. These designs eliminate the variations due to differences between the rows and the columns from the experimental error of those effects which are not confounded but, unlike the Latin Square itself, each treatment does not appear once to every row and column.

Quasi-Maximum Likelihood Estimator A set of least squares estimators for multiple equation models where the asymptotic distributional properties of the estimators are almost independent of the form of the distributions of errors in the equation, so that least squares solutions are virtually equivalent to maximum likelihood solutions.

Quasi-Normal Equations If the normal equations of least squares estimation have the parameter estimates based upon **Instrumental Variables** rather than the ordinary model variables, then the equations are said to be quasi-normal. 'Normal' in this sense has nothing to do with the Normal (Gaussian) distribution.

Quasi-Range A term proposed by Mosteller (1946) for the difference $x_s - x_r$, where $1 \leqslant r \leqslant s \leqslant n$, in the ascending order statistics of a sample of n observations. Common practice is to take the range of $n - 2i$ observations, omitting the i largest and i smallest.

Quasi-Random Sampling Under certain conditions, largely governed by the method of compiling the **Sampling Frame** or list, a systematic sample of every nth entry from a list will be equivalent for most practical purposes to a random sample. This method of sampling is sometimes referred to as quasi-random sampling.

Quenouille's Test A test proposed by Quenouille (1947) for the goodness of fit of an autoregressive model to a time series. The test was extended by Wold (1949) to the case of a moving averages model; and by Walker (1950) for an autoregressive model with error terms comprising moving averages of independent variates. Quenouille (1958) extended it further to a pair of time series or two lengths of one series.

Questionnaire A group or sequence of questions designed to elicit information upon a subject, or sequence of subjects, from an informant. [See also **Schedule.**]

Queueing Problem The problem of queues, or congestion, arises in a variety of fields where there is a service to be offered and accepted rather than a product to be made. In general the problem is concerned with the state of a system, e.g. the length of the queue or queues at a given time, the average waiting time, queue discipline and the mechanism for offering and taking the particular service. The analysis of queueing problems makes extensive use of the theory of stochastic processes.

Quintiles The set of four variate values which divide the total frequency into five equal parts. [See **Quantile**.]

Quota Sample A sample, usually of human beings, in which each investigator is instructed to collect information from an assigned number of individuals (the quota) but the individuals are left to his personal choice. In practice this choice is severely limited by 'controls', e.g. he is instructed to secure certain numbers in assigned age groups, equal numbers of the two sexes, certain numbers in particular social classes and so forth. Subject to these controls, which are designed to make the sample as representative as possible, he is not restricted to the contacting of assigned individuals as in most forms of probability sampling.

Quotient Regression A regression of the form

$$y_t = \frac{\alpha_0 + \alpha_1 x_{1t} + \ldots + \alpha_n x_{nt}}{\beta_0 + \beta_1 z_{1t} + \ldots + \beta_m z_{mt}} + \epsilon_t$$

(Wold, 1966), where x and z are regressor variables and ϵ is a random residual. The coefficients α and β are usually estimated by least squares in an iterative procedure.

R-Technique See **Q-Technique**.

Racial Likeness, Coefficient of A coefficient proposed by K. Pearson and published in 1921. It was designed for the testing of homogeneity of two multivariate distributions but was extended to the measurement of distance between them. For the latter purpose it has certain disadvantages and has been replaced by the **D^2 Statistic** of Mahalanobis.

Radix This word in its customary mathematical sense, namely as meaning a number which is the base of a numerical system or table; for example, the radix of the system of decimal numbers and of common logarithms is ten. It also occurs in an analogous statistical sense, e.g. a **Life Table** may show the numbers, surviving at different ages, of an initial number of individuals, e.g. 10,000, which is the radix of that particular tabulation.

Raikov's Theorem This theorem, first stated by Raikov (1938) shows that if x_1 and x_2 are independent and $x = x_1 + x_2$ has a Poisson distribution, then each of the random variables x_1 and x_2 has a Poisson distribution. This result can be generalised to any finite number of independent random variables.

Raising Factor Apart from its ordinary significance this term is used in the following special sense. The coefficients of a linear function of the values of the sample units used to estimate population, stratum, or higher stage unit totals are called raising, multiplying, weighting or inflation factors of the corresponding sample units. If the raising factors of all the sample units are equal, the common raising factor is called the raising factor of the sample, and the sample itself is called self-weighting. It should be noted that the raising factors depend not only on the sampling plan but also on the method of estimation.

Random This word may be taken as representing an undefined idea, or, if defined, must be expressed in terms of the concept of probability. A process of selection applied to a set of objects is said to be random if it gives to each one an equal chance of being chosen. Generally, the use of the word 'random' implies that the process under consideration is in some sense probabilistic.

Random Allocation Designs An alternative name for **Random Balance Design** which emphasises the principal of construction involving random sampling from the array of treatment combinations in a full k-factorial design.

Random Balance Design A factorial design intended to deal with the case where the number of factors exceeds the number of observations. It is constructed by selecting treatment combinations at random from the possible set, subject to constraints to ensure balance. The method was proposed by Satterthwaite (1959) and developed by other writers (e.g. Dempster, 1960).

Random Component If a magnitude consists of a number of parts compounded in some way, e.g. by addition or multiplication, any such part as is a **Variate** is a random component of the magnitude.

Random Distribution This expression is sometimes wrongly used for a probability distribution. It is also sometimes employed to denote a distribution of probability which is uniform in the range concerned, i.e. a rectangular distribution. It seems better to avoid the term altogether, or, in such expressions as 'points randomly distributed over an area' to specify clearly the law of distribution involved.

Random Effects Model An alternative name for **Model II** [See also **Variance Components**.]

Random Error An error, that is to say, a deviation of an observed from a true value, which behaves like a variate in the sense that any particular value occurs as though chosen at random from a probability distribution of such errors.

Random Event An event with a probability of occurrence determined by some probability distribution. The term is used somewhat loosely to denote either an event which may or may not happen at a given trial, such as the throwing of a 6 with an ordinary die, or an event which may or may not happen at any given moment of time such as an industrial accident to an individual.

Random Impulse Process A stochastic process describing the linear motion of a particle subject to a series of small impulses which are random.

Random Linear Graph The formation of lines joining pairs of points independently and randomly selected from a group of, say, N points.

Random Order An order of a set of objects when the ordering process is carried out in such a way that every possible order is equally probable. Tests of random order are freely used to test the hypothesis that there are systematic elements present which would prevent the observed order from being random.

Random Orthogonal Transformations A device used in multivariate analysis (see Wijsman, 1957) whereby transformations are performed with orthogonal matrices the elements of which depend on a random vector.

Random Process In a general sense the term is synonymous with the more usual and preferable 'Stochastic' **Process**. It is sometimes employed to denote a process in which the movement from one state to the next is determined by a variate which is independent of the initial and final state. It is better to denote such a process as a **Pure Random Process**.

Random Sample A sample which has been selected by a method of **Random Selection**.

Random Sampling Error **Sampling Error** in cases where the sample has been selected by a random method. It is common practice to refer to random sampling error simply as 'sampling error' where the random nature of the selective process is understood or assumed.

Random Sampling Numbers Sets of numbers used for the drawing of random samples. They are usually compiled by a process involving a chance element and in their simplest form consist of a series of digits 0 to 9 occurring (so far as can be ascertained) at random with equal probability.

Random Selection A method of selecting sample units such that each possible sample has a fixed and determinate probability of selection. Ordinary haphazard or seemingly purposeless choice is generally insufficient to guarantee randomness when carried out by human beings and devices such as tables of random sampling numbers, or analogous machines, are used to remove subjective biasses inherent in personal choice.

Random Series A series the numbers of which may be regarded as drawn at random from a fixed distribution. [See **Irregular Kollektiv**.]

Random Start In selecting a **Systematic Sample** at intervals of n from an ordered population, it is sometimes desirable to select the first sample unit by a random drawing from the first n units of that population. The sample is then said to have a random start.

Random Variable In this dictionary the word 'variate' is used to denote what is often called a 'random variable'. It is in many ways convenient to distinguish between the variable of mathematics and the variate of probability theory. [See **Variable, Variate**.]

Random Walk The path traversed by a particle which moves in steps, each step being determined by chance either in regard to direction or in regard to magnitude or both. Cases most frequently considered are those in which the particle moves on a lattice of points in one or more dimensions, and at each step is equally likely to move to any of the nearest neighbouring points. The theory of random walks has many applications, e.g. to the migration of insects, sequential sampling and, in the limit, to diffusion processes.

Randomisation A set of objects is said to be randomised when arranged in a random order; and, by a slight extension, a set of treatments applied to a set of units is said to be randomised when the treatment applied to any given unit is chosen at random from those available and not already allocated.

Randomisation Tests A test of significance based on the distribution obtained by permuting the actual observations either completely or under certain constraints.

Randomised Blocks An experimental design in which each **Block** contains a complete replication of the treatments, which are allocated to the various units within the blocks in a random manner and hence allow unbiassed estimates of error to be constructed.

Randomised Decision Function A decision function which is selected from a set of possible decision functions with the help of a chance mechanism.

Randomised Fractional Factorial Designs A class of design proposed by Ehrenfeld & Zachs (1961) which yield unbiassed estimators, valid tests and confidence intervals for the parameters of interest without the usual assumptions concerning the confounding of higher order interaction. Two methods of randomisation are used (*i*) based upon blocks of treatment combinations, a type of

Cluster Sampling; and (*ii*) random selection of treatments from every block, a form of **Stratified Sampling**.

Randomised Model A statistical model, generally an experiment design, where the treatment combinations are assigned to experimental units by some random arrangement.

Randomised Test An expression to be avoided. It occurs in the meaning of a test in which the decision to accept or reject depends on the value of a random variable. It should not be confused with **Randomisation Test**.

Range The largest minus the smallest of a set of variate values. The range is of itself an elementary measure of dispersion but, in terms of the **Mean Range** in repeated sampling, it may afford a reasonable estimate of the population standard deviation. [See also *m*th **values**.]

Range Chart A chart used in statistical quality control on which the recorded quality criterion is the range of samples. This **Control Chart** is used to maintain a check upon the variability of the quality of the particular product or processes. The range is a less sensitive criterion for changes in variability than is the sample variance but is much more easily calculated.

Rank The term occurs in statistical work in at least three contexts: (*a*) In the theory of order relations, the rank of a single observation among a set is its ordinal number when the set is ordered according to some criterion such as values of a variate borne by the individuals. (*b*) In matrix theory the term occurs in its usual mathematical sense, being the greatest number *r* of linearly independent rows or columns which can be found in it. (*c*) Derived from the previous usage, the rank of a multivariate distribution is the rank of its dispersion matrix, and is thus the number of variates which are independent in the sense of not being connected by linear equations. [See also **Singular Distribution**.]

Rank Correlation Rank correlation measures the intensity of correlation between two sets of rankings or the degree of correspondence between them. There are two principal coefficients of rank correlation: **Kendall's** τ (1938) and **Spearman's** ρ (1904). [See also ***Cograduation**.]

Rank Order Statistics Statistics based only on the rank order of the sample observations, e.g. the rank correlation coefficients. 'Rank order statistics' are distinguishable from 'order statistics', e.g. the median, the range which make use of the metric values of the observations.

Rankit A transform of quantal response data based on ranks. [See **Probit**.]

Rao's Scoring Test A large sample test of a simple hypothesis proposed by Rao (1948) using a scoring system in parameter estimation (Rao, 1965).

***Ratio (Rapporto)** Italian statistics use a number of expressions involving the word *rapporto* which are unknown in English. For example, the co-existence ratio (*rapporto di coesistenza*), which relates to the intensity of a phenomenon in two different places or of two phenomena at the same place; the duration ratio (*rapporto di durata*) which is a measure of average duration; the composition ratio (*rapporto di composizione*) measuring the relation between the intensity of a phenomenon and that of a more comprehensive phenomenon of which it forms part; the derivation ratio (*rapporto di derivazione*) which compares the intensity of a phenomenon with one which is prerequisite to it, such as births to total population, and the repetition ratio (*rapporto di repetizione*) which measures how often a phenomenon recurs in a certain time interval.

Ratio Delay Method An alternative name for the **Snap Reading Method**.

Ratio Estimator An estimator which involves the ratio of two variates, i.e. a ratio whose numerator and denominator are both subject to sampling errors. The term occurs particularly in sample survey theory. If the members of a population each bear the values of two characteristics, *x* and *y*, and the total of *x*, say *X*, is known for the population, the corresponding total of *y*, say *Y*, can be estimated by multiplying *X* by a sample ratio consisting of the sample total of *y* divided by the sample total of *x*.

Ratio Scale A graphical scale in which equal absolute variations correspond to equal proportional variations in the data. The most common form of chart employing the ratio is the logarithmic or **Semi-logarithmic Chart**.

Rational Trend A term which is, or ought to be, obsolescent. It denotes a trend which may be expressed as a mathematical function of the time, as distinct from one which has a stochastic component.

Raw Moment A moment of a frequency distribution calculated about some origin other than the arithmetic mean. The usage is not universal and some authors use this term to denote moments either about the mean or not before corrections for grouping are applied. [See **Sheppard's Corrections**.]

Raw Score The score as originally obtained in some psychological, educational or other test. [See also **T-score**, **z-score**.]

Rayleigh Distribution A χ distribution with two degrees of freedom, so called because it was considered by Rayleigh in some physical situations.

Realisation A realisation of a stochastic process $\{x_t\}$ is one of the series of values (... x_{-2}, x_{-1}, x_0, x_1, ...) to which it may give rise. The realisation may be regarded as a 'member' of the process in the same way that an individual observation is regarded as the member of a population.

In general the realisation is of infinite extent, but a finite observed section of it is also sometimes referred to as a realisation.

Records Tests Distribution free tests for trend in time series based on the breaking of record values.

An observation is called an upper/lower record if it is greater/smaller than all previous observations in the series. Foster & Stuart (1954)) have described two such tests.

Recovery of Information The standard analysis of experiments designed in the form of **Incomplete Blocks** fails to use information about treatment effects which can be obtained from comparisons among block totals. More refined methods of analysis to 'recover' this information were proposed by Yates (1940).

Rectangular Association Scheme If N is the **Design Matrix** of a partially balanced incomplete block design with three associate classes, the two modes of classification (of treatments in N) for the relation of first and second association can be superimposed, in the form of a rectangular array (Vartak, 1959).

Rectangular Distribution Strictly a continuous distribution of type

$$dF(x) = dx/k, \qquad \alpha \leqslant x \leqslant \alpha + k$$

where k is a constant. The expression is also sometimes used to denote a discontinuous distribution for which all variate values have the same probability.

Rectangular Lattice An experimental design introduced by Harshbarger (1947) as an extension of the **Square Lattice**. It is appropriate to $k(k-1)$ treatments, which are regarded as corresponding to the points of a $k \times (k-1)$ lattice.

Rectified Index Number An index number formula which is obtained by taking the geometric mean of two other index numbers of opposite bias. The two index numbers are sometimes said to be geometrically 'crossed', e.g. the **Ideal Index Number**. The object of rectification is usually to make the resulting index satisfy either, or both of, the **Factor Reversal** or the **Time Reversal Test**.

Rectifying Inspection Inspection of a product which aims at removing any defective units found and replacing them by effective units. In this way the quality of a batch of product may be considerably improved and, in any case, a batch is never rejected. This type of inspection is not applicable when the inspection test is destructive. [See also **Average Outgoing Quality Limit**.]

Rectilinear Trend An alternative name for a **Linear Trend**.

Recurrence Game A sequence of trials of an event which are conducted as a game, in that a 'reward' is received, or a 'fee' incurred, under certain recurrences.

Recurrence Time This concept occurs in the analysis of stochastic processes in two ways. In connection with a renewal process it is the time, or number of steps, between two similar (recurrent) states. With reference to a **Point Process** and in particular a stationary point process where the time origin is arbitrary, the forward recurrence is the interval from the arbitrary time origin to the next point event. The backward recurrence time is formed by direct analogy.

Recurrent Markov Chain A periodic **Markov Chain**.

Recurrent State A state k in a Markov Chain is said to be recurrent if, with probability one, the Markov Chain will eventually return to k, having started at k. A state k is said to be non-recurrent if this probability is less than one.

Some writers call a recurrent state 'persistent' and a non-recurrent state 'transient'.

'Recursive' System The word 'recursive' has been used by some writers to denote relations which are 'recurrent', presumably under the impression that the former is the correct adjectival form derived from the verb 'to recur'. A purist will avoid the word in this sense.

Wold (1953) has proposed to describe as 'recursive' systems of equations in econometrics with three properties: (1) they are recurrent in the sense that if the values of the variables are known up to time $t-1$ the equations give the values for time t; (2) the values of the variables at time t are obtainable one by one in some order or other; (3) each equation of the system expresses a unilateral causal dependence. [See **Causal Chain Model**.]

Reduced Design A device in experiment design for restricting the large size of block required by a full design. For example, in cyclic rotation experiments Patterson (1964) proposed dividing the crop sequences into groups according to comparability.

Reduced Equations A method of estimation in econometrics whereby the original equations are modified so that each endogenous variable is expressed as a function of the set of exogenous variables and, possibly, the errors. [See **Indirect Least Squares**.]

Reduced Form Method In econometrics, a method of estimating the parameters in a stochastic system which relies on the expression of the **Endogenous Variables** individually in terms of **Predetermined Variables**. [See also **Limited Information Methods**.]

Reduced Inspection See Normal Inspection.

Reduced Sample A term sometimes applied to the involuntary censoring of a sample when some items are no longer observable. [See Kaplan & Meier, 1958.]

Reduction of Data The process whereby a large number of observations are brought within manageable compass for convenience of handling and interpretation.

Reed-Münch Method A method proposed by Reed and Münch (1938) for the rapid assessment of the equivalent doses of standard and test preparations which would produce a 50 per cent quantal response. The method is strictly applicable only to **Tolerance Distributions** which are symmetrical.

Reference Period In one sense this is synonymous with **Base Period**. It may also refer to the length of time, e.g. week or year for which data are collected.

Reference Set An alternative term for **Fundamental Probability Set**.

Reflecting Barrier Certain additive or **Random Walk Processes** represent the motion of a particle in one or more dimensions and, in certain cases, limitations may be imposed on the motion in the form of barriers (constraints) which, once reached, reflect the particle and continue the motion, as distinct from absorbing it.

Refusal Rate In the sampling of human populations, the proportion of individuals who, though successfully contacted, refuse to give the information sought. The proportion is usually and preferably calculated by dividing the number of refusals by the total number of the sample which it was originally desired to achieve. Where, however, there are other causes of non-achievement, e.g. persons have died or left the area, the refusal rate is sometimes calculated as the number of refusals divided by the number of persons contacted, i.e. by the number of refusals plus the number of successful or partially successful contacts.

Regenerative Process A class of stochastic process attributed to Palm (1943) characterised by possessing regeneration points. These points are those epochs where the occurrence of the state S is sufficient to 'regenerate' the process in the sense that probabilities are no longer dependent on past history. The Markov process is a special case which requires every state to be a regeneration point.

Regressand A synonym for dependent variable in a regression relation.

Regression This term was originally used by Galton to indicate certain relationships in the theory of heredity but it has come to mean the statistical method developed to investigate those relationships.

If a variate y consists of two components, a variate and a systematic element $f(X)$ depending on a variable X, i.e. if $y = f(X) + \epsilon$ then the regression of y on X is the equation $Y = f(X)$ where it is supposed that ϵ has zero expectation. The definition remains valid if X, instead of being a single variable, refers to a set of variables $X_1, X_2, \ldots$

The general regression curve of order r is
$$\mu_{rx} = \mathcal{E}[\{y - \mathcal{E}(y\,|\,x)\}^r\,|\,x]$$
expresses dependence of central moments of y, for fixed x, upon x. If $r = 1$ we have the dependence of the mean of y for given x on the corresponding x: $\mathcal{E}(y\,|\,x) = f(x)$ for $r = 2$ we have the scedastic curve denoting dependence on the variance; $r = 3$ gives the clitic curve and $r = 4$ the kurtic curve respectively.

The most frequently considered form of $f(x)$ is a polynomial, particularly a linear function, giving the regression of y on X as $Y = \beta_0 + \beta_1 X$ or, for p variables $Y = \beta_0 + \beta_1 X_1 + \ldots + \beta_p X_p$.

Such expressions are called regression equations. The X's are called 'independent', 'predicated' variables, 'predictors' or 'regressors'. y is called the 'dependent variate', 'predictand' or 'regressand'.

Regression Coefficient The coefficient of an independent variable in a regression equation.

Regression Curve A diagrammatic exposition of a regression equation. For two variables this can be shown on a plane with the 'independent' variable X as abscissa and Y as ordinate; and for three variates it is possible to construct solid models or reduce the representation to a plane surface by use of the **Isometric Chart** and the **Stereogram**.

The term is sometimes interpreted to mean a regression equation of a degree higher than the first, the emphasis then lying on the word 'curve' as opposed to a straight line.

Regression Dependence If, for two stochastically dependent variates X, Y, $\Pr(Y \leqslant y\,|\,X = x)$ is non-increasing in x, y is said by Lehmann (1966) to be positively regression dependent. Likewise the dependence is negative regression if the probability statement is non-decreasing in x.

Regression Estimate In general, an estimate of the value of a dependent variate y obtained from substituting the known values of the 'independent' variables X in a regression equation connecting y and X.

The term has a particular application in sample surveys. If the regression of A on B may be estimated from a sample and the total of B is known for the population, the total of A may be estimated from the regression equation. It is then called a regression estimator. [See **Ratio Estimate**.]

Regression Line In general this is synonymous with

regression curve, but is sometimes and rather ambiguously used to denote a linear regression, i.e. one in which the 'dependent' variable is of the first degree only.

Regression Surface See **Regression Curve.**

Regressor A synonym for independent variable in a regression relation. [See **Regression.**]

Regret The loss function which arises in decision theory generally involves two terms the second of which represents the difference between total loss and unavoidable loss. It is this excess loss, the 'regret' which has to be minimised: see van der Waerden (1960).

Regular Best Asymptotically Normal Estimator A class of estimator proposed by Neyman (see Taylor, 1953) based upon the minimisation of a distance function of the χ^2 type.

Regular Estimator An estimator for which there hold certain regularity conditions, principally concerning the differentiability of the estimator with respect to the variate values on which it depends and of the frequency distribution with respect to its parameters. [See **Cramér-Rao Inequality.**]

Regular Group Divisible Incomplete Block Design A group divisible **Incomplete Block Design** is regular (Bose & Connor, 1952) if $r > \lambda_1$ and $rk > \lambda_2 v$ where r is number of replicates, v the product of m groups each of n treatments; λ_1 number of blocks that treatments in the same group occur and λ_2 the blocks in which treatments in different groups occur.

Rejectable Quality Level The level of quality as determined by percentage defective, for example, for which a buyer would wish to have only a low probability of accepting. This probability is the **Consumer's Risk.**

Rejection Error See **α-error; Error of First Kind.**

Rejection Line See **Acceptance Line.**

Rejection Number See **Acceptance Number.**

Rejection Region In the theory of testing hypotheses, a region of the **Sample Space** such that if a sample point falls within it the hypothesis under test is rejected.

Rejective Sampling Sampling with unequal probabilities with replacement in which the whole sample is rejected as soon as any individual is selected a second time (Hájek, 1964).

*Relative Area of Transvariation See ***Transvariation.**

Relative Efficiency (of an Estimator) A measure of comparative efficiency of two estimators of the same parameter. If estimator t_1 has, for sample size n_1, the same precision, in the sense of same sampling variance, as estimator t_2 for sample size n_2, the relative efficiency of t_1, with respect to t_2 is n_2/n_1.

Relative Efficiency (of a Sample Design) In the design of experiments this is equivalent to **Efficiency Factor.** In sample survey work the usage takes into account the cost of the survey and the relative efficiency is the ratio of the 'cost per unit of information', where **Information** is used in the sense of Fisher. A third usage concerns sampling plans and is based upon the ratio of the cost of the optimum plan to the plan in operation.

Relative Efficiency (of a Test) The ratio of sample sizes concerned with two tests of a statistical hypothesis necessary to yield the same **Power** against the same alternative hypothesis. The concept is due to Cochran (1937) and Pitman (1948). An alternative formulation due to Blomquist (1950) specifies equal slopes for the power curves at the parameter point rather than equal power.

Relative Frequency The frequency in an individual group of a frequency distribution expressed as a fraction of the total frequency.

Relative Index (Indice Relativo) In Italian usage an index is said to be relative if divided by the mean or the maximum value or other value which it may attain under certain specified hypotheses.

Relative Information A term introduced by Yates (1939) in connection with the partial confounding of experimental effects in factorial experimental designs. Where an effect is partially confounded, the relative amount of information is the ratio of the amount actually available to what would be available if there were no confounding.

Relative Potency The relationship of two estimated stimuli, one of which acts as a standard, which produce the same response. In biological assay the relative potency of a test preparation compared with a standard preparation is generally obtained from the inverse ratio of doses which result in identical responses, i.e. equally effective doses.

Relative Precision A term which is frequently used to denote the ratio of the error variances of two sample designs which are different but which are based upon the same sampling unit and the same size of sample. The usage is not universal, however, since some writers use the term **Relative Efficiency** for this concept. The relative efficiency and the relative precision are equal in the case of simple random sampling for the mean of a large population, but not necessarily otherwise.

Relative Risk A regrettable term purporting to represent the degree of association in 2×2 tables. If the relative frequencies are *a, b, c, d*, the relative risk is defined as *ad/bc* and is sometimes called the odds ratio.

Relative Variance A term sometimes used to denote the square of the coefficient of variation. [See **Variation, Coefficient of.**]

Relaxed Oscillation A time series model whereby the value of a phenomenon, although oscillating, generally increases in amplitude during a period of time through the operation of its internal forces. These forces then precipitate a 'crisis' or a 'bursting' and a return to zero value, i.e. the steadily increasing oscillations are relaxed, after which the process is repeated. A model of this nature might serve to account for the action of certain economic phenomena but the concept is more useful in the physical sciences.

Reliability This term is used in three different contexts. In connection with biological assay, Finney (1947) has defined reliability of an assay as the reciprocal of a function of the confidence interval of the estimate of potency of the stimulus.

The term is also used in factor analysis, especially in connection with the statistical analysis of psychological and educational tests. The 'reliability' of a result is conceived of as that part which is due to permanent systematic effects, and therefore persists from sample to sample, as distinct from error effects which vary from one sample to another. The term has not spread to other sciences. In a slightly more specialised sense the noun 'reliability' sometimes means a **Reliability Coefficient**.

The term is now also used in the context of the life of industrial components and equipments as the probability of survival after time t_n, that is to say, $1 - F(t)$ where $F(t)$ is the distribution function of the lifetimes. [See also **Factor Analysis.**]

Reliability Coefficient A coefficient introduced by Spearman (1910) into psychology. Its object is to assess the systematic component of a variate (test) as distinct from error components. In psychology it is usually measured by the correlation between the results of two administrations of the same test. The 'reliability' as a quantity is the complement of the error variance of the test but this usage requires care in view of the widespread use of the term in connection with industrial equipment in the form $1 - F(t)$ where $F(t)$ is the distribution function of the lifetimes. [See also **Factor Analysis**.]

Renewal Process A class of stochastic processes in which times between events are independently and identically distributed.

Renewal Theory An application of the analysis of recurrent events to problems concerning the duration of life in aggregates of physical equipment. Such aggregates are sometimes referred to as self-renewing when the failure of any unit results in its replacement.

Repeated Survey A sample survey which is performed more than once with essentially the same **Questionnaire** or **Schedule** but not necessarily with the same sample units. [See also **Fixed Sample; Sampling on Successive Occasions.**]

Repetition A term denoting the execution of a statistical inquiry at different points in space or time, usually as part of a coordinated programme, as distinct from **Replication**.

Repetitive Group Sampling Plan A set of sampling schemes introduced by Sherman (1965) for acceptance inspection based upon the proportion of defective units. According to a fixed criterion, a sample proportion indicates accept, reject, or disregard and repeat the process until a decision to accept or reject is achieved: no account is taken of the number of intermediate samples or their proportion defective.

Replacement See **Sampling with Replacement.**

Replacement Process A sequential control process for which, at various points, the system can be returned to some initial state.

Replication The execution of an experiment or survey more than once so as to increase precision and to obtain a closer estimation of sampling error. Replication should be distinguished from **Repetition** by the fact that replication of an experiment denotes repetition carried out at one place and, as far as possible, one period of time. Current usage on this point is often rather loose. [See **Duplicate Sample.**]

Representative Sample In the widest sense, a sample which is representative of a population. Some confusion arises according to whether 'representative' is regarded as meaning 'selected by some process which gives all samples an equal chance of appearing to represent the population'; or, alternatively, whether it means 'typical in respect of certain characteristics, however chosen'.

On the whole, it seems best to confine the word 'representative' to samples which turn out to be so, however chosen, rather than apply it to those chosen with the object of being representative.

Reproducibility An experiment or survey is said to be reproducible if, on **Repetition** or **Replication** under similar conditions, it gives the same results; that is to say, if the variation between experiments is small and negligible. For a similar idea in psychological tests see **Reliability**.

*Resemblance (Rassomiglianza) See ***Attraction, Index of.**

Residual A general term denoting a quantity remaining after some other quantity has been subtracted. It occurs in a variety of particular contexts. For example, if the true value of a variable is subtracted from an observed value then the difference may be called a residual; it is also frequently called an **Error**. Similarly, if a mathematical model is fitted to data, the values by which the observations differ from the model values are called residuals.

In a slightly wider and less satisfactory sense the word is used to denote a stochastic element which is associated with the 'predicated' or **'Independent' Variables** in a regression, e.g. in the linear regression

$$y = \beta X + \epsilon$$

the variate ϵ is sometimes called a residual error term, and if the value of β is estimated from the data as, say, b, the difference between an observed value of y and the so-called 'predicted' value bX is also called a residual.

Residual Sum of Squares See **Error Sum of Squares**.

Residual Treatment Effect In experiments which are continued over several consecutive periods of time on the same individual it is important to consider whether the effect of the experimental treatments administered during one period is carried over into the next and subsequent periods. Any such 'carried-over' effects are known as residual treatment effects and, if they are likely to be present, appropriate precautions have to be taken in the design of the experiment and the analysis of the results.

Residual Variance That part of the variance of a set of data which remains after the effect of certain systematic elements such as treatments is removed. It measures the variability due to unexplained causes or experimental error.

Residual Waiting Time See **Waiting Time**.

Resolvable Balanced Incomplete Block Design An incomplete block design with parameters v (treatments), b (blocks) of size $k < v$ and r (number of blocks in which every treatment occurs) is said to be α-resolvable if the blocks can be divided into t sets each of β blocks so that in each set every treatment is replicated α times. We then have the relationships

$$v\alpha = k\beta; \; r = \alpha t; \; b = \beta t.$$

Resolvable Designs An incomplete block design is said to be resolvable (Bose, 1942) if the blocks can be grouped in a way such that each group contains every treatment once, so as to form a complete replicate. This property has considerable practical advantages in allocating experimental effort or avoiding premature termination of an experiment.

Response The reaction of an individual unit to some form of stimulus. It may be reaction to a drug, as in bioassay, or the reaction to a request for information, as in sample surveys of human beings. [See **Non-response**.]

Response Error See **Non-sampling Error**.

Response Metameter The transformed measurement of the response to a given stimulus. The transformation is made, for example, in biological assay, in order to facilitate computations and diagrammatic representation. [See also **Metameter, Dose Metameter**.]

Response Surface If a response η depends upon an unknown function ϕ of k quantitative factors $\xi_1, ..., \xi_k$, the values of η for varying ξ's may be viewed as a surface in $k + 1$ dimensions. One object of experimentation is to approximate to this surface in some domain of interest, especially where η is a maximum.

Response Time Distribution Where, in biological assay, the response to the stimulus is measured in terms of the time that elapses before a given reaction appears, the different reaction times for different individuals may be put into the form of a distribution of response time.

Restricted Randomisation In complex factorial designs the randomisation of the allocation of treatments may not wholly eliminate systematic features which are felt to be undesirable. The situation may sometimes be met by imposing conditions on the randomisation which is then said to be restricted.

Restricted Chi-square Test A modification of the chi-square test proposed by Neyman (1949) where the restriction is on the alternative hypotheses as derived from an explicit model. The use of this restriction enables one to obtain higher power in some directions than in others.

Restricted Sequential Procedure A class of closed sequential procedure proposed by Armitage (1957) incorporating the principle of truncation in order to reduce the variability of sample number.

Return Period In time series, the interval of time taken by the series to return to some assigned value, usually an extreme value, as, for example, 'the return period of flooding' in a river.

Return States In the theory of Markov chains, a state which **Communicates** with itself is called a return state. A state which communicates with no state, not even itself, is called a non-return state.

Reversal Design Alternative name for **Switch-back Design**.

Reversal Test This term occurs in two quite different connections; (*a*) in certain tests of consistence for index numbers: **Factor Reversal Test** and **Time Reversal Test**; (b) in the analysis of time series, where one of the tests for random order is based upon 'reversals' in the series. A reversal occurs if, in the first differences of the series, a positive sign follows a negative sign or vice versa. A so-called 'reversal' test for randomness may be formed by considering the proportion of reversals in a given series.

Reversible Relation A relation of the type $y = f(x)$ is reversible in the functional sense if the inverse function $x = f^{-1}(y)$ exists. The relation is regarded as reversible in a causal sense if it can be interpreted with either x as the cause and y the effect or vice versa. For example, the Boyle-Mariotte law connecting the pressure, (P), volume (V) and absolute temperature (T), namely $P = cV/T$ may, in certain circumstances, be regarded as reversible with respect to P and V but not with respect to T.

*Reversion, Index of (Indice di Reversione) If two variates, measured from their respective means, take the values $(x_1, y_1) \ldots (x_n, y_n)$, the index of reversion of x on y is

$$\sum_{i=1}^{n} \pm y_i / \sum_{i=1}^{n} |x_i|$$

where the sign of y_i is $+$ or $-$ according as x_i, y_i have the same signs or not. An index of reversion of y on x may be similarly defined.

Ridit Analysis A method proposed by Bross (1958) for analysing subjectively categorised or poorly recorded measurement data. It consists of allocating scores Relative to the Identified Distribution of the data based upon a transformation to the uniform distribution rather than the Normal distribution.

Right and Wrong Cases Method A method of analysis proposed by Müller (1879) for quantal response data arising from psycho-physical experiment. Since it employed the transformation to standard normal deviates of the proportions arising from the quantal responses, it may be regarded as one of the antecedents of **Probit Analysis**.

Right Angular Design An association scheme for **Partially Balanced Incomplete Block Designs**, proposed by Tharthare (1963), defined four-associate class designs. Designs with this type of association scheme are termed right angular designs; there are $2Sl$ treatments arranged in l right-angles of equal arms of length S.

Risk This word occurs in statistics in its ordinary sense, and, apart from actuarial statistics, has one specialisation in the theory of **Decision Functions**. Where a number of possible decisions have a loss function attached, the risk is the expected cost of the experimentation plus the expected value of the loss function.

The risk function is the value of the risk taken for different decision functions.

Robbins-Munro Process A Stochastic Approximation Procedure for finding the parameter of a regression equation, or a quantile where assumptions are not made on the distribution, proposed by Robbins & Munro (1951). The concept has been developed by other writers.

Robustness Many test procedures involving probability levels depend for their exactitude on assumptions concerning the generating mechanism, e.g. that the parent variation is Normal (Gaussian). If the inferences are little affected by departure from those assumptions, e.g. if the significance points of a test vary little if the population departs quite substantially from the normality the test on the inferences are said to be robust. In a rather more general sense, a statistical procedure is described as robust if it is not very sensitive to departure from the assumptions on which it depends.

Room's Squares A type of experiment design in square form proposed by Room (1955). It consists of a square of $2n-1$ rows and columns so that in each row and column there are n symbols ($n-1$ blanks) which contain all $2n$ digits; ($n[2n-1]$) symbols.

Root Mean Square Deviation The square root of the second moment of a set of observations taken about some arbitrary origin, that is to say, the square root of the **Mean Square Deviation** or mean square error. The minimum value of the root mean square deviation occurs when the origin coincides with the arithmetic mean—it is then called the **Standard Deviation**. [See also **Variance**.]

Root Mean Square Error An alternative name for **Root Mean Square Deviation**.

*Rotation (Rotazione) In Italian usage, there is said to be functional rotation (*rotazione funzionale*) if the role of dependent and independent variables are interchanged; e.g. from $y = f(x)$ to $x = f(y)$. Statistical rotation (*rotazione statistica*) occurs if a mean or representative value is substituted for an individual value and vice versa; e.g. if the relation $M = z(\bar{f} - 10)$ where M is the age of a married man and $\bar{f}$ is the mean age of wives of men aged M, is transformed to $\bar{M} = z(f - 10)$ where f is the age of a wife and $\bar{M}$ the mean age of husbands whose wives are aged f.

Rotatable Designs If a k-factor **Response Surface** design can have the 'spherical' or nearly spherical **Variance Function** the estimated response has a constant variance at all points equidistant from the centre of the design. Such designs are rotatable designs.

Rotation Sampling A term suggested by Wilks and developed by Eckler (1955) for the situation of sampling on successive occasions with some (optimum) proportion of units common to successive pairs of samples.

Round Robin Design In the case where the same n objects have to be used repeatedly throughout an experiment we have a special form of **Resolvable Design** which may or may not be **Cyclic**. A tournament of n (even) players who meet in r rounds of $n/2$ matches is an example and known generally as the Round Robin design.

Rounding The process of approximating to a number by omitting certain of the end digits, replacing by zeros if necessary, and adjusting the last digit retained so that the resulting approximation is as near as possible to the original number. If the last digit is increased by unity the number is said to be rounded up; if decreased by unity it is rounded down. When both are under consideration the process is said to be one of rounding off.

Route Sampling A procedure similar to **Line Sampling** and used in surveys of crop acreage in districts which are well provided with roads. A route which adequately covers the area is chosen and the roadside lengths of the different crops recorded. Since the location of roads is unlikely to be random, estimates of acreage so obtained are likely to be biassed but changes in acreages may be estimated by using the same route for a number of years. The method of route sampling as a form of **Systematic Sampling** can also be applied to crop estimation.

Runs In a series of observations of attributes the occurrence of an uninterrupted series of the same attribute is called a run. In particular, a single isolated occurrence may be regarded as a run of one. In a series of variate values, a consecutive set which are monotonically increasing or decreasing are said to provide runs 'up' or 'down' respectively. The theory of runs has been developed in connection with a number of distribution free tests.

Rutherford's Contagious Distribution A discrete probability distribution proposed by Rutherford (1954) where the probability of success at any trial depends linearly upon the number of previous successes.

S-curve An alternative name for **Sigmoid Curve**.

s-test A term used by some writers for a test of significance of an observed standard deviation using **Helmert's Distribution**. It is equivalent to a χ^2 test and there seems to be no need for this separate expression.

S_B, S_U Distributions Two bounded systems of frequency distributions proposed by Johnson (1949) based on a variate transformation of type $z = \gamma + \delta f \left\{ \frac{x - \xi}{\lambda} \right\}$ where z is a unit Normal variable, γ, δ, ξ and λ are constants and f is some convenient function. The S_B distributions use the function

$$\log \left\{ \frac{x - \xi}{\xi + \lambda - x} \right\}.$$

The S_U family uses the function $\sinh^{-1} \left\{ \frac{x - \xi}{\lambda} \right\}$. [See also **Johnson's System**.]

Sack's Theorem A theorem due to Sacks (1958) which states under very general conditions that when the **Robbins-Munro Process** is used with constants $a_n = c/n$ (c a constant), the distribution of $(x_n - \theta)$ is asymptotically Normal with zero mean. The importance of this theorem is that, in conjunction with **Dvoretzky's Theorem**, it gives a theoretical basis for employing the Robbins-Monro process under a very wide range of conditions.

Sample A part of a population, or a subset from a set of units, which is provided by some process or other, usually by deliberate selection with the object of investigating the properties of the parent population or set.

Sample Census If 'census' is taken to mean the examination of each member of a population this term is self-contradictory. If, however, census refers to the kind of material collected then it is possible to use a sample.

Sample Design The usage is not uniform as regards the precise meaning of this and similar terms like 'sample plan', 'survey design', 'sampling plan' or 'sampling design'. These cover one or more parts constituting the entire planning of a sample survey inclusive of processing, etc. The term 'sampling plan' may be restricted to mean all steps taken in selecting the sample; the term 'sample design' may cover in addition the method of estimation; and 'survey design' may cover also other aspects of the survey, e.g., choice and training of interviewers, tabulation plans, etc. 'Sample design' is sometimes used in a clearly defined sense, with reference to a given frame, as the set of rules or specifications for the drawing of a sample in an unequivocal manner.

Sample Moment See Sampling Moment.

Sample Plan See Sample Design.

Sample Point A sample of n variate values $x_1, x_2, ..., x_n$ can be represented as a point or vector in an n-dimensional space, usually Euclidean, in which the values of the x's are taken as coordinates. A 'point' in this space corresponding to an observed set of sample values is the sample point. The idea generalises in a straightforward manner to p-way multivariate variation, the sample then being regarded as defining a point in pn dimensions or p vectors in n dimensions or n vectors in p dimensions.

Sample Size The number of sampling units which are to be included in the sample. In the case of a **Multi-stage Sample** this number refers to the number of units at the final stage in the sampling.

Sample Space The set of **Sample Points** corresponding to all possible samples. The permissible domain of variation of a sample point. Sometimes referred to as sample description space or **Event Space**.

Sample Statistic An expression, better avoided as redundant, which is synonymous with **'Statistic'**.

Sample Survey A **Survey** which is carried out using a sampling method, i.e. in which a portion only, and not the whole population, is surveyed.

Sample Unit This term is often synonymous with **Sampling Unit** but would be better confined to the denotation of any one of the units constituting a specified sample.

Sampling Distribution The distribution of a **Statistic** or set of statistics in all possible samples which can be chosen according to a specified sampling scheme. The expression almost always relates to a sampling scheme involving random selection, and most usually concerns the distribution of a function of a fixed number n of independent variates.

Sampling Error That part of the difference between a population value and an estimate thereof, derived from a random sample, which is due to the fact that only a sample of values is observed; as distinct from errors due to imperfect selection, bias in response or estimation, errors of observation and recording, etc. The totality of sampling errors in all possible samples of the same size generates the sampling distribution of the statistic which is being used to estimate the parent value.

Sampling Fraction The proportion of the total number of sampling units in the population, stratum, or higher stage unit within which simple random sampling, with multiple counting of sample units when sampled with replacement, is made. There are thus sampling fractions corresponding to different strata and different stages of sampling. Exactly the same definition is sometimes loosely applied to other sampling schemes, e.g. in sampling with variable probability; or multi-stage sampling (ratio of total number of ultimate units included in the sample to total units in the population). However, for general application it appears desirable to define it as the reciprocal of the **Raising Factor** of the sample when it exists, i.e. when the sample is **Self-weighting**. The term sampling ratio or rate is also used. [See also **Over-all Sampling Fraction, Variable Sampling Fraction.**]

Sampling Inspection The evaluation of the quality of material or units of a product by the inspection of a part, rather than the whole; in contradistinction to total inspection or **Screening**.

Sampling Interval See **Systematic Sampling**.

Sampling Moment A moment of a sampling distribution, as distinct from a moment of a set of sample observations, i.e. a sample moment, and a moment of the parent population, i.e. a parent moment.

Sampling on Successive Occasions The carrying out of a sampling process at successive intervals of time. Various methods of doing so are employed in sample surveys, e.g. by selection of a new sample on each occasion, by the partial replacement of the sample and by sub-sampling the initial sample.

Sampling Ratio See **Sampling Fraction**.

Sampling Structure A specification which defines a class of completely specified sample or survey designs. In problems of optimum design the optimisation is restricted to a given class of designs, and not to all conceivable possibilities.

Sampling Unit One of the units into which an aggregate is divided or regarded as divided for the purposes of sampling, each unit being regarded as individual and indivisible when the selection is made. The definition of unit may be made on some natural basis, e.g. households, persons, units of product, tickets, etc., or upon some arbitrary basis, e.g. areas defined by grid coordinates on a map. In the case of **Multi-stage Sampling** the units are different at different stages of sampling, being 'large' at the first stage and growing progressively smaller with each stage in the process of selection. The term **Sample Unit** is sometimes used in a synonymous sense; but refer to that term for a different meaning.

Sampling Variance The variance of a sampling distribution. The word 'sampling' can usually be omitted, as being defined by the context or otherwise understood. The sampling variance of a statistic is the square of its **Standard Error**.

Sampling with Replacement When a sampling unit is drawn from a finite population and is returned to that population, after its characteristic(s) have been recorded, before the next unit is drawn, the sampling is said to be 'with replacement'. In the contrary case the sampling is 'without replacement'.

A different usage occurs in sample surveys when samples are taken on successive occasions. If the same members are used for successive samples there is said to be no replacement; but if some members are retained and others are replaced by new individuals there is 'partial replacement'.

Saturated Model If, in a factorial experiment, θ_i is the binomial probability of success for the ith factor combination, then a model with i parameters is termed a saturated model.

Saturation In the factor analysis of multivariate material the correlation between a common factor and a variate is called the saturation of that particular variate. It measures the extent to which the factor 'appears' in the variate or the extent to which the variate is 'saturated' with the factor. [See **Factor Loading.**]

Scale Parameter A parameter of a frequency distribution which is functionally related to the scale of the variable, e.g. the standard deviation in a Normal distribution.

Scatter Coefficient A term proposed by Frisch (1929) to indicate a property of a multivariate distribution. It is the square root of the determinant whose elements are the intercorrelations r_{ik} between the pairs of variates; that is to say, it is the square root of the correlation determinant. For the case of two variates the scatter coefficient is the same as the **Coefficient of Alienation.**

Scatter Diagram A diagram showing the joint variation of two variates x and y. Each member is represented by a point whose coordinates on ordinary rectangular axes are the values of the variates which it bears. A set of n observations thus provides n points on the diagram and the scatter or clustering of the points exhibits the relationship between x and y.

Scedasticity A little used word denoting dispersion, especially as measured by variance. In a bivariate distribution, the graph of the variance of arrays of one variate against the corresponding values of the other variate is called a scedastic curve. [See also **Clisy, Kurtosis.**]

If the variance of one variate is the same for all fixed values of the other, the distribution is said to be homoscedastic in the first variate; in the contrary case it is heteroscedastic.

Scedastic Curve See **Regression.**

Schedule Apart from its customary connotation of 'list', this word occurs in the theory of sample surveys in the specialised sense of a group, or sequence, of questions designed to elicit information upon a subject. It is then synonymous with 'questionnaire'. Usually it is completed by an investigator on the basis of information supplied by the particular member of the population chosen for inclusion in the sample; but sometimes it is completed by that member himself, as in postal inquiries.

Scheffé's Test A conservative method (1959) of testing the significance of one or more comparisons of mean values arising in analysis of variance where the comparisons are selected by inspection as being of interest. It is similar to, and for a pair of mean values agrees with, a t-test.

Schuster Periodogram An alternative name for the unqualified term 'Periodogram' so-called because it was introduced by Sir Arthur Schuster (1894). [See also **Whittaker Periodogram.**]

Score A quantitative assessment of an individual on a scale, often related to his performance in some test, or derived from his reaction to certain stimuli. [See also **Normal Scores Test.**]

Screening Design The statistical design for a programme of experiments which has the object of selecting a promising subset of treatments for further development. The selection process can be optimum according to a number of criteria but likely to include balancing the errors of the first and second kind. The approach has been used, for example, in the fields of plant breeding, drug screening and educational selection.

Screening Inspection The complete inspection of a block of material or units of a product, and the rejection of all items or portions found defective. It is also known as Total Inspection or 100 per cent Inspection. [See also **Sampling Inspection.**]

Seasonal Variation In time series, that part of the movement which is assigned to the effect of the seasons on the year, e.g. seasonal variation in rainfall. Sometimes the term is used in a wider sense relating to oscillations generated by periodic external influences, e.g. daily variations in temperature might be described as 'seasonal'.

Second Limit Theorem A theorem which, broadly speaking, states that if the moments of a sequence of distribution functions F_n tend to the moments of a distribution function F, moments of all orders existing, then F_n tends to F, provided that the latter is uniquely determined by its moments. [See also **First Limit Theorem.**]

Second Order Stationary See **Covariance Stationary.**

Secondary Process In many types of stochastic process it is possible that the events or occurrences constituting the process may also be characterised by one or more additional random variables. Each of these may be regarded as a secondary process.

Secondary Unit In sampling, a synonym of second stage unit. [See **Multi-stage Sampling.**]

Secular Trend An alternative name for **Trend** in time series which is sometimes reserved for a trend extending over a long period of years, say, centuries, as against 'trends' extending over decades.

Selected Points, Method of A method of fitting a curve to a large number of points whereby a small number of points is selected as representative, more or less subjectively, and a curve fitted to them. The number of points chosen depends on the type of curve which it is intended to fit; for the fitting of a polynomial of degree n at least $(n+1)$ points are necessary.

Selection with Arbitrary (Variable) Probability A procedure for selecting a sample in which the probabilities of selection for the sampling units in the population are allocated in advance in a purposive but arbitrary manner. When the sample units are selected one by one, as is usually done, a different set of probabilities may be associated with each drawing. [See also **Selection with Equal Probability**.]

Selection with Equal Probability Fundamentally, selection of a single element from a set of such elements in such a way that selection probabilities of all elements are equal. There is, however, no uniform usage in respect of selection of a sample of more than one element; it has then reference to (1) the actual operation of selection, of any one of them individually and/or collectively when two or more operations are involved; (2) or the final product: the entire sample, obtained by all such operations with or without particular reference to the component sample units. Thus, for example, in stratified simple random sampling, with different sampling fractions in different strata, the entire sample is sometimes referred to as being selected with unequal probability even though the actual operation of selection within a stratum is basically with equal probability.

Selection with Probability Proportional to Size A sampling procedure under which the probability of a unit being selected is proportional to the **Size** of the unit. Generally this probability has reference to each drawing separately when sample units are selected one by one. Thus the procedure known commonly as sampling with probability proportional to size, with replacement, ensures such a probability of any particular drawing, but considered in its entirety, the series of drawings does not make the probability of inclusion in the sample of any specified unit proportional to its size unless the units are all of the same size.

Self-avoiding Random Walks A random walk in which the units have physical size and no two units may occupy the same region of space; or, on a lattice, a walk which at no point crosses or touches itself.

Self-correlation Coefficient An alternative term for a **Reliability Coefficient**, but one which is better avoided.

Self-conjugate Latin Square A Latin square which remains the same if its rows and columns are interchanged, i.e. it is symmetrical about its main diagonal.

Self-renewing Aggregate See **Renewal Theory**.

Self-weighting Sample If the **Raising Factors** of the sample units are all equal the sample is self-weighting, of course with respect to the particular linear estimator under consideration; but it may not be a self-weighting sample for another estimator. A self-weighting sample, usually in respect of the total of the entire population, is generally incorporated in a sample-design to simplify tabulation work, because the population total is easily estimated from the sample total. In two stage (multi-stage) sampling the number or proportion of second stage sample units are sometimes fixed in such a manner that the sample becomes self-weighting.

Semi-averages, Method of A particular case of the **Method of Selected Points** in which the data are divided into two equal groups and a straight line drawn through the means of the groups or two other representative points, one in each group. This method is used to provide a rapid estimate of a linear regression line.

Semi-interquartile Range An alternative name for **Quartile Deviation** that is to say, one half of the distance between the two quartiles of a sample or a distribution.

Semi-invariant In older usage this term, introduced by Thiele (1889), related to what are now called **Cumulants**. The words 'semi-invariant' or 'seminvariant' are now better confined to statistics which are independent of the origin and are multiplied by a scale factor under transformations of scale. The moments about the sample mean and the cumulants are both seminvariant in this sense and other symmetric functions of the observations exist with seminvariant properties. The term is not, but could be, used to describe statistics such as the range, which are not symmetric functions of the observations.

Semi-Latin Square An experimental design for $2k$ treatments arranged in the form of a rectangle with k rows of $2k$; each row being an arrangement of the $2k$ treatments, each pair of columns 1, 2, 3, 4, etc., being also an arrangement of the $2k$ treatments; no row or column therefore containing the same treatment more than once. It may also be regarded as a $k \times k$ Latin square with each plot split. The design has been criticised on the grounds that it leads to biassed estimates of error.

Semi-logarithmic Chart A form of graphic presentation in which one axis only is scaled in terms of logarithms. The logarithms may be based upon any suitable number although in the case of specially printed chart papers they are usually to base ten, i.e. are common logarithms.

Semi-Markov Process A stochastic process first given by Levy (1954), and Smith (1955) as a regenerative stochastic process, and Takacs (1954) to counter problems. This type of process is essentially a Markov chain with randomly distributed lengths of time in any one state.

Semi-range A statistic equal to one half of the **Range**. [See also **Mid Range**.]

Semi-Martingale See **Martingale**.

Semi-stationary Process A somewhat illogical but convenient word to describe a process which is largely stationary in the sense that its non-stationary characteristics are 'slowly' time-dependent (Priestly, 1965).

Sensitivity Data A term which is sometimes used as an alternative to **Quantal Response Data** to describe data consisting of measured reactions at various levels of a stimulus. This particular term has been largely used in connection with tests of explosives.

Sequential Analysis The analysis of material derived by a sequential method of sampling, that is to say, it is the data, not the analysis, which are sequential.

In sequential sampling the members are drawn one by one or in groups in order, and the results of the drawing at any stage decide whether sampling is to continue. The sample size is thus not fixed in advance but depends on the actual results and varies from one sample to another. The sampling terminates according to predetermined rules which are decided by the degree of precision required.

Sequential Chi-squared Test A sequential probability ratio test developed by Jackson & Bradley (1961) for the composite hypothesis concerning a vector of means μ

$$H_0 : (\mu - \mu_0)\Sigma^{-1}(\mu - \mu_0)' = \lambda_0^2$$
against $\qquad H_1 : (\mu - \mu_0)\Sigma^{-1}(\mu - \mu_0)' = \lambda_1^2$

The criterion is in the form of χ^2 and its distribution depends on a generalised hypergeometric function and known covariance matrix (Σ).

Sequential Estimation Estimation from data obtained by a sequential sampling process.

Sequential Probability Ratio Test A sequential test for a hypothesis H_0 against an alternative hypothesis H_1, due to Wald (1944). At the end of each stage in the sampling the probability ratio p_1/p_0 is computed where the suffixes 0 and 1 refer to the null and alternative hypotheses respectively and p is the known probability function of all sample members so far drawn. Then if $B < p_1/p_0 < A$ the sampling is continued another stage. But if $p_1/p_0 \leqslant B$ the null hypothesis (H_0) is accepted, and if $p_1/p_0 \geqslant A$ the null hypothesis is rejected and the alternative hypothesis (H_1) accepted. The two constants A and B are determined by reference to prescribed requirements concerning the two types of errors to be made in testing hypotheses, the rejection of H_0 when it is true and the acceptance of H_1 when it is false.

Sequential Test A test of significance for a statistical hypothesis which is carried out by using the methods of sequential analysis. An example is the **Sequential Probability Ratio Test.**

Sequential 'T²' Test A development of the univariate sequential t-test which used the method of frequency functions by Jackson & Bradley (1961). The test involved an unknown covariance matrix and made use of the confluent hypergeometric function.

Sequential Tolerance Region A **Statistical Tolerance Region** where the sample for the purpose of determining the boundaries proceeds sequentially (Saunders, 1960) and is terminated when those boundaries are unchanged after a predetermined small number of observations.

Serial Cluster A type of cluster used in India in which the actual demarcation of a cluster or listing of units constituting a cluster is avoided by means of a rule which makes use of the serial numbers already assigned to the units in the frame. [See **Entry Plot.**]

Serial Correlation The correlation between members of a time series (or space series) and those members lagging behind or leading by a fixed distance in time (or space). Thus, if the series is u_1, u_2, ... the serial correlation of order k is the correlation between the pairs (u_1, u_{1+k}), (u_2, u_{2+k}), etc. An analogous definition may be framed for a continuous series.

In this sense the serial correlation is the sample value of the parent autocorrelation. Some writers, however, use **'Autocorrelation'** to denote the correlation of members of a series with themselves whether of sample or parent, and 'serial correlation' to denote the correlations of members of two different series. [See **Lag Covariance.**]

Serial Design An experimental design (Thompson & Seal, 1964) which uses overlapping effects in time in order (a) to secure estimates of quantities for which an evaluation would otherwise be impossible and (b) to secure a basis for continuing the experiment for a long period. [See also **Evolutionary Operation.**]

Serial Sampling Inspection Schemes These schemes assume that batches of items produced sequentially in industrial situations will be positively correlated. Schemes can be constructed by several methods which include those derived from Bayes' theorem applied after setting up a stochastic process representing the system and thereby obtaining a sentencing rule for particular batches.

Serial Variation A statistic for studying short term variations in time series proposed by Jowett (1952). It is the mean semi-squared difference

$$d_{(x)s} = \sum_{t=1}^{n-s} \tfrac{1}{2}(x_t - x_{t+s})^2/(n-s)$$

and is a simple linear function of the **Serial Correlation Coefficient.**

Serially Balanced Sequence A design of experiment proposed by Finney & Outhwaite (1955) for the situation where a single experimental subject has to receive a number of treatments during a period of time. The treatments are arranged in a series of complete blocks so that residual effects of any treatment occurs the same number of times in association with the direct effect. A somewhat similar procedure for the Latin Square design was suggested by Bradley (1958).

*Series-Seriation (Serie-Seriazione) In Italian usage, series (*serie*) in its widest sense is a succession of numbers referred to any variable. If the numbers express statistical data, the series is called *serie statistica*. In a narrower sense, a statistical series is a succession of statistical data referred to qualitative values, while a succession of statistical data referred to quantitative values is called seriation (*seriazione*).

Two series or seriations are said to be parallel (*serie o seriazioni parallele*) if the frequency of the values of one of them is a constant multiple of the frequency values of the other.

Series in the narrower sense are classified as ordered (*serie ordinate*) and not ordered (*serie non ordinate*) the first ones presenting and the second ones not presenting a natural order of succession. The ordered series may be further subdistinguished as rectilinear (*serie rettilinee*), if they present also two extreme values (as in the grades of a hierarchy) and periodic or cyclical (*serie cicliche*) if—except for a convention—they do not present extreme values (as in the days of the weeks). Among the non-ordered series a particular type is that of the unconnected series (*serie sconnesse*) whose qualitative values can be arranged in any order.

From another point of view series in the narrower sense may be classified into **Time Series** (*serie temporali o storiche*); geographical (*serie territoriali*) and qualitative (*serie qualitative*).

*Series (Serie) In Italian usage 'series' refers to data arranged according to the values of a variable character, the serial quality residing in the arrangement of these values, not (as in the English 'series') in a temporal or spatial arrangement of individuals. The Italian 'series' is thus more akin to the English 'distribution'. Ordered data are referred to as *serie ordinata*, but order is usually to be understood if *serie* is unqualified. Two series are said to be parallel (*serie parallele*) when each term of one is a constant multiple of the corresponding term of the other. A geographical series (*serie territoriale*) is one for which the defining variable is expressed in terms of location, e.g. births in a given period classified by nation. Discrete series (*serie sconnesse*) are those for which the values of the defining variable are not naturally related one to the next, as, for example, persons arranged by name in alphabetical order, when the groups depend on the conventional order of the alphabet.

Series Queues A queuing system wherein each arrival unit is served in turn by facilities 1, 2, ..., k which are parts of one system. A somewhat different concept is expressed by the term **'Tandem Queues'**.

*Seriola The Italian equivalent of sub-series.

Shape Parameter A term used in the past to denote a parameter of a frequency distribution which is associated with **Skewness** or **Kurtosis**. The term is to be avoided since it has been shown that, contrary to earlier belief, the ordinary measures of skewness and kurtosis based on moments are not necessarily good representations of shape.

Shapiro-Wilk Test An analysis of variance-type test of Normality for a complete sample (1965) where the test statistic is the ratio of the square of a linear combination of the sample order statistics to the usual estimate of variance.

Sheppard's Corrections The calculation of moments from a grouped frequency distribution introduces certain errors as a result of assuming that frequencies are concentrated at the central values of the class intervals. Sheppard (1897, 1907) proposed a set of corrected moments ($\bar{\mu}$) which, for moments about the mean are given by

$$\bar{\mu}_2 = \mu_2 - \tfrac{1}{12}h^2,$$
$$\bar{\mu}_4 = \mu_4 - \tfrac{1}{2}\mu_2 h^2 + \tfrac{7}{240}h^4, \text{ etc.,}$$

where h is the grouping interval.

Similar corrections have been given by various authors to cover factorial moments, the multivariate case, discontinuous variation and cumulants. [See also **Correction for Grouping**.]

Sherman's Test Statistic A test proposed (1950) of the null hypothesis that data from an observed series of events arise from a **Poisson Process** against the alternative hypothesis that the data arise from a **Renewal Process**. [See also **Moran's Test Statistic**.]

Shewhart Control Chart See **Control Chart**.

Shock and Error Model A system of equations which contains both stochastic elements associated with specific variables (errors in variables) and elements associated with specific equations in the system, i.e. shocks (errors in equations). [See also **Errors in Equations; Errors in Variables**.]

Shock Model In econometric analysis, a system of equations which contain random disturbances, as opposed to one in which the variables are subject to **Errors of Observation**. [See also **Errors in Equations**.]

Short Term Fluctuation A fluctuation in a time series which has a short duration; a continuing set of such fluctuations. 'Short' for this purpose is a somewhat arbitrarily defined expression. [See **Trend**.]

Shortest Confidence Intervals See **Most Selective Confidence Intervals.**

Shot Noise See **Noise.**

Sigmoid Curve A curve lying between two horizontal asymptotes representing a function which increases monotonically and has a point of inflexion somewhere near a point half-way between them; hence a curve somewhat resembling a letter S. In statistical work this type of curve is met in connection with, *inter alia*, the distribution function of unimodal distributions, growth curves, such as the **Logistic**, and a particular dose-response relationship in biological assay.

Sign Test A test of significance depending on the signs of certain quantities and not on their magnitude; for example, one possible test for trend in a time series is based on the ratio of positive to negative signs of the first differences.

Signed Rank Test A class of distribution free tests of the hypothesis $H_0 : d = 0$ where the d_i are the differences within a set of paired observations. Each difference has a sign and a rank order and the test uses the sum of the differences having regard to their sign. The first test of this kind was proposed by Wilcoxon (1945).

Significance An effect is said to be significant if the value of the statistic used to test it lies outside acceptable limits, that is to say, if the hypothesis that the effect is not present is rejected. A test of significance is one which, by use of a test statistic, purports to provide a test of the hypothesis that the effect is absent. By extension the critical values of the statistics are themselves called significant. [See **Levels of Significance.**]

Significance Level See **Level of Significance.**

Similar Action The name given to the action of mixtures of stimuli, e.g. the toxic effect of a mixture of poisons, when the stimuli are statistically independent and additive. The effect of a mixture is then predictable from the relative proportions of the constituents and the known response of each.

Similar Regions In the theory of testing hypothesis a region in the sample space is said to be similar to another if a correspondence can be set up between them such that the probability of a sample point falling in a part of one is proportional to the probability that a sample point falls into the corresponding part of the other. For example, the distribution of n independent Normal variates with zero mean and unit variance is spherically symmetric in the sample space. It is possible to set up a correspondence between the whole space and the surface of a hypersphere of unit radius centred at the origin. The probability that a sample point falls in any cone with vertex at the origin is proportional to the probability that a point on the sphere falls in the area which that cone cuts off on it; the surface of the sphere is thus similar to the sample space.

Similarity Index If two individuals each bear the values of p (0, 1) variables, and in m cases they both exhibit, or do not exhibit the same variate value, the ratio m/p is a similarity index. The complementary quantity $1 - (m/p)$ is a dissimilarity index, and is used particularly in Cluster Analysis as a distance function.

*Simple Abnormal Curve See ***Abnormal Curve.**

Simple Hypothesis A statistical hypothesis which completely specifies the distribution function of the variates concerned in the hypothesis. [See also **Composite Hypothesis.**]

Simple Lattice Design See **Square Lattice.**

Simple Random Sampling Sampling in which every member of the population has an equal chance of being chosen and successive drawings are independent as, for example, in sampling with replacement.

Simple Sample A random sample is said to be simple when the probabilities of selection of members are all equal and are constant throughout the drawing.

Simple Structure See **Structure.**

Simple Table A table which shows only classifications according to one variate or at the most two variates. The tabulation of an ordinary univariate frequency distribution is a common example of a simple table. The expression is, perhaps, better avoided in a special technical sense.

Simplex Centroid Design A design of experiments with mixtures proposed by Scheffé (1963) wherein if there are m components and all mixtures are of equal proportions, the design involves observations on all subsets of mixtures using from 1 to m components.

Simplex Designs A simplex is the n-dimensional analogue of the triangle in two, or the tetrahedron in three, dimensions. In experimental design the simplex usually consists of the $n + 1$ vertices which are thought of as equidistant, like an equilateral triangle, surrounding the domain of interest; the object being to reach an optimal point by some such method as steepest ascents. If the design is applied to investigating the composition of multi-component systems, the sum of proportions of components being unity, it is sometimes described as a simple lattice.

The superposition of two or more simplexes to provide certain kinds of rotatable designs is known as a simplex sum design (Box & Behnken, 1960).

Simplex Method An **Algorithm** to solve Linear **Programming** problems due to Dantzig (1949). The linear constraints, in general, define a feasible region within which the optimum solution must lie. This is a simplex, and the method, in brief, seeks a point on the simplex and finds its way to the optimum by traversing a path along the edges of the simplex.

Simulation Model A model of a dynamic system which is too complicated for explicit analytical solution, but whose behaviour can be simulated in a variety of conditions by starting from suitable numerical initial circumstances.

Simulator A physical system which is analogous to a model under study as, for instance, an electric network in which the elements are in correspondence with those of an economic model. The variables of interest in the model appear as physical variables such as voltages and currents and may be studied by an examination of the physical variables in the simulator.

Simultaneous Confidence Intervals The setting of confidence intervals for several parameters which are simultaneously under estimate. The problems of locating the parameters separately in confidence intervals when they are not independent is possibly unsolvable without some further assumption, but they may jointly be located in a confidence region. [See **Joint Prediction Interval**.] Similar conditions apply to Simultaneous Tolerance Intervals and Simultaneous Discrimination Intervals.

Simultaneous Discrimination Intervals See **Simultaneous Confidence Intervals**.

Simultaneous Equations Model A model representing a stochastic situation in which the relations between the variates are expressed by a set of simultaneous equations containing them. This interdependent system was first proposed by Haavelmo (1943).

Simultaneous Estimation The estimation of two or more parameters on one and the same occasion from the same data.

Simultaneous Tolerance Interval See **Simultaneous Confidence Intervals**.

Simultaneous Variance Ratio Test A test of equality of variances of $(k+1)$, $k \geq 2$, univariate Normal populations, proposed by Gnanadesikan (1959), for the case where one variance is chosen as the standard and the other k variances compared with it. The alternative hypothesis in this case is that not all the other k variances are equal to the standard; this distinguishes the test from the **Maximum F-ratio test**.

Single Factor Theory A representation of multivariate data, introduced into factor analysis by Spearman (1904),

in which there is only one single common factor. There is some ambiguity of terminology since the 'Single (Common) Factor Method' is equivalent to the two factor method of Spearman which uses one common factor and one specific or unique factor for each test. [See also **General Factor; Hierarchy**.]

Single Sampling A type of sampling inspection where the decision to accept or reject the hypothesis that the material concerned accords with some specification is taken after the inspection of a single sample. [See also **Double Sampling**.]

Single Sampling Plan In quality control, a procedure which provides for the drawing of one sample, on the basis of which a lot is accepted or rejected; as distinct from a double sampling plan, which may allow for indecision on the first sample to be resolved by a further sample.

Single Tail Test An alternative term for a **One Sided Test**.

Singly Linked Block Design A class of incomplete block design proposed by Youden (1951) in which every pair of blocks has one treatment in common. For example, for ten treatments in five blocks of four plots each the design is as follows:

1	1	2	3	4
2	5	5	6	7
3	6	8	8	9
4	7	9	10	10

It is a particular case of the **Triangular Design**.

Singular Distribution A multivariate distribution is singular if the rank of the correlation matrix or, equivalently, the dispersion matrix is less than the number p of variates. It is then possible to express the frequency in terms of fewer than p variates, linearly related to the original set.

Singular Weighing Design A **Weighing Design** where the matrix $S = \mathbf{X}'\mathbf{X}$ is singular; $X = \{x_{ij}\}$ is the design matrix (Raghavarao, 1964).

Sinusoidal Limit Theorem A theorem stated by Slutzky (1927) to the effect that if a random series x_t is subject to n iterated summations of pairs of items, followed by the calculation of the mth differences and, if the ratio m/n is kept constant, any arbitrary section of the resulting series will tend (with probability 1 as $n \to \infty$) to a sine curve of period $T = 2\pi/\text{arc cos } r$, where $r = (1-m/n)/(1+m/n)$. The result has subsequently been generalised.

Six Point Assay One of the general class of designs for **Symmetrical Parallel Line Assays**. The six points are grouped into three pairs corresponding to low, medium and high doses of the standard and test preparations or stimuli.

Size This is a very elastic term, e.g. (1) size of a sample means the total number of sample units in the sample, usually with multiple counting of repeated sample units in sampling with replacement. In multi-stage sampling the size must have reference to the stage of sampling, but it sometimes stands for the size of the ultimate sample, i.e., the total number of ultimate sample units taken up for detailed inquiry; (2) sometimes the size of a stratum is used to mean the number of units constituting the stratum; and similarly the size of a primary unit stands definitely for the number of second stage units constituting the primary unit; but (3) the size of a stratum of sampling units, primary or otherwise, etc., may sometimes be measured in several ways. Thus, the size of a village may stand for its population, area, or something similar.

The total size of a sample means the sum of the sizes (2) or (3) of the sample units constituting the sample; this should be distinguished from size (1).

Size of a Region In the theory of testing statistical hypotheses, the size of a **Critical Region** is a measure of probability and is the same as the probability of an α-error or **Error of the First Kind**. For composite hypotheses it has sometimes been used to denote the limits of the α-error where no **Similar Regions** exist.

Size (of a Test) The size of the test of a statistical hypothesis (H_0) is the probability (α) that the hypothesis is rejected when it is true.

Skew Correlation A term denoting correlation in bivariate distributions which are asymmetrical according to at least one variate. The modern tendency is to regard the correlation coefficient as a doubtful measure of relationship in such cases and the term is obsolescent.

Skew Distribution A distribution which is not symmetrical; a distribution for which a measure of **Skewness** has some value other than zero.

Skew Regression An obsolete term for **Curvilinear Regression**.

Skewness An older and less preferable term for asymmetry, in relation to a frequency distribution; a measure of that asymmetry. The concept of asymmetry is easily defined, a measure of asymmetry less easily so.

If a unimodal distribution has a longer tail extending towards lower values of the variate it is said to have negative skewness; in the contrary case, positive skewness.

Skip Free Process A term proposed by Keilson (1962) for a class of random walk where in passing from x_1 to $x_2 > x_1$ all intervening states must be encountered at least once. As stated the process would be positive skip free, the opposite mode, negative skip free, requires the passage from x_2 to $x_1 < x_2$.

Slippage Test A significance test of k samples in which the hypothesis is one of homogeneity in the means, as against the alternative that one member or set of members has 'slipped' away from the others. For example, where the samples are observations on an industrial process at successive points of time and it is suspected that, owing to tool wear, the magnitude of the variable is systematically moving away from the intended value.

Slope Ratio Assay A general class of biological assay where the dose response lines for the standard test stimuli are not in the form of two parallel regression lines but of two lines with different slopes intersecting the ordinate corresponding to zero dose of the stimuli. The relative potency of these stimuli is obtained by taking the ratio of the estimated regression coefficients, i.e. the ratio of the two slopes, hence the same 'slope ratio' assay.

The slope ratio assay generally employs an odd number of points, and is called a $(2k+1)$-point design. This compares with the $2k$-point design of the parallel line assay, although this general class of design can be adapted for the slope ratio assay by omitting the test at the common zero dose.

Slutzky Process A synonym for the **Moving Average Process**.

Slutzky's Theorem A theorem derived by Slutzky (1925) concerning convergence in probability of rational function of variates. If $\alpha_n, \beta_n, \ldots, \theta_n$ are variates converging in probability to the constants $a, b, \ldots, t$ then any rational function $\phi(\alpha_n, \beta_n, \ldots, \theta_n)$ convergence probability to the constant $\phi(a, b, \ldots, t)$ provided that the latter is finite. The theorem is true, more generally, for a continuous function ϕ. It implies that **Convergence in Probability** is invariant under continuous functional transformations.

Slutzky-Yule Effect An effect in the averaging of random series studied independently by Slutzky and Yule. If a moving average be applied to such a series the averaged series contains undulations of an apparently systematic kind. Further averaging enhances the effect and under certain types of repeated moving average the resulting series approaches a sine wave. [See **Sinusoidal Limit Theorem**.]

Small Numbers, Law of A term suggested by von Bortkiewicz (1898) to describe the behaviour of rare events obeying a **Poisson Distribution**. The term is not antithetical to the Law of Large Numbers, and in fact is itself related to the behaviour of large numbers in which only small proportions are events of the kind under study. It is probably better avoided.

Smirnov Tests See **Kolmogorov-Smirnov Test, Cramér-von Mises Test**.

Smooth Regression Analysis A concept proposed by Watson (1964) extending the basic principles of smoothing frequency or probability density functions to the bivariate case of regression analysis.

Smooth Test A test of goodness of fit between data and hypothesis in which the alternate hypotheses are regarded as moving away from the null hypothesis 'smoothly' in the parameters, i.e. with high continuity and differentiability in them. This kind of test was proposed by Neyman (1937). It has the important property of taking into account the nature and the order of the signs of deviations between observation and expectation as well as the size of these deviations.

Smoothing The process of removing fluctuations in an ordered series so that the result shall be 'smooth' in the sense that the first differences are regular and higher order differences small. Although smoothing can be carried out by freehand methods, it is usual to make use of moving averages or the fitting of curves by least squares procedures. In fact the concept is closely allied to that of **Trend Fitting**. [See also **Error Reducing Power**.]

Smoothing Power A term used in connection with the smoothing of a series. The 'smoothness' of a series may be tested by examining the order as well as the size of differences between successive observations. There are a number of bases upon which a smoothing index can be constructed: the conventional one involving the use of differences of the third order. [See also **Error Reducing Power**.]

Snap Reading Method A simple method of time sampling proposed by Tippett (1935) for determining the proportions of time a system spends in various states. The term derives from 'snap (shot = time) reading'.

Snedecor's F-Distribution A popular term for the **Variance-Ratio Distribution**, so-called by Snedecor after the initial of R. A. Fisher's surname.

Snowball Sampling A form of sampling proposed by Goodman (1961) whereby a random sample of n individuals is drawn from a finite population. Each individual is asked to name k further individuals; and so on for s-stages. The initial sample solution and its size govern the kind of analysis and inferences that can be drawn from the data.

Spatial Point Process A **Stochastic Point Process** in two dimensions usually represented on a plane.

Spatial Systematic Sample A systematic sample taken in two dimensions; also termed 'plane sampling' by Quenouille (1949).

Spearman-Brown Formula A formula for the estimation of the **Reliability Coefficient** of a psychological or educational test which is n times as long as a basic test for which the reliabilities are known. If r_1 is the reliability of a test of unit length, the reliability of a test of length n (not necessarily integral) is $nr_1/\{1+(n-1)r_1\}$. [See also **Split Half Method**.]

Spearman Estimator The estimator, non-parametric, of the mean effective dose relevant to the Spearman-Kärber Method of bio-assay or other stimulus experiment. If the doses (k) are equally spaced the mean is $x_k + \frac{1}{2}d - d\Sigma p_i$ where d is the common difference between doses and p_i the individual dose/stimuli responses.

Spearman's Footrule A coefficient of rank correlation proposed by Spearman (1906) and defined as follows: Given two rankings a_i, b_i ($i = 1, 2, ..., n$) and defining $d_i = a_i - b_i$, the coefficient R is given by

$$R = 1 - \frac{3 \sum\limits_{i=1}^{n} |d_i|}{n^2 - 1}$$

The employment of absolute values of the differences and other reasons have prevented the coefficient from coming into general theoretical use.

Spearman's ρ A coefficient of rank correlation proposed by Spearman (1906). If the two rankings are a_i, b_i, and we define $d_i = a_i - b_i$, $i = 1, 2, ..., n$, the coefficient is given by

$$\rho = 1 - \frac{6 \sum\limits_{i=1}^{n} d_i^2}{n^3 - n}.$$

It is also the product moment correlation between the rank numbers a and b.

Spearman-Kärber Method A method for estimating **Equivalent Doses** of stimuli which generate **Quantal Responses**. In general, this method estimates the average logarithmic tolerance, i.e. the mean effective dose, rather than the **Median Effective Dose** and requires an unlimited range of doses for its successful application.

Spearman Two Factor Theorem An alternative term for the basic result underlying the two factor theory, due largely to the work of Spearman. [See also **Single Factor Theory**.]

Species of Latin Square In the enumeration of Latin squares by combinatorial methods certain distinguishable types appear from which other squares may be obtained by permutation of the letters or of rows or of columns and also by interchange of the three categories. These types have been called species.

Specific Factor See **Common Factor**.

Specific Rate A rate which is based upon some homogeneous sub-groups of a population instead of the whole population. For example, death rates may be specific for age, that is to say, may be calculated separately for a number of age groups of the population.

Specification Bias A term suggested by T. W. Anderson and Hurwicz (1947) for the bias which arises from incorrect specification of the model under analysis, e.g. by the use of a model with **Errors in Variables** instead of one with **Errors in Equations**.

Specificity In multivariate, and particularly in factor analysis, the specificity of a variate is the proportion of its total variance attributable to a specific factor. [See **Common Factor, Factor Analysis**.]

Spectral Average The average over a **Spectral Density Function** $f(w)$ of the form

$$J(A) = \int A(w)f(w)dw$$

where $A(w)$ is a suitably chosen function. It occurs particularly in the smoothing of the spectral density function and $A(w)$ is sometimes known as the **Spectral Window**. Special forms of such a window are known by the name. of Daniell, Bartlett, Blackman-Tukey and Parzen.

Spectral Density The derivative of the **Spectral Function**.

Spectral Distribution Function See **Spectral Function**.

Spectral Function A necessary and sufficient condition for $\rho(\tau)$, $\tau = 0, 1, 2 ...$, to be an autocorrelation function of a discrete stationary stochastic process is that it is expressible in the form:

$$\rho(\tau) = \frac{1}{\pi} \int_0^\pi \cos \tau w \, dF(w)$$

where $F(w)$ is a non-decreasing function called the spectral distribution function with $F(0) = 0$, $F(\pi) = \pi$. For a continuous process the corresponding condition is that:

$$\rho(\tau) = \int_0^\pi \cos \tau w \, dF(w)$$

with $F(0) = 0$, $F(\infty) = 1$. Conversely we have

$$F(w) = w + 2 \sum_{j=1}^\infty \frac{\rho_j}{j} \sin jw, \quad 0 \leqslant w \leqslant \pi,$$

for the discontinuous process and

$$F(w) = \frac{2}{\pi} \int_0^\infty \rho(x) \frac{\sin xw}{x} \, dx, \quad 0 \leqslant w \leqslant \infty,$$

for the continuous case. The function $F(w)$ is variously called the spectral function, integrated function, power spectrum or integrated power spectrum; the first appearing to be the simplest usage. Similarly $\dfrac{dF(w)}{dw}$ is called the spectral density.

Both spectral function and spectral density can be defined, without invoking the concept of autocorrelation, in terms of the intensities given by harmonic analysis. [See Periodogram.] It is, however, usual to consider the spectral density as a function of the frequency w and the periodogram intensity as a function of the period $2\pi/w$, at least in graphical representation.

Spectral Weight Function A weight function used in the estimation of the spectral density; individual specifications have been proposed, for example, by Daniell (1946), Bartlett (1948), Blackman & Tukey (1959) and Parzen (1961). The function is sometimes known as a 'spectral window'.

Spectral Window See **Spectral Weight Function**.

Spectrum A term which is applied by physical analogy (a) to the graphical representation of the spectral function; (b) to the graphical representation of the spectral density; (c) to the spectral function itself; (d) to the spectral density function itself.

Usage varies, but it would seem convenient to speak of 'spectral function' and 'spectral density function' for the mathematical functions; of the 'integrated spectrum' as the graph of the spectral function against frequency as abscissa; of the 'spectrum' as the graph of the spectral density against frequency as abscissa; and of 'periodogram' as the graph of spectral density against period as abscissa.

As for ordinary frequency functions, the spectral density may not exist, the spectral values condensing at certain points to provide a line spectral 'density' function. When it is desired to distinguish the cases the expressions 'discontinuous spectrum', 'discontinuous spectral frequency function' might be used.

The concept of spectrum for a single series can be extended to the relations between a pair of series. If the **Cross-correlations** between series 1 and 2 are $\rho_{(12)s}$, $s = -\infty, ..., \infty$, the cross-spectral density is

$$w_{12}(\alpha) = \sum_{-\infty}^\infty \rho_{(12)s} \exp(is\alpha) = c(\alpha) + iq(\alpha)$$

where $c(\alpha)$ is the **Co-spectrum** or co-spectral density and $q(\alpha)$ the **Quadrature Spectrum** or quadrature spectral density. The sum of squares $c^2 + q^2$ is called the **Amplitude** and, if $w_1(\alpha)$ and $w_2(\alpha)$ are spectral densities of the two series, the quantity

$$C(\alpha) = \frac{c^2(\alpha) + q^2(\alpha)}{w_1(\alpha) + w_2(\alpha)}$$

is called the **Coherence**. The quantity, plotted against α as abscissa,

$$\arctan \{q(\alpha)/c(\alpha)\}$$

is the phase diagram and the quantity $\{w_1(\alpha)/w_2(\alpha)\}C(\alpha)$, similarly plotted is called the gain diagram.

Spencer Formula A moving average-type graduation formula by Spencer (1904) in two forms; one using 15 terms and the other 21 terms, both reproducing cubic trends. The 15-point formula has weights $[4]^2 [5] [-3, 3, 4, 3, -3]/320$ and the 21-point formula the expression $[5]^2 [7] [-1, 0, 1, 2, 1, 0, -1]/350$, where $[r]$ means a simple moving average of r.

Spent Waiting Time See **Waiting Time**.

Spherical Normal Distribution A tri-variate Normal distribution with zero correlation between pairs of variates, zero means and unit variance, i.e. standardised.

Spherical Variance Function See **Variance Function**.

Spitzer's Identity An identity by Spitzer (1956) connecting a sequence of distributions (U_n) from random walks and distributions (V_n) of unrestricted sums of variates. The distributions (U_n) are those of the maximum distance achieved by a particle during n steps.

Splicing In an index number it may become necessary at certain times to make provision for the appearance of new items or the disappearance of items previously in use e.g. in price index numbers, when commodities go off the market. The method of effecting the change is known as splicing. For example, if the index at period k based on period o is I_{ok} and a change then occurs in the content of the index; if a new index for period l on period k as base is I'_{kl}, then one form of spliced index relating period k to period o is $I_{ok} \times I'_{kl}$ divided by 100 if necessary. [See also **Chain Index**.]

Split Half Method A method used, mainly in psychology, to estimate the reliability of a test. Two scores are obtained from the same test, either from alternate items, the so-called odd-even technique, or from parallel sections of items. The correlation between the halves is usually raised to the reliability expected for the test as a whole by the **Spearman-Brown Formula**.

An analogous use of this term occurs in connection with the design of sample surveys. If there is a question which permits of two formulations, the sample can be split into two halves, and one version given to each half. In this way it is possible to determine the appropriate wording of the question or the more general interpretation of the replies.

Split Plot Confounding Confounding in a design embodying split plots. There are two different kinds of confounding possible (a) the effects of whole plots may be confounded just as they would be if no splitting were present; (b) interactions between the factors represented in the split plots and certain differences between whole plots may be confounded. The object is to ensure that a limited number of important comparisons can be made within plots while less important comparisons are made between plots.

Split Plot Design An experimental method in which additional or subsidiary treatments are introduced by dividing each plot into two or more portions. For example, the division of experimental plots into halves enables an additional factor or treatment to be included at two levels.

Split Test Method An alternative term for the **Split Half Method**.

Spread See **Concordant Sample**.

Spurious Correlation A term proposed by K. Pearson (1897) for the case where correlation is found to be present between ratios or indices in spite of the original values being random observations on uncorrelated variates. More generally, correlation may be described as spurious if it is induced by the method of handling the data and is not present in the original material. It is to be distinguished from **Illusory Correlation**.

Square Contingency See **Contingency**.

Square Lattice An experimental design for testing treatments which are a perfect square, say k^2, in number. If the treatments in arbitrary order are denoted by the number pairs (i, j), $i, j = 1, 2, ..., k$, they may be arrayed in a square:

$$
\begin{array}{cccc}
(1, 1) & (1, 2) & \cdots & (1, k) \\
(2, 1) & (2, 2) & \cdots & (2, k) \\
\cdot & \cdot & \cdots & \cdot \\
(k, 1) & (k, 2) & \cdots & (k, k).
\end{array}
$$

Various designs can be constructed from this array; for example (trivially) k blocks of k by selecting rows; a set of $2k$ blocks by selecting rows and columns (a simple lattice design); a set of $3k$ blocks by taking in addition the members corresponding to identical letters in a $k \times k$ Latin square superposed on this square (a triple lattice design), and so on. There are in general $(k+1)k$ blocks of k providing orthogonal comparisons.

From the point of view of factorial experiments the k^2 treatments are regarded as the combination of two factors each at k levels. [See **Quasifactorial Design**.]

If k is prime it is possible to find $k+1$ replicates of the square such that the $k^2 - 1$ degrees of freedom assignable to treatment effects are divided into $k+1$ orthogonal sets of $k-1$ degrees of freedom. If each of these is confounded with the rows of one replicate and the columns of another, the $k+1$ replicates are called a completely balanced (or balanced) lattice square; every treatment then occurs with every other in one row and in one column. For non-prime k such a design may not exist but certain designs possessing a kind of balance may sometimes be found. If fewer than the $k+1$ replicates are employed the design is said to be partially balanced.

Square Root Transformation A variate transformation which is used to 'stabilise the variance' of sample data drawn from Poisson populations; that is to say, to give variates whose variance will be nearly independent of their means. The square root transformation bears the same relation to the Poisson distribution as the **Arc Sine Transformation** to the binomial distribution. It avoids the difficulty of testing homogeneity in variance analysis where variances in different classes differ; but in cases other than homogeneity tests it distorts the model. [See **Stabilisation of Variance**.]

Squariance A term proposed by Pitman (1938) in place of the phrase 'sum of squares about the mean' for the sake of brevity. [See also **Deviance**.]

Stability Test A rough but convenient test for binomial variation in data which accrue over a period. The cumulative proportion of successes is plotted against the number of observations. If the variation is binomial (Bernoullian) the graph of successive points 'settles down' to a straight line, or nearly so, with diminishing fluctuation about such a line. The test consists of a judgment by eye of the appearance of this effect.

Stabilisation of Variance The process of transforming a variate whose distribution is dependent on a parameter, in order to make the variance of the transformed variate as insensitive as possible to the values of the parameter. The transformation is used mainly in order to provide tests of significance, the same test being then approximately valid over a fairly wide range of parameter values; or to ensure, in tests of homogeneity of means in variance-analysis, that the variances of the variates are approximately equal, as is required by standard tests. Unrestricted enthusiasm in the use of stabilisation transformations is undesirable, as they may distort the model under test.

Stable Process (Distribution) An alternative term for **Stationary** (stochastic) **Process**. The term is to be avoided because of confusion with a 'loi stable' in the sense used by French authors. A distribution law is said to be stable if the convolution of two variates which follow it is also distributed in the same form: e.g. the sum of two independent Normal variates is Normal and hence the Normal law is stable in this sense.

Stacy's Distribution A form of **Generalised Γ-distribution**.

Staircase Design An experiment design proposed by Graybill & Pruitt (1958) extending the basic **Randomised Block** to the case where all blocks do not contain the same number of experimental units.

Staircase Distribution A picturesque but unnecessary name for a discontinuous distribution based upon the fact that its distribution function is a step function.

Staircase Method An alternative name for the '**Up and Down' Method** especially when used in connection with fatigue tests.

Standard Deviation The most widely used measure of dispersion of a frequency distribution. It is equal to the positive square root of the **Variance**. The standard deviation should not be confused with the **Root Mean-Square Deviation**.

Standard Equation This term is sometimes used (*a*) to denote the expression of a frequency distribution when given in standard form and (*b*) as an alternative term to **Normal Equation**, i.e. one of the sets of equations necessary for the estimation of constants by the method of least squares. Neither of the usages is very widespread.

Standard Error The positive square root of the variance of the sampling distribution of a statistic.

Standard Error of Estimate An expression for the standard deviation of the observed values about a regression line, i.e. an estimate of the variation likely to be encountered in making predictions from the regression equation. For example, in simple linear regression of y on x the standard error of estimate of y is given by $\sigma_y(1-r^2)^{\frac{1}{2}}$ where σ_y^2 is the variance of y and r is the correlation between y and x.

Standard Latin Square A Latin square in which the first row and column are in the natural order of letters, A, B, C, …, or numbers, 1, 2, 3, …. All Latin squares of a given order can be obtained by permutations of the rows and columns of the standard Latin squares of that order. For example, the 576 squares of order 4 are obtained in this way from 4 standard squares.

Standard Measure If x is a variate with mean μ and standard deviation σ the transformed variate $y = (x-\mu)/\sigma$ is said to be in standard measure. It has zero mean and unit standard deviation.

Standard Population The population in a given period or area which can be used as a basis for comparison with that at another period or in another area, e.g. in constructing standardised rates of birth or death.

Standard Score An alternative name for z-**score**. [See also **T-score**.]

Standardised Mortality Ratio An index number in the **Paasche Form** used in the analysis of vital statistics. The ratios of the age specific death rates for the given year to similar rates for the base or standard year are weighted by the 'expected' deaths in the given year. [See also **Comparative Mortality Figure**.]

Standardised Deviate The value of a deviate reduced to standardised form (zero mean, unit variance) by subtracting the parent mean and then dividing by the parent standard deviation.

Standardisation of the sample values is often carried out by a similar process with the sample mean and standard deviation, where the parent values are unknown.

Standardised Regression Coefficients A regression is usually expressed as $y = \sum_{j=1}^{p} \beta_j x_j$ with unstandardised variables. If the x's are in a standard form (zero mean, unit variance) the β's are then called standardised regression coefficients and are equivalent to correlations if y is also standardised. The same terminology may be used if the x's are reduced to unit variance but not necessarily zero mean; the coefficient β_0 absorbing the difference.

Standardised Variate A variate in **Standard Measure**.

Stationary Distribution A phrase sometimes used to denote a distribution, e.g. of a human population which remains constant in time. There is some danger of confusion with a stationary stochastic process and perhaps the expression is better avoided.

Stationary Population See **Stationary Distribution**.

Stationary Process A stochastic process $\{x_t\}$ is said to be strictly stationary if the multivariate distribution of $x_{t_1+h}, x_{t_2+h}, \ldots, x_{t_n+h}$ is independent of h for any finite set of parameter values $t_{1+h}, \ldots, t_{n+h}, t_1, t_2, \ldots, t_n$.

The process is said to be stationary in the wide sense if the mean and variance exist and are independent of t.

Statistic A summary value calculated from a sample of observations, usually but not necessarily as an estimator of some population parameter; a function of sample values.

Statistical Decision Function See **Decision Function**.

Statistical Population See **Population**.

Statistical Tolerance Limit An upper, and lower, value of a variate following a given distribution form between which, it is asserted with confidence β, a proportion γ of the population will lie. The basis is a sample of n independent observations and various forms of the interval cover the proportion 'at least' (β-content), 'average probability content at most' (β-expectation). These statistical tolerance limits may also be in a distribution free context.

Statistical Tolerance Region A generalisation of the concept of a statistical tolerance interval between an upper and lower **Statistical Tolerance Limit** to the case of two or more dimensions.

Statistically Equivalent Block A term proposed by Tukey (1947) for the multivariate analogue of the **Statistical Tolerance Interval** based upon order statistics.

Statistics Numerical data relating to an aggregate of individuals; the science of collecting, analysing and interpreting such data.

Steepest Ascent, Method of A method introduced by Box & Wilson (1951) to find the maximum value of a **Response Surface**, i.e. to select from experimental data the treatment or set of treatments which are, in some sense, best. The method uses an initial two level factorial experiment to determine the direction near the starting point in which the response surface rises most steeply. It then proceeds in a standard fashion until a three level experiment is used to investigate the shape of the response surface near the optimum.

Stein's Two Sample Procedure A method of sampling proposed by Stein (1945) to overcome a difficulty in sampling from a Normal population that, for fixed sample size n, no similar test of the mean μ exists which is independent of the variance σ^2. The method chooses a sample of fixed size n_0 and then proceeds to a further sample of $n-n_0$, n being determined by the observations of the first sample.

Steiner's Triple Systems In 1853, Steiner proposed the problem of arranging w different objects in triplets such that every pair occurs in one and only one triplet. It is a special case of a **Balanced Incomplete Block Design**. In generalisation, a balanced incomplete block design arranging v elements in m blocks of size k so that each set of β elements occurs exactly once is known as a Steiner system.

Step Down Procedure A procedure (see Ray, 1958) whereby differences are examined in one observed variable first and then in another observed variable after eliminating by regression the effect of the first, and so on. The variables are, in some sense, arranged in a descending order of importance.

Stephan's Iterative Process An iterative process proposed by Stephan (1942) for adjusting sample frequency tables when the expected marginal totals are known.

Stepwise Regression A method of selecting the 'best' set of regressor variables for a regression equation. It proceeds by introducing the variables one at a time (stepwise forward) or by beginning with the whole set and rejecting them one at a time (stepwise backwards). The criterion for accepting or rejecting a variable usually depends on the extent to which it affects the multiple correlation coefficient or, equivalently, the residual variance. Some machine programs proceed by considering more than one variable at a time, or by reconsidering some which have already been accepted or rejected.

Stereogram A general class of diagram which purports to show a three dimensional figure on a plane surface. In particular, the name is given to the three dimensional form of the **Histogram**, namely the diagram showing the frequencies of a bivariate distribution.

'STER' Distribution This distribution is one where, given a non-negative integral valued random variable y having a probability function $P_y(y)$, the random variable whose probability function is defined by sums which are Successively Truncated from what starts out to be the Expectation of the Reciprocal of the (zero-truncated) random variable Y.

Stevens-Craig Distribution The number of colours represented in a random sample of size n drawn with replacement from an urn containing ka balls, a being of each of k different colours has a Stevens-Craig distribution with parameters k and n and probability function

$$\Pr(x) = \frac{1}{k^n} \left(\frac{k}{x} \right) \Delta^x 0^n = \frac{1}{k^n} k^{(x)} s_n^x$$

where $x = 1, 2, ..., k$; $k = 1, 2, ...$; and $n = 1, 2, ...$. This is a particular case of the multivariate hypergeometric distribution.

Stirling Distribution A Stirling distribution of the first type is an m-fold convolution of the **Logarithmic Series Distribution**. The name appears to be derived from the use of Stirling numbers of the first kind in the numerator of the density function.

Stochastic The adjective 'stochastic' implies the presence of a random variable; e.g. stochastic variation is variation in which at least one of the elements is a variate and a stochastic process is one wherein the system incorporates an element of randomness as opposed to a deterministic system.

The word derives from Greek στόχος, a target, and a στοχαστιχης was a person who forecast a future event in the sense of aiming at the truth. In this sense it occurs in sixteenth-century English writers. Bernoulli in the *Ars Conjectandi* (1719) refers to the *'ars conjectandi sive stochastice'*. The word passed out of usage until revived in the twentieth century.

Stochastic Approximation Procedure A non-parametric method of iterative estimation for a functional or regression relationship which incorporates random elements (see Derman, 1956). It may be contrasted with the Newton-Raphson method of approximation and compared with the **'Up and down Method'**. An extension to take account of time changes was made by Dupač (1965). [See also **Kiefer-Wolfowitz Process, Robbins-Munro Process.**]

Stochastic Comparison, of Tests If two or more allowable tests of significance of a single hypothesis are made on the one set of data the critical levels obtained may be regarded as random variables. In this sense Bahadur (1960) referred to the stochastic comparison of tests: when applied to a comparison of two tests the concept was first explicitly suggested by Anderson & Goodman (1957).

Stochastic Continuity See **Stochastic Process**.

Stochastic Convergence One of several types of convergence concepts in Probability. A sequence $\{x_n\}$ of variates is said to converge stochastically to a variate x if

$$\lim_{n \to \infty} \Pr\{ | x_n - x | > \epsilon \} = 0$$

for every $\epsilon > 0$. This is also known as convergence in probability and convergence in measure.

If x_n and y_n are variates and, as n tends to infinity

$$\Pr\{ | x_n - y_n | > \epsilon \}$$

tends to zero x_n is said to converge stochastically to y_n; but by convention the case where x_n and y_n do not converge is usually excluded.

Stochastic Dependence The relationship between variates which are not independent (**Independence**); as contrasted with mathematical dependence, which is a relationship between variables. In statistical usage either form is often referred to as 'dependence', the meaning generally being clear from the context.

Stochastic Differentiability See **Stochastic Process**.

Stochastic Disturbance A disturbance which possesses a probability distribution. [See also **Shock Model, Error Model, Shock and Error Model.**]

Stochastic Integrability See **Stochastic Process**.

Stochastic Kernel A sophisticated way of referring to a probability distribution dependent on a parameter.

A stochastic kernel K is a function of two variables, a point and a set: such that $k(x\Gamma)$ is

(*i*) for a fixed x, a probability distribution in Γ, and

(*ii*) for any interval Γ, a particular closed set of continuous functions in x.

If the probability distribution in (*i*) is a **Defective Distribution** the kernel is referred to as a sub-stochastic kernel.

Stochastic Matrix In general, a matrix of stochastic elements. It is sometimes, but regrettably, used to denote a matrix of transition probabilities in stochastic processes, the matrix being called stochastic if the sum of the entries in any row is equal to 1. If, in addition, the sum of the entries in any column is equal to 1 the matrix is said to be doubly stochastic.

Stochastic Model A **Model** which incorporates some stochastic elements.

Stochastic Process A family of variates $\{x_t\}$ where t assumes values in a certain range T. In most practical cases x_t is the observation at time t and T is a time range but t may also refer to distribution in space and may be considered for discontinuous or continuous values.

A stochastic process $\{x_t\}$ is said to be stochastically

continuous if, for values $t, t+h_1, t+h_2, ...,$ with h_n tending to zero as n tends to infinity

$$\lim_{n \to \infty} x_{t+h_n}$$

exists in the sense of **Stochastic Convergence** and is equal to x_t.

Likewise if

$$\lim \frac{x_{t+h_n} - x_t}{h_n}$$

exists in the sense of stochastic convergence the process is said to be stochastically differentiable. And if the process exists in $a \leqslant t \leqslant b$ and the Riemann integral

$$\int_a^b x_t dt$$

exists in the sense of stochastic convergence the process is said to be stochastically integrable.

Stochastic Programming An advanced form of **Linear Programming** where the parameters of the objective function, derived from the constraints, are not fixed but comprise sample items from a given distribution of values.

Stochastic Transitivity The condition that, given two probabilities Π_{ij} and Π_{jk} where Π_{ij} is the probability that a binary random variable x_{ij} has unit value, $\Pi_{ij} \geqslant \frac{1}{2}$ and $\Pi_{jk} \geqslant \frac{1}{2}$ imply $\Pi_{ik} \geqslant \frac{1}{2}$. If the final term is replaced by $\max(\Pi_{ij}, \Pi_{jk})$ we have a more stringent condition known as strong stochastic transitivity.

Stochastic Variable An alternative name for a **Variate** or random variable.

Stochastically Larger or Smaller A variate with distribution function $F(x)$ is said to be stochastically larger than a variate y with distribution function $G(y)$ if $F(x) \leqslant G(x)$ for all x and $F(x) < G(x)$ for some x. In this case y is said to be stochastically smaller than x.

Stopping Rule A procedure in sequential sampling for dividing a sample space into two regions; one in which further observations are taken, and the other in which sampling is terminated.

Strata Chart A chart upon which two or more time series are plotted with the vertical scales arranged so that the curves do not cross. The bands, or strata, between successive curves may be distinctively coloured or hatched. This kind of chart is valuable in connection with the presentation of time series data in which a total can be broken down into its constituent parts.

Strategy In the theory of games, a schedule giving the possible courses of action open to an individual according to the state of the game and possibly to previous action by his opponents. If the strategy lays down a single course of action for each situation it is said to be pure. If there are choices which are determined by a chance mechanism the strategy is mixed.

In the theory of sampling an estimation procedure is sometimes referred to as a strategy; for example, Pathak (1967).

Stratification The division of a population into parts, known as strata; especially for the purpose of drawing a sample, an assigned proportion of the sample then being selected from each stratum. The process of stratification may be undertaken on a geographical basis, e.g. by dividing up the sampled area into sub-areas on a map; or by reference to some other quality of the population, e.g. by dividing the persons in a town into two strata according to sex or into three strata according to whether they belong to upper, middle or lower income groups.

The term stratum is sometimes used to denote any division of the population for which a separate estimate is desired, i.e. in the sense of a **Domain of Study**. It is also used sometimes to denote any division of the population for which neither separate estimates nor actual separate sample selection is made, e.g. see the use of (sub-) strata in multiple stratification without control of sub-strata; or, e.g. the use of strata in stratification after selection when it is used to improve the estimate pertaining to the entire population.

Stratification after Selection It sometimes happens that the proportional numbers lying in certain strata are known but that it is impossible to identify in advance the stratum to which a chosen member belongs. The sample selection then has to be made without reference to the strata, e.g. by simple random sampling. The resulting sample may, however, be stratified after selection and treated as an ordinary stratified sample. The procedure is almost as efficient as sampling with a **Uniform Sampling Fraction**.

Stratified Sample A sample selected from a population which has been stratified, part of the sample coming from each stratum.

Stratum See **Stratification**.

Strength, of a Test If for a pair of tests of hypotheses with errors of first and second kind equal to (α, β) and (α', β') we have $\alpha < \alpha'$ and $\beta \leqslant \beta'$ or $\alpha \leqslant \alpha'$ and $\beta < \beta'$ the first test is said to be stronger than the second test. If $\alpha < \alpha'$ and $\beta > \beta'$ or vice versa, the tests are not comparable in this particular sense.

Strictly Stationary Process See **Stationary Process**.

Strong Law of Large Numbers Let $\{x_i\}$ $i = 1, 2, ...,$ be a sequence of variates with expectations μ_i. In its classical form the Strong Law of Large Numbers gives conditions under which

$$\frac{1}{n} \sum_{i=1}^{n} (x_i - \mu_i) \to 0$$

with probability unity.

The Weak Law gives conditions under which

$$\Pr\left\{ \left| \frac{1}{n} \sum_{i=1}^{n} (x_i - \mu_i) \right| > \epsilon \right\} \to 0$$

for any given $\epsilon > 0$.

Modern versions of both laws are concerned with conditions under which these statements hold for more general normalising constants.

Strongly Consistent Estimator An estimator which converges strongly to its limit in the probabilistic sense. [See also **Strong Convergence, Weak Convergence.**]

Strongly Distribution Free A concept very difficult to define in verbal terms, introduced by Birnbaum & Rubin (1954) in connection with a study of the relationship between order statistics, symmetric functions and distribution free properties.

Structural Equation An expression which is usually employed to denote a mathematical relation among variables entering into the specification of a model and hence expressing its structure. In statistics and economics the variables are frequently stochastic in character, and the term has been used to differentiate between such relations and those of a functional kind in the ordinary mathematical sense.

Structural Parameters The parameters appearing in structural equations. But see **Partially Consistent Observations.**

Structure The structure of a model is the pattern of relationship between its constituent variables as distinct from their values or coefficients associated with them. In factor analysis the structure expresses the pattern of relationship between the variates and the underlying common factors. In the special case where each variate does not depend on all the factors common to the system the structure is called simple.

An equation appearing in the explicit formulation of a model is called a structural equation; the estimation of any parameters appearing in it is called (not very happily) structural estimation; a coefficient in such an equation is called a structural coefficient. Strictly speaking, perhaps, the adjective 'structure' should be applied only to those variables which appear in the system more than once and hence knit the structure together; but this requirement is not always observed. [See **Factor Pattern, Partially Consistent Observations.**]

Studentisation The process of removing complications due to the existence of an unknown parent scale parameter by constructing a statistic whose sampling distribution is independent of it; especially by dividing a statistic which is of a certain degree in the observations by another statistic of the same degree. The expression is derived from the *nom de plume* of W. S. Gosset, who

first introduced the process in 1907 by discussing the distribution of the mean divided by the sample standard deviation. [See also **'Student's' Hypothesis.**]

Studentised Maximum Absolute Deviate Given a sample of independent value $x_1, x_2, \ldots, x_k$ this quantity is defined as

$$d = \max_{t = 1, 2, \ldots, k} \frac{|x_i - \bar{x}|}{s}$$

where s is the sample standard deviation. If the x's are Normal the significance points of d can be tabulated.

Studentised Range The range of a sample of n divided by the sample standard deviation, usually based on $n-1$ degrees of freedom.

'Student's' Distribution See *t*-distribution.

'Student's' Hypothesis A Composite Hypothesis asserting that the mean of a sample drawn from a Normal population has a certain value or lies in a certain range. For this hypothesis the *t*-test based on 'Student's' distribution has certain optimal properties.

Sturges' Rule An empirical rule for assessing the desirable number of frequency groups into which the distribution of observed data should be classified. If N is the number of items and k the number of groups, then:

$$k = 1 + 3 \cdot 3 \log_{10} N.$$

For example, a distribution of 100 items should have not less than eight frequency groups according to this particular rule.

Subexponential Distribution A property of a distribution function proposed by Tukey (1958) and defined as follows: the distribution is subexponential to the right if

$$\frac{F(z+h) - F(h)}{1 - F(h)} = 1 - \frac{1 - F(z+h)}{1 - F(h)}$$

is monotonically decreasing for fixed $z > 0$ as h increases. The property of being subexponential to the left, or in both directions, is defined analogously.

Sub-group Confounded In certain kinds of experimental design the comparisons made among the observations may be regarded as a group in the mathematical sense. When certain sets of high order interactions are **Confounded** they form a sub-group, also in the mathematical sense. The design is then said to have a sub-group of comparisons confounded.

Subjective Probability This expression can be used in at least three senses: (*i*) to describe the intensity of belief in a proposition held by an individual (not necessarily quantifiable); (*ii*) to denote a theory of (quantifiable) probability founded on such intensity of belief with the help of axioms; (*iii*) to describe any theory which is not objective in the sense of being based either on generally accepted axioms or on observable frequencies. [See **Psychological Probabilities.**]

Submartingale   See **Martingale**.

Subnormal Dispersion   A term proposed by Lexis (1877) to denote the case where the **Lexis Ratio** is less than unity. Lexis referred to data giving rise to such a situation as 'constrained' (*gebunden*). [See also **Poisson Variation, Supernormal Dispersion**.]

Subsample   A sample of a sample. It may be selected by the same method as was used in selecting the original sample, but need not be so. [See also **Multi-phase Sampling**.]

Subsampling   This term is used in two different senses; one related to multi-stage sampling and the other to multi-phase sampling. In multi-stage sampling the process of selecting sample units, say, at the second stage from any selected first stage unit is called subsampling of the first stage unit. In **Multi-phase Sampling** that part, say, of the first-phase sample which is taken up for the second phase inquiry is said to constitute a subsample of the first phase sample, and the process of selection at the second phase may be called subsampling.

Substitute F-ratio   For the purpose of variance analysis the usual mean square estimators of the variance can be replaced, with some gain of convenience but loss of efficiency, by estimators based upon the range; and the ratio of two independent estimates by the ratio of an estimator based upon a single range to an independent estimator based upon mean range. Such ratios are known as substitute F-ratios; the usual F-test for the ratio of two independent estimators of variance then no longer applies and a different test has to be employed.

Substitute *t*-ratio   A modified form of 'Student's' t in which the numerator and denominator of the t-ratio are replaced by more easily calculated statistics such as mean ranges. [See **Substitute F-ratio**.]

Substitution   In sampling inquiries it is sometimes difficult to make contact with, or obtain information from, a particular member of the sample. In such cases it is sometimes the practice to substitute a more convenientyl examined member of the population in order to maintain the size of sample. Any such substitution should, however, be carried out upon a strictly controlled plan in order to avoid bias.

Successive Difference Statistic   A group of statistics related to observations which have an order in space or, more usually, in time. They depend on the difference between successive members of the series, or the differences of those differences. [See **Variate Difference Method**.]

Sufficiency   A property of an estimator defined by R. A. Fisher (1921). An estimator t is said to be sufficient for a parameter θ if the distribution of a sample $x_1, x_2, ..., x_n$

given t does not depend on θ. The distribution of t then contains all the information in the sample relevant to the estimation of θ and a knowledge of t and its sampling distribution is 'sufficient' to give that information.

Generally a set of estimators or statistics $t_1, ..., t_k$ are 'jointly sufficient' for parameters $\theta_1, ..., \theta_l$ if the distribution of sample values given $t_1, ..., t_k$ does not depend on these θ's. If $k > l$ the set is exhaustive and if $k = l$ the set is sufficient (Fisher, 1956). [See also **Minimal Sufficient Statistic**.]

Sukhatme *d*-Statistic   A little used expression for the quantity d occurring in the **Behren's-Fisher Test**.

Superefficiency   A term denoting the efficiency of an estimator which is more efficient, i.e. has smaller variance than a maximum likelihood estimator. This super-efficiency may exist only for some values of the parameters under estimate: the group of such values being termed the set of superefficiency.

Superfluous Variable   In regression analysis, an independent variable which does not add anything to the goodness of fit of a regression line to data. In '**Bunch Map**' **Analysis** a variable is deemed to be superfluous if its inclusion in the analysis does not make the 'bunch' tighter.

Supernormal Dispersion   See **Lexis Variation**.

Superposed Process   A stochastic process consisting of the superimposition of p individual processes.

Superposed Variation   Variation which is additive to the variation under discussion but is not part of the generative scheme, e.g. errors of observation; as contrasted with variation, like the error terms in an **Autoregressive Equation** where the occurrence of any particular value is followed by its incorporation into the motion of the system.

Supersaturated Design   A supersaturated design (Satterthwaite, 1959) is a factorial design in n observations in which the number of factors is $n + 1$. If the design matrix is constructed at random this results in one of the **Random Balance Designs**. In order to avoid certain disadvantages, Booth & Cox (1962) proposed a method of nearly orthogonal systematic designs.

Supplementary Information   In sample survey design, information about the sampling units which is supplementary to the characteristics under investigation in the survey. Such information may be used for stratification, for the determination of the probabilities of selection, or in estimation. When used in estimation it provides the basis for estimators based on ratios or regression. For example, in a survey of business firms, supplementary information on, say, gross turnover provided by a previous census may be used either in the sample design or to improve the efficiency of sample estimates.

Supplemented Balance A feature of experiment design, originally proposed by Hoblyn, Pearce & Freeman (1954) and developed by Pearce (1960), whereby one particular treatment (control) has additional replications.

Surprise Index A device proposed by Weaver (1948) to provide some basis for the realisation of an event (E_i) governed by a set of probabilities (E_j). If there is a simple statistical hypothesis (H), the surprise index associated with E_i is

$$\lambda_i = \frac{\mathscr{E}(p_j | H)}{p_i} = \frac{\Sigma p_j^2}{p_i} \, (i, j = 1, 2, 3, \ldots)$$

A generalisation to the multivariate Normal distribution was given by Good (1954).

Survey An examination of an aggregate of units, usually human beings or economic or social institutions. Strictly speaking, perhaps, 'survey' should relate to the whole population under consideration and to material collected in considerable detail. However, it is often used to denote a **Sample Survey**, i.e. an examination of a sample made in order to draw conclusions about the whole.

Survey Design See **Sample Design**.

Survivor Function In a **Renewal Process**, if x denotes the identically distributed times between events in the process, e.g. life times and $F(x)$ is the distribution function associated with x, the survivor function $R(x)$ is defined as the complement of $F(x)$. This is a quite general concept associated with the analysis of the life of human, biological or material units.

Switchback Design An experiment design which relates to a three period sequence for two treatments rather than a two, or multiple of two, period sequence; i.e. ABA, rather than AB. Denoting periods as suffices, the use of $A_1 B_2 A_3$ and $B_1 A_2 B_3$, for example, would deal with the case where trend effects contribute to experimental error independently of treatment effects.

Symmetric Sampling A symmetric sampling procedure is one where all elements in the population are treated on the same footing during the drawing procedure. For example, sampling with probability proportional to size is not symmetric.

Symmetrical Distribution A frequency distribution for which the variate values equidistant from a central value are equally frequent. In a symmetrical distribution all odd order moments about the mean and all odd order cumulants, where they exist, are zero.

Symmetrical Factorial Design A factorial design is said to be symmetrical if, in the experiment to which it relates, the number of levels of each factor is the same. In the contrary case it is said to be asymmetrical.

Symmetrical Unequal Block Arrangement A device proposed by Kishen (1940-41) to overcome the difficulty in incomplete block design if naturally defined blocks comprising different numbers of plots. The symmetric unequal block arrangements preserve the property of complete balance.

Symmetrical Test See **Double Tailed Test**.

*Symmetry (Simmetria) In Italian usage symmetry, in relation to a frequency curve, coincides with English usage. Asymmetry (*asimmetria*) is measured by the **Dissimilarity Index** between the positive deviations from the median and the absolute values of the negative deviations. A symmetrical frequency curve is said to be inversely symmetrical (*inversamente simmetrica*) because the two halves can be brought into coincidence only by rotation outside the plane; any curve for which coincidence can be achieved by rotation in the plane would be called directly symmetrical (*direttamente simmetrica*).

An analogous measure, said to be of dissymmetry (*dissimmetria*) is obtained by taking the measure of asymmetry with regard to sign. One such index is mean −mode/mean deviation, which resembles one of Pearson's measures of **Skewness**.

'Sympathy' Effect A term used in connection with the sampling of human populations to describe the situation where a member of the sample responds to a question from the investigator in the way which is believed to please the investigator, rather than giving an accurate reply. The respondent is often not conscious of deliberate falsehood, which usually adds to the difficulty of countering the bias to which the effect gives rise. [See also **'Vanity' Effect**.]

Systematic This word is frequently used in statistics in contrast to 'random' or 'stochastic'. Thus, a variate y consisting of a constant m plus a variate x with zero mean is sometimes said to have a systematic component m and a stochastic component x, although it might equally well be regarded as a stochastic component y with mean m. Similarly, an error variate is said to be a systematic error if it has a non-zero mean; and a sampling process is 'systematic' if it is not random.

The usage is convenient but occasionally gives rise to difficulty. Many processes embody both 'systematic' and 'stochastic' elements and should not properly be described by either adjective alone; e.g. the so-called systematic sampling of a list may begin from a randomly chosen point, and a random sample may be chosen from systematically determined strata. The basic difficulty is that even a random event may be the outcome of a systematic procedure and it is not to be resolved by substituting some other word for 'systematic'.

Systematic Design An experimental design laid out without any randomisation. The term is difficult to define exactly because in one sense every design is systematic; it usually refers to a situation where experimental observations are taken at regular intervals in time or space. [See **Systematic**.]

Systematic Error As opposed to a random error, an error which is in some sense biassed, that is to say, has a distribution with mean, or some equally acceptable measure of location, not at zero.

Systematic Sample A sample which is obtained by some systematic method, as opposed to random choice; for example, sampling from a list by taking individuals at equally spaced intervals, called the sampling intervals, or sampling from an area by determining a pattern of points on a map.

Systematic Square Early attempts at creating experimental designs for the elimination of variability in two directions orthogonal to each other did not generally use the principle of randomisation. Thus the allocation of treatment to the rows and columns proceeded in some 'systematic' way. For example, a square design could be laid out as follows with systematic arrangement in the N.W.-S.E. diagonals:

A	B	C	D
D	A	B	C
C	D	A	B
B	C	D	A

[See also **Knut Vik Square**.]

Systematic Statistic A term proposed by Mosteller (1946) for a statistic consisting of a linear combination of order statistics. There appears to be nothing in order statistics or in linearity of combination to warrant the restriction of the word 'systematic' to such quantities, although it is useful to have a word for them.

Systematic Variation A term used in two slightly different senses: (*i*) to denote variation which is deterministic, as opposed to stochastic, and which can therefore be represented by a deterministic mathematical expression; and (*ii*) to denote variation in observations resulting from experimental or other situations as a result of factors which are not under statistical control. [See also **Assignable Variation**.]

***t*-distribution** This distribution, originally due to 'Student' (1908), is usually written in the form, as modified by R. A. Fisher (1925):

$$dF = \frac{\Gamma\{\frac{1}{2}(\nu+1)\}}{\Gamma(\frac{1}{2}\nu)\sqrt{(\nu\pi)}} \left(1+\frac{t^2}{\nu}\right)^{-\frac{1}{2}(\nu+1)} dt, -\infty \leqslant t \leqslant \infty,$$

where ν is called the number of degrees of freedom. The distribution is, among other things, that of the ratio of a sample mean, measured from the parent mean, to a sample variance, multiplied by a constant, in samples from a Normal population. It is thus independent of the parent scale parameter and can be used to set confidence intervals to the mean independently of the parent variance. [See also **Studentisation**.]

***T*-distribution** An alternative name for **Hotelling's T-distribution**.

***T*-score** A variate value obtained by a method of rescaling marks, or scores, in a test, proposed by McCall (1923). The method is essentially one of transforming the scores into deviates of a Normal distribution which has a mean of 50 and standard deviation of 10 units. Hence the range of the T-score of 0 to 100 is equivalent to one of 5 times the standard deviation on each side of the mean in a Normal distribution. [See *z*-score.]

***t*-test** A test based on the distribution known as 'Student's'. [See *t*-distribution.]

***T*-test** There are three tests of significance which may be encountered under this name. One is the test using **Hotelling's T-distribution**. Another is a rank order test of trend in a time series introduced by Mann (1945) and the third is a non-parametric test for comparing two variances proposed by Sukhatme (1957). In this latter case the statistic is

$$T = \frac{1}{mn} \sum_{i=1}^{m} \sum_{j=1}^{n} \psi(x_i, y_j)$$

which is a modified form of the **Wilcoxon-Mann-Whitney Statistic**; $\psi(x_i, y_j) = 1$ if $y_i < x_i < o$ or $o < x_i < y_j$ and zero otherwise.

Tail Area (of a Distribution) The portion of the area under a **Frequency Curve** which lies between the start of the distribution and some ordinate lying between the start and the mode; or symmetrically, between some ordinate lying between the mode and the end of the variate range and the end of the distribution. The term is usually applied only to distributions which 'tail off' at their extremes, i.e. have frequency functions tending to zero.

Takacs Process A **Markov Process** in continuous time obtained by Takacs (1955) by considering a single server queueing system with Poisson input and a general service time distribution. If $X(t)$ is defined to be the time which a customer who arrived at time t would have to wait until the commencement of service, $X(t)$ is called the Takacs or 'virtual waiting time' process.

Tandem Queues A situation where the output from the service stage of one queueing system is the direct input to a service stage of a second system (Reich, 1957). This is not the same concept as **Series Queues**.

Tandem Tests A term proposed by Abramson (1966) for a sequence of two Wald-type **Sequential Probability**

Ratio Tests the first carried out on one of two variables X_1 or X_2 and the second, which may depend upon the outcome of the first, performed on the other variable.

*Tantiles (Tantili) In Italian usage the $n-1$ partition values of a variable which divide the total amount of an extensive magnitude into a given number n of equal amounts. If $n = 2$, the central tantile is called dividing value (**Valore Divisorio**). [See **Quantiles**.]

Tchebychev-Hermite Polynomials Polynomials based upon derivatives of the **Normal Distribution**. If the distribution is represented by

$$\alpha(x) = \frac{1}{\sqrt{(2\pi)}} e^{-\frac{1}{2}x^2}$$

then the polynomial of order r, $H_r(x)$, is defined by:

$$\left(\frac{-d}{dx}\right)^r \alpha(x) = H_r(x)\alpha(x).$$

These polynomials have important orthogonal properties. It appears that they were originally derived by Laplace but are known in statistical work as Hermite or Tchebychev-Hermite polynomials. The first four are $H_1 = x$, $H_2 = x^2-1$, $H_3 = x^3-3x$, $H_4 = x^4-6x^2+3$.

Tchebychev Inequality If $g(x)$ is a non-negative function of a variate x, Tchebychev's (1874) inequality states that for every $k > 0$

$$\Pr\{g(x) > k\} \leqslant \mathscr{E}\{g(x)\}/k.$$

If $g(x) = (x-m)^2$, m being the mean of x and $k = t^2\sigma^2$, σ^2 being the variance of x, this reduces to the **Bienaymé-Tchebychev Inequality**.

More general inequalities of a similar kind involving moments higher than the second are sometimes known as inequalities 'of the Tchebychev type'. A multivariate inequality was given by Olkin & Pratt (1958).

Temporally Continuous Process An expression sometimes used to denote a stochastic process which is dependent upon a continuous time parameter. The terminology is not recommended.

Temporally Homogeneous Process A stochastic process for which the **Transition Probabilities** are the same for any time interval of length t.

Terminal Decision In sampling schemes of a sequential type, a decision which involves terminating the sampling process. For example, under a single sampling scheme for acceptance inspection there are two possible decisions which are both terminal decisions: to accept or reject the lot under inspection. If the scheme made provision for a third type of decision, to continue sampling, this third type would not be terminal.

Terry's Test A rank order test, proposed by Terry (1952), that a set of n observations come from an unknown population against the alternative of a set of

Normal populations with means $d_i\xi$, n, $i = 1, 2, \ldots$, when the d's are known constants.

Test Coefficient In factor analysis, a synonym for **Factor Loading**.

Test of Normality A test of a set of observations to see whether they could have arisen by random sampling from a Normal population. Such tests may be carried out by a comparison of the sample distribution function with a Normal distribution function. Certain other tests are said to be tests of Normality when, in fact, they are only tests of agreement of certain sample statistics with the values of the corresponding population parameters; for example, a test of the sample moment ratio $b_1 = m_3/m_2^{3/2}$, where m_r is the rth sample-moment against the normal value of zero, or $b_2(= m_4/m_2^2)$ against the normal value 3 are spoken of as tests of normality.

Test Statistic A function of a sample of observations which provides a basis for testing a statistical hypothesis.

Tetrachoric Correlation An estimate of the parameter ρ, equivalent to the product moment correlation between two Normally distributed variates, obtained from the information contained in a two by two table or double dichotomy of their bivariate distribution. The term is almost entirely confined to a particular estimator of rather complicated form developed by K. Pearson.

Tetrachoric Function A function which is related to the **Tchebychev-Hermite Polynomials** and which is used in the calculation of the **Tetrachoric Correlation** coefficient. The function of order r may be defined as:

$$\tau_r = \frac{(-1)^{r-1}D^{r-1}\alpha(x)}{(r!)^{\frac{1}{2}}} = \frac{H_{r-1}(x)\alpha(x)}{(r!)^{\frac{1}{2}}}$$

where $D^{r-1}\alpha(x)$ is the $(r-1)$th derivative of $\alpha(x) = e^{-\frac{1}{2}x^2}\sqrt{(2\pi)}$, the standard Normal distribution, and $H_{r-1}(x)$ is the Tchebychev-Hermite polynomial of order $r-1$.

Tetrad Difference See **Hierarchy**.

Theoretical Frequencies The frequencies which would fall into assigned ranges of the variate if some theoretical distribution law were exactly followed, as distinct from the actual frequencies which may be observed in a sample.

Theoretical Variable A somewhat undesirable expression used to denote a variable or variate introduced into structural relations but not directly observable; as distinct from a 'true' variable or variate, which is observable except for errors of observation and an 'observable' variable or variate which can be directly observed. [See also **Latent Variable**.]

Thomas Distribution This is equivalent to the **Double Poisson Distribution**. Consider two independent Poisson random variables x_1 and x_2 with parameters λ_1 and λ_2

respectively. If the random variable x_1 is generalised by the random variable $1 + x_2$ then we have a Thomas (1949) distribution with parameters λ_1 and λ_2.

Thompson's Rule A studentised rejection rule for outlying observations proposed by Thompson (1935). The criterion, determined from F- or t-tables, provides the basis for rejecting all observations whose studentised residual is larger than the critical value. This rule is not particularly suitable for a single outlier, or two outliers which lie at opposite extremes of the sample observations.

Three Dimensional Lattice A general class of lattice design of which the **Cubic Lattice** is a particular case. For example, 120 treatments could be laid out as a $4 \times 5 \times 6$ three-dimensional lattice. This would be regarded as equivalent to a factorial design with three factors at 4, 5 and 6 levels.

Three point Assay The simplest case of the general class of $(2k+1)$ **Point Slope Ratio Assays**. This design is useful only where the validity of the assay is known *a priori*, since in the analysis of the results no degrees of freedom are available for tests of validity.

Three Series Theorem A theorem due to Kolmogorov (1928) concerning the sums of mutually independent variates. Let $\{x_k\}$ be a sequence of independent random variates and let $\{a_k\}$ be a bounded sequence of positive numbers. Define
$$y_k = \begin{cases} x_k \ if \ |x_k| \leqslant a_k \\ 0 \ \ if \ |x_k| > a_k. \end{cases}$$
Then the series Σx_k converges with probability unity if, and only if, all the following three series converge:

(i) $\Sigma \Pr\{x_k \neq y_k\}$,

(ii) $\Sigma \mathscr{E}\{y_k\}$,

(iii) $\Sigma \mathscr{E}[y_k - \mathscr{E}\{y_k\}]^2$.

Ticket Sampling A method of selecting a random sample in which the characteristics of each member of the population are noted on a separate card and the required number of cards are drawn from the resulting pack. Also known as Lottery Sampling.

Tied Double Change Over Design An experimental design proposed by Federer & Ferris (1956) for situations where treatments (t) are applied in sequence and the effects persist into the succeeding period. Both direct and residual effects may be estimated and the design makes use of $(t-1)$ orthogonal Latin Squares in its construction.

Tied Ranks When a set of objects have to be ranked it may happen that certain of them are indistinguishable as regards their order and are therefore placed together in a group. To complete the ranking equal rank numbers are allotted to each member of the group, which are then said to be 'tied' and to exhibit 'tied ranks'. The most common method is to allocate to each member the mean of the ranks which the tied members would have if they were ordered. This is called the mid-rank method.

Tightened Inspection See **Normal Inspection**.

Tilling A technique used in confluence or **Bunch Map Analysis** for setting out systematically all the elementary regressions which occur in all possible subsets of variables in a regression equation.

Time Antithesis An index number formula derived from another formula by interchanging the subscripts denoting the base period and the given period and then taking the reciprocal. [See also **Factor Antithesis**.]

Time Comparability Factor In the analysis of vital statistics it sometimes happens that the standard population used to construct index numbers of mortality becomes out of date. In order to make comparisons between periods during which this has occurred an adjusting factor known as the Time Comparability Factor is used. A common form for this factor is derived as the average of the age specific death rates for the new time period weighted by the mean populations at the specific ages in the base period, divided by a similar average of rates weighted by the mean populations at the specific ages in the new time period. [See also **Area Comparability Factor**.]

Time Lag The difference in time by which one observation lags behind or is later than another. [See also **Lag, Lag Covariance**.]

Time Reversal Test One of the criteria proposed by Irving Fisher for a 'good' index number. The time reversal test is satisfied when an index number satisfies the following relationship:
$$I_{on}I_{no} = 1$$
where the base and given periods are designated by 'o' and 'n' respectively. The advantage of index numbers obeying this test is that the comparison of two periods is symmetric and consistent results are obtained whichever is regarded as the base.

Time Series A time series is a set of ordered observations on a quantitative characteristic of an individual or collective phenomenon taken at different points of time. Although it is not essential, it is common for these points to be equidistant in time. The essential quality of the series is the order of the observations according to the time variable, as distinct from those which are not ordered at all, e.g. in a random sample chosen simultaneously or are ordered according to their internal properties e.g. a set arranged in order of magnitude.

Tolerance Distribution The distribution among a number of individuals of the critical level of intensity at which

153

a stimulus will just produce a reaction in each individual. Although the distribution of these tolerances may be skew, it is often possible to make it approximately Normal by a simple transformation such as the logarithmic.

Tolerance Factor In quality control, the difference between the upper and lower tolerance limits divided by some measure of the variability of the product, usually the standard deviation. Sometimes one half of this quantity is taken as the tolerance factor, especially where the distribution of the variate under measurement is symmetrical.

Tolerance Limits In quality control, the limiting values between which measurements must lie if an article is to be acceptable, as distinct from **Confidence Limits**. [See also **Statistical Tolerance Limits**.]

Tolerance Number of Defects An expression which in better English would be 'tolerable number of defects'. It is obtained by multiplying the **Lot Tolerance Per Cent** or fraction **Defective** by the size of the **Lot** or batch submitted for inspection.

Total Correlation The **Zero Order Correlation** between two variates, i.e. the correlation between the original data rather than between residuals after some common variation has been abstracted.

Total Determination, Coefficient of In regression analysis, the square of the coefficients of multiple correlation: R^2. It represents the proportion of the total variance of the dependent variate which is accounted for by the variation of the independent variables in the multiple correlation.

In this respect it is a generalisation of the **Coefficient of Determination** and is sometimes called the coefficient of multiple determination. Similarly, a coefficient of total non-determination, or multiple non-determination, can be written:

$$K = \sqrt{(1 - R^2)}.$$

Total Inspection See **Screening Inspection**.

Total Regression A regression coefficient of zero order, i.e. a coefficient which involves only one dependent and one independent variate. The expression is probably better avoided.

Traffic Intensity A critical ratio of importance in the analysis of congestion problems; it is generally given as

$$\rho = \frac{\text{mean rate of arrival (at a queue)}}{\text{mean rate of service}}$$

but may also be stated as the mean service time divided by the mean inter-arrival time.

Transfer Function Alternatively called the **Frequency Response Function**.

Transformation Set of Latin Squares If the rows, columns and letters of a Latin Square are permuted, the resulting set of Latin squares is known as the transformation set. In the case of squares of certain sizes, e.g. 6×6 squares, not all the squares of a transformation set will be different.

Transient State Alternative name for **Non-recurrent State**.

Transition Probability In the theory of stochastic processes, the transition probability is the **Conditional Probability**, that a system in state E_j will be in state E_k at some designated later time.

Translation Parameter A parameter of location.

*Transvariation (Transvariazione) In Italian usage, given two groups of quantities, there is said to exist transvariation between them with respect to a certain mean if, among all the possible differences between the values of one group and those of the other, some have opposite signs to that of the difference between means chosen to represent the two groups. Such differences are called transvariations. Twice their sum divided by its maximum possible value measures the intensity of transvariation (*intensità di transvariazione*); this maximum being taken as the sum of absolute values of all possible differences between the terms of the two groups measured from the respective arithmetic means.

If the groups are represented by two frequency curves $f_1(x)$ and $f_2(x)$, s_1 is the area between $f_1(x)$, the x-axis and the ordinates at the extreme values; similarly for s_2; and s is the area common to s_1 and s_2, the relative area of transvariation (*area relativa di transvariazione*) is $2s/(s_1 + s_2)$. When $s_1 = s_2$ this quantity is called the transvariation ratio (*rapporto di transvariazione*).

If two distributions D_1 and D_2 have respective means M_1 and M_2 $(M_1 - M_2 > 0)$ a value of the variate x is said to be discriminatory (*valore discriminativo di transvariazione*) if it minimises the error committed in supposing that all values of D_2 are less than x and all values of D_1 are greater than x. If D_1 and D_2 have the same total frequency this is called the critical value (*valore critico di transvariazione*).

The concept has been extended to more than two groups.

Treatment In experimentation, a stimulus which is applied in order to observe the effect on the experimental situation, or to compare its effect with those of other treatments. In practice, 'treatment' may refer to a physical substance, a procedure or anything which is capable of controlled application according to the requirements of the experiment.

Treatment Mean Square A mean square in a **Variance**

Analysis assignable to differences among the effects of one of the experimental treatments.

Trend A long term movement in an ordered series, say a time series, which may be regarded, together with the oscillation and random component, as generating the observed values. An essential feature of the concept of trend is that it is smooth over periods that are long in relation to the unit of time for which the series is recorded. 'Long' for this purpose is somewhat arbitrarily defined so that a movement which is a trend for one purpose may not be so for another; e.g. a systematic movement in climatic conditions over a century would be regarded as a trend for most purposes, but might be part of an oscillatory movement taking place over geological periods of time.

In practice trend is usually represented by some smooth mathematical function (analytic trend) such as polynomials in the time variable or logistic form; but graduation procedures by **Moving Averages** are also common.

Trend Fitting The general process of representing the trend component of a time series. A trend may be represented by a particular curve form, e.g. the logistic, or by a particular form of the general class of polynomial in time, or by a **Moving Average**. [See also **Variate Difference Method**.]

Trial In probability theory a 'trial' is a deliberate attempt to generate an event which is supposed to be happening under a probabilistic scheme; e.g. the tossing of a coin is a 'trial', the outcome being one of two possible events, a head or a tail.

More generally, a 'trial' is any controlled experiment with an outcome of an uncertain kind.

Triangle Test A test in which three objects, two of which are alike are presented to a judge who attempts to select the dissimilar object. [See also **Duo-trio Test**.]

Triangular Association Scheme This type of association scheme for **Partially Balanced Incomplete Block Designs** states that, for the $n(n-1)/2$ treatments, there exists sets (S_j) such that (i) each S_j consists of $n-1$ treatments, (ii) any treatment is in precisely two sets, (iii) any two sets S_i, S_j have exactly one common treatment.

Triangular Design A class of experimental design in which $\frac{1}{2}n(n-1)$ treatments are arranged in n incomplete blocks of $n-1$ according to a pattern which may be illustrated as follows for the case $n = 4$.

X	1	2	3
1	X	4	5
2	4	X	6
3	5	6	X

The diagonals of a 4 by 4 table are eliminated and the six treatments filled in as shown. Each row then constitutes an incomplete block.

The phrase 'triangular design', due to R. C. Bose, is also applied to more general incomplete block designs based on the above scheme. [See also **Linked Blocks**.]

Triangular Distribution A frequency distribution which, when graphed as a frequency polygon, has a triangular shape. It may be written in the form

$$dF = \frac{2}{a(a+b)}(a+x)dx, \quad -a \leqslant x \leqslant 0;$$
$$= \frac{2}{b(a+b)}(b-x)dx, \quad 0 \leqslant x \leqslant b.$$

Triangular Multiply Linked Block Design For a **Triangular Design** to be multiply linked, an extension of singly and doubly linked,
either $r = 2\lambda_1 - \lambda_2$ with $b = n$
or $r = (n-3)\lambda_2 - (n-4)\lambda_1$ and $b = \frac{1}{2}(n-1)(n-2)$
with the letters denoting replicates (r), blocks (b) and number of treatments (n) in a group.

Triangular (Singly or Doubly) Linked Blocks A subtype of **Triangular Designs**.

Trimming A procedure for the rejection of observations suggested by Tukey (1962), for removing equal numbers of lowest and highest observaions in order to improve estimation of a location parameter from a simple random sample. The motive is to guard against 'wild shots' or a long tailed distribution.

Trinomial Distribution The **Multinomial Distribution** for three exhaustive and mutually exclusive probabilities.

Triple Comparisons The statistical model for **Paired Comparisons** can be extended to the $\binom{t}{3}$ triple comparisons that may be formed from t treatments. The **Asymptotic Relative Efficiency** of triple comparisons, compared with paired comparisons, is $1 \cdot 5$ and $9/(4\pi)$ when compared with ordinary analysis of variance.

Triple Lattice See **Square Lattice**. Generally for any lattice design, if three replications are selected from those possible under the design the resultant is said to be triple.

Trough An observation in a discontinuous time series which is lower than each of the two neighbouring observations; or, in the continuous case, a point where the series has a minimum.

True Mean An alternative, although little used, term for the mean of a population, as distinct from the mean of a sample.

True Regression An expression which is sometimes used in reference to a sample to denote the regression of one variate upon another which would have been obtained if there had been no errors of observation in the independent variable. The usage is open to misunderstanding owing to possible confusion with the 'parent' regression, as distinct from sample regression, and is probably better avoided.

Truncation   A truncated distribution is one formed from another distribution by cutting off and ignoring the part lying to the right or left of a fixed variate value. A truncated sample is likewise obtained by ignoring all values greater than or less than a fixed value.

In this sense truncation is to be distinguished from **Censoring**. The word also occurs in a different sense to denote the cessation of a sampling process. For example, in **Sequential Analysis** the successive drawing of members of a sample may have to be stopped before a decision has been reached under the terms of the sequential scheme. This cutting off with respect to time may be called truncation but is different from cutting off with respect to a variate value. [See also **Cut Off**.]

Tukey Statistic   A multiple comparison test of mean values (1951, 1952) arising in analysis of variance based upon the **Studentised Range**. It is a step by step procedure but when all the sample sizes are equal can be modified to a simultaneous procedure. [See **Gabriel's Test**.]

Tukey's Gap Test   A test for comparing individual mean values in analysis of variance suggested by Tukey (1949). It is based upon excessive gaps occurring between individual or groups of mean values.

Tukey's q-test   A test of significance for the equality of mean values based upon the studentised range (q) of the observations and the range of those mean values.

Tukey's Quick Test   A simple and compact test for comparing two samples proposed by Tukey (1959). It is based upon the overlap of the sample values, i.e. upon counts of the number of observations in one sample which are greater than those in the other sample and vice versa. The test is quick to execute and the critical values for the common percentage points are easy to memorise.

Turning Point   In an ordered series, an observation which is a **Peak** or a **Trough**. When several contiguous values are equal, and greater than or less than the neighbouring values, a convention is required to determine which is regarded as the turning point, e.g. the middle one may be chosen.

Two-by-Two Frequency Table   A term for the presentation in tabular form of data subject to **Double Dichotomy**. If each member of a set of n can bear or not bear an attribute A and an attribute B, such a table might be of the form

	Bearing A	Not Bearing A	Totals
Bearing B	a	b	$a+b$
Not Bearing B	c	d	$c+d$
Totals	$a+c$	$b+d$	$n=a+b+c+d$

Here, for example, b is the number of members bearing B but not bearing A.

Two-Factor Theory   See **Single-Factor Theory**.

Two-phase Sampling   See **Multiphase Sampling**.

Two Sided Test   An alternative name for **Double Tailed Test**.

Two Stage Sample   A simple case of a **Multi-stage Sample**. In this case the population to be sampled is first classified into primary units, each of which consists of a collection of the basic sampling unit, the secondary unit. A sample of these primary units is taken, constituting the first stage, and these are then subsampled with respect to their secondary units: this constitutes the second stage.

Two Way Classification   The classification of a set of observations according to two criteria of classification as, for example, in a double dichotomy or a **Correlation Table**.

Type   A word occurring in several specialised usages in statistics, e.g. in relation to types of frequency function and types of critical region. In older literature two other usages, both obsolescent, are found:
- (a) in relation to a value such as the mean or the median, which is taken as 'typical' of a frequency distribution;
- (b) in relation to the central values of the group of a bivariate frequency table, an array of x-values in a group of y centred at the value y_0 being said to be of type y_0.

Type Bias   Bias which may be imparted to an index number as a result of using a particular type of average on the constituent series. The actual existence of bias depends on the nature of the series to which the averaging is applied and also, of course, on what is meant by 'bias' in the context under discussion.

Type One Counter Model   See **Counter Model, Type I**.

Type Two Counter Model   See **Counter Model, Type II**.

Type A Distribution   A form of compound Poisson distribution proposed by Neyman (1933) for use as a **Contagious Distribution**. It is to be distinguished from the Gram-Charlier type A expansion of a frequency function.

The frequency of the discontinuous variate x (which takes values 0, 1, 2, ..., ∞) at j is the coefficient of t^j in

$$\exp\{-m_1(1-e^{m_2(t-1)})\}.$$

Other forms are also in use.

Type B Distribution A distribution formed by the limit of the sum of a large number of independent Neyman Type A distributions wherein the second parameter has a uniform distribution.

Type C Distribution A distribution formed in a manner similar to the Neyman Type B with a beta distribution for the second parameter.

Type I Distribution One of the three main types of the Pearson system of frequency distributions. In general it is a unimodal distribution with limited range. With the origin at the mode, it was written by Pearson in the form

$$dF = k\left(1+\frac{x}{a_1}\right)^{m_1}\left(1-\frac{x}{a_2}\right)^{m_2} dx, \ -a_1 \leqslant x \leqslant a_2;$$
$$m_1, m_2 > -1.$$

where $m_1/a_1 = m_2/a_2$. With a suitable choice of origin and scale, it is equivalent to the beta distribution:

$$dF = \frac{1}{B(p, q)} x^{p-1}(1-x)^{q-1}dx, \ 0 \leqslant x \leqslant 1; p, q > 0.$$

Type II Distribution A particular case of the Type I distribution. It may be written:

$$dF = k\left(1-\frac{x^2}{a^2}\right)^m dx, \ -a \leqslant x \leqslant a; m \geqslant -1;$$

and is symmetrical, usually platykurtic in shape and with limited range. An alternative way of expressing the distribution is

$$y = \frac{1}{B(p, p)} x^{p-1}(1-x)^{p-1}, \ 0 \leqslant x \leqslant 1; p > 0.$$

In the special case when $m = 0$ or $p = 1$ it becomes the rectangular distribution.

Type III Distribution This distribution, in the Pearson system of frequency distributions, has unlimited range in one direction and it is generally unimodal. It may be written:

$$dF = k\left(1+\frac{x}{a}\right)^{\gamma a}e^{-\gamma x}dx, \ -a \leqslant x \leqslant \infty; \gamma > 0, a > 0.$$

or, more simply, by choice of a suitable origin and scale:

$$dF = \frac{1}{\Gamma(p)}x^{p-1}e^{-x}dx, \ 0 \leqslant x \leqslant \infty; p > 0.$$

The **Chi-squared Distribution** is of this form.

Type IV Distribution One of the three main types of the Pearson system of frequency distributions. Its general shape is that of a unimodal skew distribution with unlimited range in both directions. It may be written:

$$dF = k\left(1+\frac{x^2}{a^2}\right)^{-m}e^{-\mu\tan^{-1}(x/a)}dx, \ -\infty \leqslant x \leqslant \infty;$$
$$a > 0, \ \mu > 0.$$

Type V Distribution A unimodal distribution of special

type in the Pearson system with origin at the start of the distribution. It is usually written in the form:

$$dF = kx^{-p}e^{-\gamma/x}dx, \ 0 \leqslant x \leqslant \infty; \gamma > 0, p > 1.$$

A transformation of type $y = \gamma/x$ turns it into a **Type III Distribution**.

Type VI Distribution The third of the three main types in the Pearson system of frequency curves. It is generally unimodal and skew with unlimited range in one direction and may be written as

$$dF = kx^{-q_2}(x-a)^{q_2}dx, \ a \leqslant x \leqslant \infty; q_1 > q_2 - 1.$$

By the substitution $y = a/x$ this distribution can be reduced to the **Type I Form**.

Type VII Distribution A unimodal symmetrical distribution of special kind in the Pearson system. It has unlimited range in both directions and may be written:

$$dF = k\left(1+\frac{x^2}{a^2}\right)^{-m}dx, \ -\infty \leqslant x \leqslant \infty; m > \tfrac{1}{2}.$$

The *t*-distribution is a special case of this type.

Type VIII Distribution A member of the less important group of Pearson curves. It may be written:

$$dF = k\left(1+\frac{x}{a}\right)^{-m}dx, \ -a \leqslant x \leqslant 0; 0 \leqslant m \leqslant 1.$$

Type IX Distribution A less important member of the Pearson system of distribution curves which may be written:

$$dF = k\left(1+\frac{x}{a}\right)^m dx, \ -a \leqslant x \leqslant 0; m > -1.$$

Type X Distribution A distribution of the Pearsonian system which is the same as the **Exponential Distribution**.

Type XI Distribution A J-shaped distribution in the Pearson system which may be written:

$$dF = kx^{-m}dx, \ b \leqslant x \leqslant \infty; m > 0.$$

The start of the distribution is at an ordinate $x = b$. [See **Pareto Distribution**.]

Type XII Distribution A special distribution in the Pearson system which has a twisted J-shape and constitutes a particular case of the Type I distribution. The form is

$$dF = \left(\frac{1+\dfrac{x}{a_1}}{1-\dfrac{x}{a_2}}\right)^m dx, \ -a_1 \leqslant x \leqslant a_2; |m| > 1.$$

Type I Error An alternative term for α-error or **Error of the First Kind**. [See also **Producer's Risk**.]

Type II Error An alternative term for β-error or **Error of the Second Kind**. [See also **Consumer's Risk**.]

Type I and II Probabilities See Type I Sampling.

Type A Region In the theory of testing statistical hypotheses, a locally unbiassed critical region for testing a simple hypothesis specifying one parameter. Regions of this kind are obtained by maximising the curvature of the power curve at $\theta = \theta_0$ subject to conditions of local unbiassedness and control of errors of the first kind.

If a Type A region does not suffer from the restriction of being merely locally unbiassed, but is, in effect, unbiassed everywhere, then it is known as a Type A_1 region.

Type B Region An extension of the concept of the **Type A Region** to the case of a composite hypothesis.

Type C Region An extension of the concept of the unbiassed critical region of Type A proposed by **Neyman and Pearson** to cover a simple hypothesis specifying two parameters. A critical region of this class must be of a given **Size**, unbiassed and of best local power in the neighbourhood of the null values of the parameters, say θ_1^0, θ_2^0. The exact determination of Type C regions rests upon knowledge of the errors of the second kind and in order to overcome the general absence of this information Isaacson (1951) proposed the region of Type D.

Type D Region An unbiassed critical region proposed by Isaacson (1951) for testing simple hypothesis specifying the values of several parameters. The Type D region, which is a generalisation of the Type A region, is one which maximises the curvature of the power surface subject to conditions of size and unbiassedness.

Type 'E' Region A development of the **Type D Region** but stated to be of doubtful existence for a wide class of multivariate tests (Giri & Kiefer, 1964).

Type I Sampling A term sometimes used in Bayesian analysis to denote ordinary sampling from a given population. If this population is itself the result of sampling from a super-population the term Type II sampling is applied to the selection of the population.

The probabilities that govern the two types of sampling are called Type I and Type II probabilities respectively. The hierarchy can be extended to Types III, IV, etc.

Type II Sampling See **Type I Sampling**.

Type A Series A term introduced by Charlier to denote the expansion of a continuous frequency function as a series of derivatives of the Normal frequency function. It is more usually known as the Gram-Charlier series. [See also **Edgeworth Series**.]

Type B Series A term introduced by Charlier to denote the expansion of a frequency function in terms of derivatives of a Poisson distribution. [See **Gram-Charlier Series, Type B**.]

Type C Series An expansion of a frequency function proposed by Charlier as an alternative to his Type A. The latter can give rise to negative frequencies in the tails of the distribution and Type C purports to remove this anomaly. It has not come into general use. [See **Gram-Charlier Series, Type C**.]

*Typical Characteristic (Carattere Tipico) See ***Characteristic**. [See also **Mode**.]

*Typical Period (Periodo Tipo) The period of time to which, in weighted index numbers, the weights of the individual indices are referred. It may or may not coincide with the base period.

*Typical Year (Anno Tipo) See **Base Period**.

U-shaped Distribution A frequency distribution shaped more or less like a letter U, though not necessarily symmetrical, i.e. with the maximum frequencies at the two extremes of the range of the variate.

'U' Statistics A class of statistic occurring in the construction of rank order tests for the two sample problem: Wilcoxon (1945) and Mann & Whitney (1947). They are of the form

$$U = \sum_{\lambda=1}^{n_1} \sum_{\eta=1}^{n_2} z_{\lambda\eta}$$

where

$$z = 1 \text{ if } x_\lambda < y_\eta$$
$$= 0 \quad x_\lambda \geqslant y_\eta$$

and $(x_1, ..., x_{n_1})$, $(y_1, ... y_{n_2})$ are ordered observations from the two samples.

U_N^2 Test A goodness of fit test, analogous to the Cramér-von Mises Test based on the W_N^2 Statistic, introduced by Watson (1961, 1962) and developed by Pearson & Stephens (1962). The statistic is as follows:

$$U_N^2 = N \int_{-\infty}^{\infty} \left\{ F_N(x) - F(x) - \int_{-\infty}^{\infty} [F_N(y) - F(y)] dF(y) \right\}^2 dF(x)$$

and the test may be extended to the case of $U_{M,N}^2$ where the two samples are not necessarily the same size.

Ultimate Cluster The aggregate of ultimate or final stage units included in a primary unit.

Unadjusted Moment A moment of a frequency distribution before any adjustment is made for the effect of grouping the observations, e.g. before the application of **Sheppard's Corrections**. [See also **Raw Moment**.]

Unbiassed Confidence Intervals A system of confidence intervals which is selective in the sense of covering the true value of the parameter with the assigned probability $1-\alpha$ and covers other values as little as possible is said to be unbiassed if $\Pr\{\delta \epsilon \theta \mid \theta\} = 1-\alpha \geqslant \Pr\{\delta \epsilon \theta \mid \theta'\}$ where δ is a typical interval: θ is the true value and θ' an alternative. This usage of 'unbiassed' is analogous to that for tests of significance.

Unbiassed Critical Region See **Critical Region**.

Unbiassed Error An error which may be regarded as a member drawn at random from an error population with zero mean. Thus in the long run positive and negative errors tend to cancel out in the sense of having a mean which tends to zero.

Unbiassed Estimating Equation An equation for the estimation of a parameter in which the terms are unbiassed estimators of the corresponding parent values. It does not follow that the estimator of the parameter is then unbiassed itself. For example, if the estimator t of a parameter θ is given by $a - bt = 0$, where a and b are variates, a and b may be unbiassed and hence the equation is unbiassed, but the ratio a/b may still give a biassed estimator of θ.

Unbiassed Estimator An estimator, say t, which, for all sample sizes, has its expected value equal to the parameter, say θ, under estimate, i.e. $\mathscr{E}(t) = \theta$. For example, the sample variance is not an unbiassed estimator of the population variance σ^2: since

$$\mathscr{E}\left[\frac{1}{n}\sum_{i=1}^{n}(x_i - \bar{x})^2\right] = \frac{n-1}{n}\sigma^2.$$

On the other hand the statistic $\frac{1}{n-1}\Sigma(x_i - x)^2$ is an unbiassed estimator of the population variance since

$$\mathscr{E}\left[\frac{1}{n-1}\Sigma(x_i - \bar{x})^2\right] = \sigma^2.$$

An estimator which is not unbiassed is called biassed. If it is unbiassed for all distributions, as in the case of the estimator for the variance given above, it is absolutely unbiassed. If it is unbiassed and linear in the observations it is an unbiassed linear estimator.

If the estimator tends to be unbiassed as the sample number increases it is asymptotically unbiassed. The first estimator of σ^2 mentioned above, for example, is so.

Unbiassed Sample A sample drawn and recorded by a method which is free from bias. This implies not only freedom from bias in the method of selection e.g. random sampling, but freedom from any bias of procedure, e.g. wrong definition, non-response, design of questions, interviewer bias, etc. An unbiassed sample in these respects should be distinguished from unbiassed estimating processes which may be employed upon the data.

Unbiassed Test A test of significance may be unbiassed in a manner different from the use of that term in estimation. If w is a critical region of size α for $H_0: \theta = \theta_0$ against the simple alternative of $H_1: \theta = \theta_1$ is such that its power

$$\Pr\{\mathbf{x} \in w \mid \theta_1\} \geqslant \alpha$$

it is said to give an unbiassed test if H_0 against H_1. An important class of unbiassed tests are those which are also **Uniformly Most Powerful**.

Unequal Subclasses See **Disproportionate Subclass Numbers**.

Uniform Distribution An alternative term for the **Rectangular Distribution**.

Uniform Sampling Fraction If a sample is selected from a population which has been grouped into strata, in such a way that the number of units selected from each stratum is proportional to the total number of units in that stratum, the sample is said to have been selected with a uniform sampling fraction.

Uniform Spectrum A spectrum in which all ordinates have the same value. It may arise from **'White Noise'**, a term given by engineers to a sequence of small stochastic elements which are independent from one instant to the next.

Uniformity Trial An experiment, or set of trials, in which each experimental unit receives exactly the same treatment. The object, *inter alia*, may be to estimate some standard characteristic of that treatment or to investigate some aspects of the experimental technique, e.g. plot size or layout in agricultural trials.

Uniformly Best Constant Risk (U.B.C.R.) Estimator Of the class of constant risk estimators, i.e. those for which the **Risk Function** is constant, the estimator that minimises the expected risk with reference to an *a priori* distribution is termed a uniformly best constant risk estimator. It is usual to obtain this restricted class of estimators indirectly through the theorem that any constant risk estimator which is also a minimax estimator is also a uniformly best constant risk estimator. [See also **Minimax Principle**.]

Uniformly Best Distance Power (U.B.D.P.) Test If the alternative hypotheses H_1, H_2, ..., to a null hypothesis H_0 can be specified by parameters with continuous variation then it is possible to determine a region w_0 so that it yields the same **Power** to a given sub-set of hypotheses, which are said to be equidistant from the null hypothesis. A test of H_0 based upon the region w_0 as acceptance region is a uniformly best distance power test if the size of the region with respect to H_0 is α (the level of significance) and for any specified alternative hypothesis the power is not less than the power for any other region of the same kind.

Uniformly Better Decision Function The merit of a decision function may be judged by reference to its **Risk Function**. A decision function δ_1 is said to be a uniformly better decision function than δ_2 if the risk function for δ_1 is never greater than the corresponding function for δ_2 and is smaller for some values.

Uniformly Minimum Risk A characteristic of some

multiple decision procedures, where the risk, in relation to a low function is minimised over a range of hypotheses.

Uniformly Most Powerful (U.M.P.) Test A test of a hypothesis against a family of alternative hypotheses which is most powerful for each of the alternative hypotheses. In most cases a uniformly most powerful test only exists when the alternative hypotheses are restricted in some fashion; for instance, if the hypothesis is that some parameter $\theta = 0$, the alternatives might be $\theta > 0$ or $\theta < 0$ but not both. If the test is uniformly most powerful for either of these alternative sets it is said to be the uniformly most powerful one sided test.

Unimodal An adjective describing a frequency distribution which has a **Single Mode.**

Union Intersection Principle A method proposed by Roy (1958) for denoting regions of acceptance and rejection in constructing tests of significance in multivariate analysis.

Unique Factor In factor analysis this term sometimes occurs in the sense of **Specific Factor** and should be avoided in that sense. More usually, in psychology, it refers to a clearly identifiable trait which forms a factor common to the tests under discussion.

Uniqueness See **Factor Analysis.** The uniqueness of a variate (test) is the complement of the communality.

Unit Stage Sampling See **Multi-stage Sampling.**

Unit Normal Variate A variate which is Normally distributed with zero mean and unit standard deviation. It is often written $N(x; 0, 1)$ or $N(0, 1)$.

Unitary Sampling Sampling in which the ultimate units are directly chosen, as contrasted with a multi-stage sampling where primary groups are first chosen.

Unitemporal Model See **Dynamic Model.**

Univariate Distribution A distribution of one variate only as contrasted with bivariate, trivariate, ..., multivariate distributions.

Universe An alternative term for **Population** derived from the 'universe of discourse' of classical logic.

Unlikelihood Ratio A type of nonsequential multiple decision rule proposed by Hall (1958) where the sample size is a minimum subject to upper bounds on the probabilities of making incorrect decisions.

Unreduced Designs A particular kind of **Balanced Incomplete Block Design** derived by taking all possible sets of k-treatments from the full set t, each set of k forming a block. The design is completed by permuting the blocks and arranging the treatments randomly.

Unreliability See **Reliability.**

Unrestricted Random Sample A sample which is drawn from a population by a random method without any restriction; that is to say, all possible samples have the same chance of being selected.

Unweighted Mean A mean of a set of observations in which no weights are attached to them, except in the trivial sense that each has weight unity.

Unweighted Means Method (in variance analysis) In variance analysis, a simple method for the analysis of a set of results for which the sub-group frequencies are unequal. It involves taking the mean values for each subclass and carrying out an ordinary variance analysis on those means.

Up and down Method A method of estimating the 50 per cent response point of quantal response data. It is essentially a unit sequential process of testing. If the first object to be tested reacts to a given stimulus the next is subjected to a decreased stimulus. If this reacts then the level is again reduced but if it fails to react the object is retested at the previous high level. This progressive testing at levels of the stimulus which are put 'up and down' according to each result accounts for the name of the method.

The method has been developed for responses less than 50 per cent by calculating the proportion of zeros rather than positive responses, and for responses greater than 50 per cent in Wetherill *et al.* (1966).

Up Cross See **Down Cross.**

Upper Control Limit See **Control Chart.**

Upper Quartile See **Quartile.**

Upward Bias See **Downward Bias.**

Uspensky's Inequality An inequality developed from the **Bienaymé-Tchebychev Inequality** for use in cases where large positive deviations from the mean are important. It may be written

$$\mathrm{Pr}(x - \bar{x} \leqslant t) \leqslant \sigma^2/(\sigma^2 + t^2) \qquad \text{if } t < 0$$
$$\geqslant 1 - \{\sigma^2/(\sigma^2 + t^2)\} \text{ if } t \geqslant 0.$$

This can also be applied to the sum of random variables.

V_N Test A **Goodness-of-fit** test adapted from the **Kolmogorov-Smirnov Test** by Kuiper (1960) using the statistic

$$V_N = \sup_{-\infty < x < \infty} [F_N(x) - F(x)] - \inf_{-\infty < x < \infty} [F_N(x) - F(x)].$$

It was extended to cover the ordinary two-sample case by Maag & Stephens (1968). [See also W_N^2 **Test,** U_N^2 **Test.**]

Validation A procedure which provides, by reference to independent sources, evidence that an inquiry is free from bias or otherwise conforms to its declared purpose. In statistics it is usually applied to a sample investigation with the object of showing that the sample is reasonably representative of the population and that the information collected is accurate. For example, a sample of human beings is partly validated by comparing, *inter alia*, its sex and age constitution with the known figures for the population from which it was chosen; except, of course, where the sample was deliberately chosen to secure concordance in these respects, as in quota sampling.

Validity is to be contrasted with consistency, which is concerned with the internal agreement of data or procedures among themselves.

*Valore Poziore An Italian term with no English counterpart. It means that value of a variate which, when multiplied by its frequency, yields a maximum.

Value Index An index number formed from the ratio of aggregate values in the given period to the aggregate values in the base period. Strictly speaking this is not an **Index Number** as ordinarily understood but a value relative. [See also **Price Relative**.]

Van Der Waerden's Test A distribution free test (1952, 1953) which is sensitive to differences in location for two samples which are from otherwise identical populations. The test statistic is

$$V = \sum_{i=1}^{n} \Phi^{-1}\left(\frac{R_i}{N+1}\right)$$

where Φ^{-1} are inverse Normal scores, N is the combined sample size and R_i the ranks of x_i in the combined sample.

'Vanity' Effect A form of bias encountered in surveys of human populations which can be introduced into the results through distorted responses by the individuals being questioned. The person questioned fails to give an accurate reply but makes instead an assertion more pleasing to his personal vanity.

*Variability (Mutabilità) See ***Modality**.

Variable Generally, any quantity which varies. More precisely, a variable in the mathematical sense, i.e. a quantity which may take any one of a specified set of values. It is convenient to apply the same word to denote non-measurable characteristics, e.g. 'sex' is a variable in this sense since any human individual may take one of two 'values', male or female.

It is useful, but far from being the general practice, to distinguish between a variable as so defined and a random variable or **Variate**.

Variable Lot Size Plan A sampling plan for acceptance inspection which combines the useful features of a lot by lot plan with the flexibility of **Continuous Plans**. It is intended to deal with continuous production; an essential feature is that screening of rejected 'lots' does not have to be done immediately.

Variable Sampling Fraction If from a stratified population a simple random sample is selected from each stratum in such a way that the proportion of units sampled in each stratum varies from stratum to stratum, the sample is said to be selected with variable sampling fraction. Applicability of the term to other sampling schemes rests upon the general definition of **Sampling Fraction**.

Variables Inspection **Acceptance Inspection** where the criteria for classifying or judging a sample submitted for inspection are quantitative rather than qualitative. In this sense 'variable' relates to a measurable quantity, as distinct from an attribute, and is not used in the broader sense noted in the definition of **Variable**.

Variance The variance (Fisher, 1918) is the second moment of a frequency distribution taken about the arithmetic mean as the origin namely

$$\int_{-\infty}^{\infty} (x - \mu_1')^2 dF$$

where μ_1' is the mean and F the distribution function. It is a quadratic mean in the sense that it is the mean of the squares of variations from the arithmetic mean. It may also be regarded as one half of the mean square of differences of all possible pairs of variate values.

Variance Analysis The total variation displayed by a set of observations, as measured by the sums of squares of deviations from the mean, may in certain circumstances be separated into components associated with defined sources of variation used as criteria of classification for the observations. Such an analysis is called an analysis of variance, although in the strict sense it is an analysis of sums of squares. Many standard situations can be reduced to the variance analysis form.

Variance Component One of the objects of variance analysis is to split up the sum of squares of observations about their mean into portions which can be assigned to variation between the classes or sub-classes according to which the data are classified. If the variables defining the classes are 'fixed', that is to say, if all the classes under consideration actually appear, these constituent parts of the sums of squares indicate through mean-squares the magnitude of class differences, and the extent to which they differ from the residual mean square affords a test of the hypothesis that such differences are governing the situation.

A second generating model often considered in variance analysis regards the classificatory variables observed as themselves variates, i.e. as samples chosen from a wider classificaton. The expected values of the mean squares derived from the variance analysis can then be used to

161

obtain estimates of the variances of the classifying variates. For example, in a two way classification with r rows and c columns and k members in each cell, one possible model expressing additive row and column effects is that the observations x are given by

$x_{ij} = a_i + b_j + (ab)_{ij} + e_{ij}, i = 1, 2, ..., r; j = 1, 2, ..., s.$

The expected mean squares in a variance analysis are

Rows: $\sigma_e{}^2 + k\sigma_{ab}{}^2 + kc\sigma_a{}^2$

Columns: $\sigma_e{}^2 + k\sigma_{ab}{}^2 + kr\sigma_b{}^2$

Interaction: $\sigma_e{}^2 + k\sigma_{ab}{}^2$

Residual: $\sigma_e{}^2.$

where $\sigma_a{}^2$, for example, is the variance of a_i. The various σ^2 are called the variance components of x and are usually estimated by equating the estimated and observed mean squares.

Variance-Covariance Matrix See **Covariance Matrix**.

Variance Function A concept involved in the design of multi-factor response surface experiments (see Box & Hunter, 1957). This function provides a standardised measure of the precision of the estimated response at any point in the space of the variables. In the absence of prior information, the variance function is concerned only with Σx^2 and the design is said to be **Rotatable** and have a spherical variance function.

Variance Ratio Distribution The distribution of the ratio of two independent quantities each of which is distributed like a variance in Normal samples, i.e. in the **Type III**, χ^2 or **Gamma Form**. The distribution, due to R. A. Fisher, may be put in the form

$$dF = \frac{\nu_1{}^{\frac{1}{2}\nu_1}\nu_2{}^{\frac{1}{2}\nu_2}\Gamma\{\frac{1}{2}(\nu_1+\nu_2)\}}{\Gamma(\frac{1}{2}\nu_1)\,\Gamma(\frac{1}{2}\nu_2)}\;\frac{F^{\frac{1}{2}\nu_1-1}}{\left(\frac{\nu_1}{\nu_2}F+1\right)^{\frac{1}{2}(\nu_1+\nu_2)}}\;dF$$

where ν_1 and ν_2 are the degrees of freedom of numerator and denominator of the ratio $F = s_1{}^2/s_2{}^2$. The distribution was first studied by Fisher in a transformed form (see z-**distribution**), the ratio F being so denominated by Snedecor from the first letter of the discoverer's name. The distribution is a simple transform of the **Type I** or **Beta Distribution**.

Variance Ratio Test A test based on the ratio of two independent statistics, each of which is distributed as the variance in samples from Normal populations with the same parent variance. Usually the statistics themselves are quadratic estimators of the parent variance. The test is widely employed in variance analysis to test the homogeneity of a set of means.

Variate In contradistinction to a **Variable**, a variate is a quantity which may take any of the values of a specified set with a specified relative frequency or probability. The variate is therefore often known as a random variable. It is to be regarded as defined, not merely by a set of permissible values like an ordinary mathematical variable, but by an associated frequency (probability) function expressing how often those values appear in the situation under discussion.

Variate Difference Method A method of analysis of time series which consist of a systematic and a random component. It is based essentially on the consideration that if the systematic part of a series can be represented by a polynomial, then successive differencing will eliminate this element and hence allow of the isolation of the random element, or at least the estimation of its variance.

Variate Transformation The transformation of one variate into another, usually by a mathematical equation connecting them. The object is, as a rule, to transform the distribution function of one variate exactly or approximately into a distribution function of known form and properties.

Variation, Coefficient of The standard deviation of a distribution divided by the arithmetic mean; sometimes multiplied by 100. It was proposed by K. Pearson (1895) for the purpose of comparing the variabilities of frequency distributions, but is sensitive to errors in the means and is of limited use.

Variation Flow Analysis A technique of evaluating the transfer of variations in stock, where the product from several machines at one processing stage is fed randomly to the several machines of the succeeding stage.

*Variazione The Italian equivalent of a standardised deviation, i.e. a value measured from the arithmetic mean and divided by the standard deviation.

Varimax See **Factor Rotation**.

Variogram A graphical representation similar to the **Correlogram** which shows the **Serial Variations Function** as ordinate.

Vector Alienation Coefficient See **Vector Correlation Coefficient**.

Vector Correlation Coefficient A generalisation of the product moment correlation between two variates for the purpose of measuring the relation between a p-way vector variate and a q-way vector variate. If the dispersion matrices of the vectors are respectively v_1 and v_2 and the covariance matrix of one set with the other is v_3 the vector correlation coefficient is defined as $= \{|v_3|/|v_1|\ |v_2|\}^{\frac{1}{2}}$.

Similarly, if the covariance matrix of all $(p+q)$ variates together is v, the vector alienation coefficient is $\{|v|/|v_1|\ |v_2|\}^{\frac{1}{2}}$.

Venn Diagram A graphical method of representing operations on sets; of great use in illustrating problems in probability.

Virtual Waiting Time Process An alternative name, derived from the particular variable considered, for the **Takacs Process.**

Von Mises Distribution The circular Normal distribution as derived by von Mises (1918).

Von Neumann's Ratio The ratio of the **Mean Square Successive Difference** to the variance of a series was proposed by von Neumann (1944) as a statistic for testing the independence of successive observations in an ordered series for which the underlying distribution is Normal. In large samples from a random series the distribution of the ratio tends to be Normal with mean 2 and variance $4(n-2)/(n^2-1)$ where n is the number of observations. The use of the ratio to test independence in a series of observations is equivalent to the use of the older **Abbe-Helmert Criterion.**

'W' Statistic The test statistic for a rank sum two sample test on dispersions proposed by Ansari & Bradley (1960). It is generally written as $W = \Sigma R(Z)$ which is, effectively, the sum of ranks associated with one sample of observations after the combined samples have been ranked.

W_N^2 Test See **Cramér-von Mises Test.**

'W' Test for Normality An analysis of variance type test for complete samples proposed by Shapiro & Wilk (1965):

$$W = \left(\sum_{i=1}^{n} a_i y_i \right)^2 \bigg/ \sum_{i=1}^{n} (y_i - \bar{y})^2$$

where the $\{a_i\}$ are normalised best linear unbiassed coefficients. The test statistic is scale and origin invariant and ranges between $na_1^2/(n-1)$ and unity.

WAGR Test A sequential test procedure for $H_0:p_0+p_1$ where the proportion is derived from observations of a Normally distributed variate greater than a given value. The designation is derived from the names Wald, Arnold, Goldberg and Rushton.

Waiting Line The American equivalent of the English meaning of the English word 'queue'. [See also **Queueing Problem.**]

Waiting Time In the theory of queues this term is self explanatory, but, in **Renewal Theory** residual waiting time is the period from a given point in time (t) to the next renewal point. Spent waiting time is the time elapsed since the last renewal. The sum of the two waiting times is the length of recurrence interval. In the theory of **Random Walks** residual waiting time is called 'point of first entry' or 'hitting point' for the interval t, ∞.

Wald-Wolfowitz Test A large-sample distribution free test of randomness based upon serial covariance proposed by Wald and Wolfowitz (1943). For a series of observations $x_1, \ldots, x_n$ measured about their mean the test statistic is:

$$R_k = \sum_{t=1}^{n} x_t x_{t+k},$$

with $x_{n+j} = x_j$, i.e. the formula is **Circular.**

Wald's Classification Statistic A statistic suggested by Wald in 1944 which is effectively the same as Fisher's discriminant function of 1936. [See **Discriminant Function.**]

Walker Probability Function A function derived by Sir Gilbert Walker (1914) in connection with tests of significance for the ordinates of a periodogram. Developing a result due to Schuster (1898) Walker stated that the probability that one value of the intensity (S^2), of the $m = \frac{1}{2}n$ independent values in a Fourier sequence of intensities, shall not exceed $4\sigma^2 k/n$ is $1-(1-e^{-k})^m$. The value of this function for arguments k and m is known as the Walker (Probability) Function.

Waring Distribution See **Factorial Distribution.**

Watson's 'U' Statistic A statistic proposed by Watson for tests of the hypothesis that two independent random samples (m, n) come from the same unknown population. It may be written $U^2 = mn(m+n)^{-2}\Sigma(d_i - \bar{d})^2$ where d_i is the difference between the pooled sample distribution functions at the ith point. Hence $d_i = n_i/n - m_i/m$ and $\bar{d} = \Sigma d_i/(m+n)$.

Wedge Plans An alternative name for **Closed Sequential** t-tests.

Weibull Distribution A distribution of the general form

$$f(x) = \frac{k}{v-\epsilon} \left\{ \frac{x-\epsilon}{v-\epsilon} \right\}^{k-1} \exp \left\{ -\left(\frac{x-\epsilon}{v-\epsilon} \right)^k \right\}$$

with $x, v > \epsilon; k > 1$, proposed by Weibull (1939) to describe data arising from life and fatigue tests. It was later derived as a model for this kind of data as the third asymptotic distribution of extreme values.

Weighing Design An experiment design, due to Hotelling (1945), for the efficient weighing of N objects using a two-pan balance. Various measures of efficiency have been proposed each being optimum for a set of conditions on the experiment.

Weight The importance of an object in relation to a set of objects to which it belongs; a numerical coefficient attached to an observation, frequently by multiplication, in order that it shall assume a desired degree of importance in a function of all the observations of the set.

Weight Bias Bias, usually in an index number, due to the use of incorrect or undesirable weights. Since the true value of the complete quantity which an index purports to measure is not in general capable of direct measurement, bias in this sense is to some extent an arbitrary quantity.

Weight Function A non-negative function used for weighting purposes; especially in the theory of decision functions, where the word is often used synonymously with 'Loss Function'.

Weighted Average An average of quantities to which have been attached a series of **Weights** in order to make proper allowance for their relative importance. For example, a weighted arithmetic mean of $x_1, x_2, ..., x_n$ with weights $w_1, w_2, ..., w_n$ is given by

$$\sum_{j=1}^{n} w_j x_j \Big/ \sum_{j=1}^{n} w_j.$$

Weighted Battery A group of educational or psychological tests wherein the relative importance of each test is determined by attaching a weight to the score obtained in that test.

Weighted Index Number An index number in which the component items are weighted according to some system of weights reflecting their relative importance. In one sense nearly all index numbers are weighted by implication; for example, an index number of prices amalgamates prices per unit of quantity and the size of these units may vary from one commodity to another in such a way as to constitute weighting. It is, however, usual to describe an index as 'weighted' only when weighting coefficients enter explicitly into its definition and calculation.

Weighting Coefficient The coefficient attached to an observation as its **Weight** in a procedure involving weighting. [See also **Raising Factor**.]

White Noise By analogy with the continuous energy distribution in white light from an incandescent body, a **Covariance Stationary Stochastic Process** which has equal power in all frequency intervals over a wide frequency range is called white noise. The sequence $\{X(t), t \geqslant 0\}$ is said to be a white noise process if it possesses a constant **Spectral Density Function**.

Whittaker Periodogram A form of **Periodogram** defined by Sir Edmund Whittaker. If a series is formed into groups of m consecutive terms arrayed one under another (cf. Buys-Ballot Table), the corresponding ordinate η^2 of the periodogram is the variance of the column sums divided by the variance of the series. The periodogram graphs η as ordinate against m as abscissa.

Wide Sense Stationary See **Covariance Stationary Process**.

Wiener-Hopf Technique A method used in the analysis of autocorrelation functions $R(\tau)$ of time series in connection with mean square error estimation. For this to be minimised, a weighting function attached to the auto-correlation function has to obey a certain integral equation.

Wiener Process See **Brownian Motion Process**.

Wiener-Khintchine Theorem An expression occasionally found to denote the theorem that the **Covariance Function** of a stationary stochastic process is positive definite.

Wilcoxon Signed Rank Test A distribution free test of the difference between two treatments using matched samples. If the differences $|x_i - y_i|$ of n pairs of observations are ranked according to size, and each rank given the sign of the original difference, the sum of positive ranks is the test statistic proposed by Wilcoxon (1945) and developed by other writers.

Wilcoxon's Test In ranked material, a test of the common origin of two samples proposed by Wilcoxon (1945). Like **Kendall's Tau**, it is based on the signs of the differences in rank numbers.

Wilks' Criterion A criterion of general use in multivariate analysis for testing hypotheses concerning multivariate Normal populations, especially hypotheses of homogeneity in means or dispersions. The criterion essentially depends on the ratio of the determinants of two matrices of sums of squares and products; the numerator corresponding to a sum-within-classes and the denominator to a total sum. It occurs in various forms.

The Λ criterion was derived by Wilks in 1932 and has subsequently been extended by him and other authors. He also derived a test of significance of the criterion, sometimes known as Wilks' test.

Wilks' Internal Scatter A term introduced by Wilks (1932) for the generalised sample variance; that is to say for $|u_{ij}|$ where the elements are those of the sample variance-covariance matrix.

Wilks-Lawley U_l Statistic This is used in multivariate analysis of variance to test certain hypotheses.

It is defined as $U_l = \prod_{i=1}^{l} \theta_i$ where θ_i ($i = 1, 2, ..., l$) are the roots of the determinantal equation $|\mathbf{A} - \theta(\mathbf{A} + \mathbf{C}|$ $= 0$ and l the number of independent gamma variables. $\mathbf{A}$ and $\mathbf{C}$ are independent sum of product matrices based on sample observations with n_1 and n_2 degrees of freedom respectively and $l = \min(p, n_1)$. This statistic can be used to test (a) equality of two dispersion matrices, (b) equality of the p-dimensional mean vectors and (c) the independence between the p-set and a q-set of variates.

Wilks-Rosenbaum Tests These are distribution free tests of location and dispersion proposed by Rosenbaum (1953, 1954) based upon the concept of **Statistical Tolerance Limits** due to Wilks (1942).

Wilson-Hilferty Transformation A transformation of χ^2 proposed by Wilson and Hilferty (1931) for the purpose of ascertaining its distribution function approximately from the Normal distribution. If the number of degrees of freedom is v, the transformed quantity $(\chi^2/v)^{\frac{1}{3}}$ is distributed approximately normally with mean $1-2/9v$ and variance $2/9v$.

Window See **Spectral Weight Function**.

Winsorised Estimation A method of estimating the mean of a relatively small sample of observations by using **Linear Systematic Statistics** and replacing extreme observations by those next in magnitude. This process is associated with the name of C. P. Winsor who proposed this approach and it can apply to symmetric or non-symmetric censoring and adjustment.

Wishart Distribution The joint distribution of variances and covariances in samples from a p-variate Normal population, given by Wishart in 1928. If n is the sample number; a_{ij} the sample covariance of the ith and jth variates; A_{ij} the corresponding parent covariance; $|a|$ the determinant of the matrix (a_{ij}); (A^{ij}) the matrix inverse to (A_{ij}) whose determinant is $|A|$; the Wishart distribution may be written:

$$dF = \frac{(\frac{1}{2}n)^{\frac{1}{2}p(n-1)}|A|^{\frac{1}{2}(n-1)}|a|^{\frac{1}{2}(n-p-2)}}{\pi^{\frac{1}{4}p(p-1)}\prod\limits_{k=1}^{p}\Gamma\{\frac{1}{2}(n-k)\}}$$

$$\exp\left(-\tfrac{1}{2}n\sum\limits_{i,j=1}^{p}A^{ij}a_{ij}\right)\prod\limits_{i<j}^{p}da_{ij}.$$

Within-group Variance See **Intraclass Variance**.

Wold's Decomposition Theorem A theorem which asserts that any univariate non-deterministic stochastic process can be decomposed into a one sided moving average process and a deterministic process.

Wold's Markov Process of Intervals A generalisation of the **Renewal Process** due to Wold (1948) which assumes that the sequence of intervals between events in the process x_1, x_2, ..., forms a time homogeneous Markov sequence.

Working Probit The iterative calculations for the maximum likelihood estimation of a **Probit Regression Line** are usually performed by finding the weighted linear regression of the working probit on the dose metameter. The working probit is a quantity compounded of the **Empirical Probit** and the **Expected Probit**, with which it

coincides if the empirical value lies exactly on the provisional line.

Working Mean An alternative term for **Arbitrary Origin**.

Yates' Correction An adjustment proposed by Yates (1934) in the calculation on χ^2 for 2×2 table. It consists of subtracting $\frac{1}{2}$ from one cell in the table and adjusting the other cells so that row and column totals remain constant and working on the value of χ^2 computed from the resulting table. The general effect is to bring the distribution based on discontinuous frequencies nearer to the continuous χ^2 distribution from which the published tables for testing χ^2 are derived.

Youden Square An experimental design proposed by Youden in 1937. It is not a square, and would be better known as a 'Youden design'. For example, a design for seven treatments could be laid out as follows:

```
A  B  C  D  E  F  G
B  C  D  E  F  G  A
D  E  F  G  A  B  C
```

which may be regarded as three rows of a Latin square— hence the alternative name of **Incomplete Latin Square**. The above design is read downwards as comprising seven blocks of three; in the whole design every treatment and every pair of treatments occur equally often, and the arrangement provides for the analysis of positional effects within blocks.

Yule Distribution A distribution proposed by Yule (1925) for biological species investigations of the form

$$f(i) = AB(i, \rho+1)$$

where A and ρ are constants and $B(i, \rho+1)$ is a beta function. Its use was extended by Simon (1955).

An entirely different form of distribution is also associated with Yule (1924) for which see **Factorial Distribution**.

Yule Process A stochastic birth process used by Yule in 1924. It is essentially equivalent to the **Furry Process**. The name is sometimes applied to an **Autoregressive Process** of the second order.

Yule's Equation A name sometimes given to an auto-regressive equation of the second order, e.g.

$$u_t + au_{t-1} + \beta u_{t-2} = \epsilon_t.$$

Yule's Hyperbolic Distribution A modified form of the **Yule** (1925) **Distribution** proposed by Powell (1955) in connection with the generation time of bacteria:

$$f(t) = \frac{\lambda}{t}\left[\left\{1-e^{-t\psi/l_2}\right\}^g - \left\{1-e^{-t\psi/l_1}\right\}^g\right]$$

where $\psi = \Gamma_1(g+1) - \Gamma_1(1)$, g is the number of entities to be duplicated, t the generation time; λ, l_1 and l_2 are constants relating to a segment of a rectangular hyperbola used in the analysis.

Z-chart A form of graphic presentation of a time series consisting of three lines which usually take the shape of a letter 'Z'. The lower line is a plot of original data in the form of a time series; the centre line is a cumulative total; the upper line consists of a moving total of the original data.

z-distribution The distribution of a logarithmic transformation of a variance ratio due to R. A. Fisher. If there are two independent estimates of the population variance s_1^2 and s_2^2 based upon n_1 and n_2 degrees of freedom the function is defined as

$$z = \tfrac{1}{2} \log_e \frac{s_1^2}{s_2^2}.$$

Fisher chose the transformation of the variance ratio to z to simplify interpolation in tables of significance points. [See also **Beta Distribution, Variance Ratio Distribution**.]

z-score A term used by some writers in connection with educational and psychological testing as an alternative to standardised scores. A z-score for an observation is the score expressed as a deviation from the sample mean value in units of the sample standard deviation.

z-test A significance test based upon the **z-distribution**. In most cases it is tantamount to a **Variance Ratio Test**; but also is used as an approximation to tests with more complicated distributions, in which case variance ratios may not be involved.

z-transformation See **Fisher's Transformation**.

Zelen's Inequality A one sided inequality of the Tchebychev type stated explicitly by Zelen (1954)

$$Pr[\xi - \mathscr{E}(\xi) \geqslant t\sigma] \leqslant \left[1 + t^2 + \frac{(t^2 - t\alpha_3 - 1)^2}{\alpha_4 - \alpha_3^2 - 1} \right]^{-1}$$

for $t \geqslant \tfrac{1}{2}[\alpha_3 + (\alpha_3^2 + 4)^{\frac{1}{2}}]$, where α_3 and α_4 are standardised third and fourth moments.

When t is at its lower limit the inequality reduces to the one-sided version of the **Bienaymé-Tchebychev Inequality**. At its upper limit, it reduces to **Cantelli's Inequality**.

Zero Sum Game A game played by a number of persons in which the winner takes all the stakes provided by the losers so that the algebraic sum of gains at any stage is zero. It has been argued that many decision problems may be viewed as zero sum games between two persons.

Zeta Distribution See **Discrete Pareto Distribution**.

Zipf's Law A general law proposed by Zipf (1949) to approximate to the distribution of individuals bearing variate values which are either integral or grouped in intervals of equal width. The simplest form is $f(z) = k/x^p$ $0 < \alpha < x$ where p is some constant. A special case occurs when $p = 1$ and x takes the range 1, 2, 3, ..., and hence can be regarded as a rank; in this case the product of the frequency and its rank is a constant.

There are various later generalisations of this simple form. See, for instance, **Waring's Distribution**.

The term 'harmonic distribution' has been used for Zipf's distribution: this should be avoided as inaccurate and likely to lead to confusion with the distribution of harmonics in spectral analysis.

Zonal Polynomial A polynomial of importance in problems of multivariate noncentral distributions developed by James (1960). If we have a symmetric matrix S, its zonal polynomials are certain homogeneous symmetric polynomials in the characteristic roots of S.

Zonal Sampling A term used mainly by Indian statisticians to indicate sampling by zones; zone in this context denoting a stratum determined on a geographical basis.

Zone of Indifference See **Zone of Preference**.

Zone of Preference In connection with the test of an hypothesis, the zone of indifference is defined as the region in the sample space, if any, which is left after the removal of a region of acceptance and a region of rejection. These latter two together are sometimes called a zone of preference.

It is more customary, and seems better practice, to use the word 'region' instead of 'zone'.

Printed in Great Britain by
T. & A. CONSTABLE LTD.
Edinburgh